All-Russian Scientific Conference with International Participation on Contemporary Issues of Hydrogeology, Engineering Geology and Hydrogeoecology in Eurasia

IOP Conference Series: Earth and Environmental Science
Volume 33

Tomsk, Russia
23 – 27 November 2015

Editor:

S.L. Shvartsev

ISBN: 978-1-5108-2274-0
ISSN: 1755-1307

Printed from e-media with permission by:

Curran Associates, Inc.
57 Morehouse Lane
Red Hook, NY 12571

Some format issues inherent in the e-media version may also appear in this print version.

Printed by Curran Associates, Inc. (2016)

For permission requests, please contact the Institute of Physics
at the address below.

Institute of Physics
Dirac House, Temple Back
Bristol BS1 6BE UK

Phone: 44 1 17 929 7481
Fax: 44 1 17 920 0979

techtracking@iop.org

Additional copies of this publication are available from:

Curran Associates, Inc.
57 Morehouse Lane
Red Hook, NY 12571 USA
Phone: 845-758-0400
Fax: 845-758-2633
Email: curran@proceedings.com
Web: www.proceedings.com

Table of contents

Volume 33

All-Russian Scientific Conference with International Participation on Contemporary Issues of Hydrogeology, Engineering Geology and Hydrogeoecology in Eurasia
23–27 November 2015, Tomsk, Russia

Accepted papers received: 16 March 2016
Published online: 15 April 2016

Preface

Papers

Hydrogeochemistry

Geoecology

Well drilling

Geology, Hydrogeology and Engineering Geology

The results of the International Conference on "Modern Problems of Hydrogeology, Geology Engineering and Hydrogeoecology of Eurasia" are published in these Proceedings. The conference was held in National Research Tomsk Polytechnic University and dedicated to the 85th anniversary of Hydrogeology, Engineering Geology and Hydrogeoecology Department. The Department was one of the first in Siberia and the first beyond the Urals and was established in 1930 due to the intensive mining industry, natural resources development, population growth, underexplored hydrogeological and geo-engineering conditions, and absence of qualified engineering staff in Siberia.

Despite the challenges, the department was intensively developing and trained highly in demand qualified engineers. Since then about 3000 engineers, 500 Bachelors (since 1993), more than 100 Masters, 160 Candidates of Science and 30 Doctors of Science have received a degree at the department. Today, among the alumni of the department there are heads of big companies, university rectors, directors of scientific organizations, members of academies and expeditions and recipients of national awards.

The research fields of the department, as well as teaching strategies and techniques applied for training qualified engineering and research staff constituted the basis for the conference's scientific fields. Precisely, the conference covered issues in such areas as hydrogeology, groundwater geochemistry, water body contamination, migration of chemical elements, application of up-to-date modeling techniques for understanding the principles of different water types' formation, groundwater mapping methods and drilling, etc.

The conference reports revealed that the groundwater of Eurasia, including both the top hydrodynamic zone and deep water horizons (up to 3-4 km), are being intensively studied. The main focus was laid to the oil and gas production regions. It was stated that the mineralization of such water reaches 500 g/l. Such strong brines are typical within the Siberian platform. It is an interesting fact that not only Cl-Na brines, but also Cl-Ca ones, the genesis of which is rather controversial, can be met within this territory.

A great number of the reports demonstrated that water-host rock non-equilibrium is typical within Siberia. It implies not only low-salinity water but also strong brines. The character of the phenomenon being not always clear, the participants proposed different viewpoints on the issue.

Modeling of hydro-geochemical processes in the water-rock system was the focus of a significant number of the reports. Also, these processes are studied via isotope analysis of

IOP Conf. Series: Earth and Environmental Science **33** (2016) 011001 doi:10.1088/1755-1315/33/1/011001

different chemical types of water with dissolved salts and gases (isotopes of O, H, N, C, Sr, Cl, Br, He, S, etc.).

The greatest attention of most reporters was paid to such critical geo-engineering issues as urban underground areas development, urban agglomeration infrastructure, territory zoning and classification, hazardous geocryological processes along main oil and gas pipelines, conditions of geo-environmental components, mining activities and ground base modeling.

The modern hydrogeology and engineering geology focused on geoecological conditions and environmental protection, which has determined the scope of the conference. The participants discussed the environmental impact caused by the mining companies (water and soil contamination and flora impact), ecological conditions in urban areas, the environmental impacts of waste water effluents, geological factors connected with the population health and drinking water quality, as well as the solutions for environmental crises.

Youth Research School was organized within the conference. Four lectures were delivered by the professors of different universities and institutes: 1) S.L. Shvartsev (TPU) "Aquagenetic Theory of Environment Evolution"; 2) B.N. Ryzhenko (VIGAC RAS) "Geochemical Processes Modeling within the Water-Rock-Gas System to Solve Fundamental and Engineering Problems"; 3) O.N. Gryaznov (USMU) "Engineering-Geological Zones of the Ural"; (4) S.O. Grinevskiy (MSU) "Geohydrological Modeling in Groundwater Resource Estimation". Besides, there were 35 student reports from 9 Russian Universities. Some of the reports were published in the Proceedings.

S.L. Shvartsev

IOP Conf. Series: Earth and Environmental Science **33** (2016) 011002 doi:10.1088/1755-1315/33/1/011002

Program committee

Committee chairman: V.I. Osipov – Academician

Vice-chairmen: A.R. Kurchikov – Corresponding member, S.L. Shvartsev – Doctor of Sciences (Geology and Mineralogy)

Committee secretariat: M.N. Kolpakova – PhD (geological and mineralogical science), O.G. Tokarenko – PhD (geological and mineralogical science)

Committee members: V.G. Rumynin – Corresponding member, S.V. Alekseev – Dr. of Science, V.K. Popov – Dr. of Science, S.P. Poznyakov – Dr. of Science E.M. Dutova – Dr. of Science, D.S. Pokrovskiy – Dr. of Science, V.N. Kataev – Dr. of Science, A.V. Kirukhin – Dr. of Science, V.M. Matusevich – Dr. of Science V.E. Olkhovatenko – Dr. of Science, V.V. Pendin – Dr. of Science, A.M. Plusnin – Dr. of Science, L.P. Rikhvanov – Dr. of Science, B.N. Ryzhenko – Dr. of Science, O.G. Savichev – Dr. of Science, L.A. Strokova – Dr. of Science, V.M. Shvets – Dr. of Science

Tributes to colleagues

Editorial board of the collection (journal) thanks the organizers of the International Conference "Modern problems of hydrogeology, geology engineering and hydrogeoecology of Eurasia":

S.L. Shvartsev – Plenary session co-Chairman, Dr. of Science, Professor, TPU, Tomsk, Russia
G.N. Kopylova – Plenary session co-Chairman, Dr. of Science, Kamchatka Branch of the Geophysical Service RAS, Petropavlovsk-Kamchatsky, Russia
V.G. Kondrat'ev – Plenary session co-Chairman, Dr. of Science, Transbaikal State University
E.M. Dutova – Plenary session co-Chairman, Dr. of Science, Professor, TPU, Tomsk, Russia
E.Yu. Pasechnik – Plenary Section Secretary, PhD, associate Professor, TPU, Tomsk, Russia
O.E. Lepokurova – Plenary Section Secretary, PhD, associate Professor, TPU, Tomsk, Russia
S.O. Grinevskiy – 1th Section co-Chairman, Dr. of Science, Professor, MSU, Moscow, Russia
D.S. Pokrovskiy – 1th Section co-Chairman, Dr. of Science, Tomsk State University of Architecture and Building, Tomsk, Russia
E.A. Soldatova – 1th Section Secretary, PhD, assistant, TPU, Tomsk, Russia
O.N. Grjaznov – 2th Section co-Chairman, Dr. of Science, Professor, Ural State Mining University, Yekaterinburg, Russia
V.E. Olkhovatenko – 2th Section co-Chairman, Dr. of Science, Professor, Tomsk State University of Architecture and Building, Tomsk, Russia
V.V. Kramarenko – 2th Section Secretary, PhD, associate Professor, TPU, Tomsk, Russia
S.V. Alekseev – 3th Section co-Chairman, Dr. of Science, Professor, Institute of Earth's Crust SB RAS, Tomsk, Russia
V.K. Popov – 3th Section co-Chairman, Dr. of Science, Professor, TPU, Tomsk, Russia
N.V. Guseva – 3th Section Secretary, PhD, associate Professor, TPU, Tomsk, Russia
L.V. Zamana – 4th Section co-Chairman, PhD, Institute of Natural Resources, Ecology and Cryology SB RAS, Chita, Russia
O.G. Savichev – 4th Section co-Chairman, Dr. of Science, Professor, TPU, Tomsk, Russia
A.A. Hvaschevskaya – 4th Section Secretary, PhD, associate Professor, TPU, Tomsk, Russia
B.N. Ryzhenko – Youth School Section co-Chairman, Dr. of Science, Vernadsky Institute of Geochemistry and Analytical Chemistry, Moscow, Russia
L.A. Strokova – Youth School Section co-Chairman, Dr. of Science, Professor, TPU, Tomsk, Russia
Yu.G. Kopylova – Youth School Section co-Chairman, phd, TPU, Tomsk, Russia
O.G. Tokarenko – Secretary of the Conference, PhD, associate Professor, TPU, Tomsk, Russia
M.N. Kolpakova – Secretary of the Conference, PhD, assistant, TPU, Tomsk, Russia
I.G. Sharkov – Department for development of publication activities, TPU, Tomsk, Russia

Peer review statement

All papers published in this volume of *IOP Conference Series: Earth and Environmental Science* have been peer reviewed through processes administered by the proceedings Editors. Reviews were conducted by expert referees to the professional and scientific standards expected of a proceedings journal published by IOP Publishing.

Underground water in Sukhodol Bay area (South Primorye)

G Chelnokov[1] and Yu Kolubaeva[2]

[1] Far East Geological Institute, Far Eastern Branch RAS, Vladivostok, Russia
[2] Trofimuk Institute of Petroleum Geology and Geophysics of Siberian branch RAS, Tomsk Division, Tomsk, Russia

E-mail: [1]geowater@mail.ru, [2]kolubaeva@inbox.ru

Abstract. The paper describes the specific chemical groundwater composition in Sukhodol Bay (eastern Ussuri Bay). The equilibrium between groundwater and carbonate-aluminosilicate minerals in water-bearing rocks was calculated. It was established that a fraction of the studied water was calcite saturated. However, dolomite and magnesium undersaturated groundwater is non-equilibrium with primary aluminosilicates.

1. Introduction

The significance of this research is relevant to the composition formation of highly calcium-chloride / calcium-sodium chloride concentrated water. In this case, this research was never conducted in the south Far East, Russia. However, groundwater occurrences with high salt concentration were found during prospect drilling of coal, hydrocarbon, limestone, mineral water, etc. [1].

The target of this research includes not only the investigation of the hydrogeochemical properties but also the evaluation of the highly-mineralized groundwater -primary rock-forming mineral equilibrium within Sukhodol Bay area (Figure1).

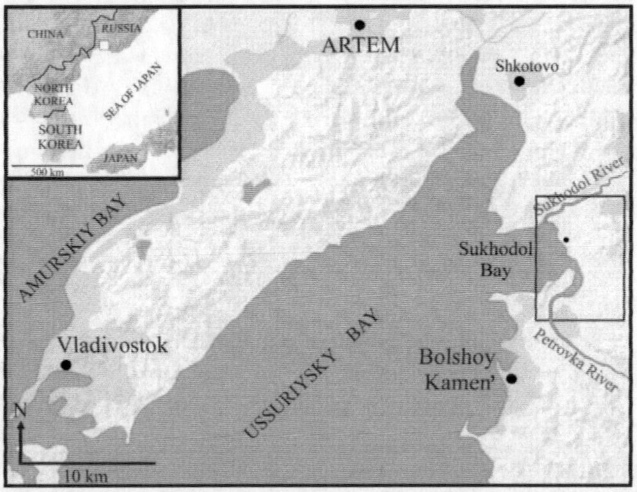

Figure 1. Map of study area.

The investigation of the groundwater in Sukhodol Bay area was launched in 2007. The first results describing the groundwater geochemistry were reflected in [2]. This paper embraces unique research data of the last few years (2014 and 2015) in chemical composition of groundwater and coastal marine water (150m from the well).

Macro-micro component analyses were conducted in the approved testing laboratory within the framework of Far East Geological Institute, Far Eastern Branch RAS. The major cations and anions were determined via ion chromatography liquid method (HPLC-10AVp, SHIMADZU). Bromine was analyzed in the Central approved chemical testing laboratory JSC "Primorgeology."

2. Geological background

Investigated area is located within Partisan depression, South Primorye trough (downside area Khanka massif) [3].

The most ancient formations within this region are Mid- Upper Triassic sediments embracing dark gray aleurolites with limestone-marl lenses and concretions, fine-grained sandstone and Ammonoidea layered. Upper Jurassic sediments include gray quartz-feldspar sandstones, sandstones with embedded coal lenses, limestone- siltstones. Cretaceous sediments are interlayered tuff siltstone, sandstone, weathered and weakly-fractured aleurolites, as well as fractured argillites. Recent sediments are found throughout the coastal area of Sukhodol estuary (liman) and subside to the coastline for about 500m. They are divided into two genetic types: (1) recent alluvial-marine sediments occurring immediately at subsurface and including dark gray fine-grained silt sand and peat bands The thickness is 1,5-6.0m. (2) alluvial Upper Quaternary sediments underlying alluvial-marine sediments embracing gravel with block masses and sand inclusions and sandy-clayey gravel boulders. The thickness is up to 10m.

3. Study area

The study area is characterized by complex hydrogeological conditions and varied hydrogeochemical cross-section. Hydrogeologically, the study area is located in the central part of South Primorye province, Primorye composite artesian basin. Based on the geological structure and hydrogeological conditions, the following aquifer types were defined:

1) Upper Quaternary aquifer – recent marine and alluvial-marine sediments ($a\text{-}mQ_{III\text{-}IV}$) are well-represented in Sukhodol bay mouth and embraces sands, sand loam and pebbles interbedded with silt. Static water level in wells is 1.3–1.7 m., well flow rates from 1.4 to 2.69 liter/sec., filtration coefficient – from 7.4 to 15.2 m/a day. Water here is predominately fresh, hydrocarbonate and hydrocarbonate chloride but, rarely calcium chloride or of mixed cation composition.

2) Upper Cretaceous aquifer- Korkian sediments. The hydrogeological conditions of this aquifer is associated with the litho-facies and structure-tectonic features of water-bearing rocks. Fault water is confined not only to zones of weathered Cretaceous rocks but also to zones of tectonic fracturing and intrusive contact disturbance. Static water level in wells is 0.8 m, well flow rates from 0.6–0.7 liter/sec. under depression level of 15–30 m. filtration coefficient for lithologically different rocks varies; for sandstones and conglomerates it is averagely 0.1–1.5 m/a day, but sometimes, 5–10 m/a day. Wells of up to 15m deep drilled at 0-500m from the coastal zone and 1km. from Sukhodol and Petrov Rivers revealed low-medium salt water (0.6–8.5 gr./liter); whereas drilled 15-300m wells revealed high calcium chloride sodium-calcium chloride mineralized water and brine water (20–110 gr/liter).

4. Results and discussion

Water chemical composition is described in table 1. The following types of water were studied: fresh, salt and brine water, as well as marine water. Based on the chemical water composition data depicted in table 1 and previous published data [2], three hydrogeochemical zones were identified in the cross-section:

1. fresh water with mineralization of up to 1 gr/liter, mixed anion-cation composition (calcium sulphide-hydrocarbonate, calcium-sodium chloride-hydrocarbonate). Non-artesian water extends to a depth of 15m. and is confined to Upper Quaternary-Recent marine and alluvial-marine sediments. Probably, there are traces of the water- marine sediment interaction in this zone;

IOP Conf. Series: Earth and Environmental Science **33** (2016) 012001 doi:10.1088/1755-1315/33/1/012001

2. calcium sodium-chloride brine water with mineralization of 50gr/liter was found in the depth interval of 15-250 m;
3. calcium chloride, calcium sodium and sodium-chloride salt water with mineralization of 27 to 35 gr/liter was found in the depth interval of 250-300m.

Brine and salt water zones (2 and 3) confined to Upper Cretaceous terrigenous sediments have pressure head (+1.4 m). Water-bearing rocks embrace aleurolites, tuff aleurolites, fine-grained sandstones and argillites. The difference between 2nd and 3rd zones involves mineralization and macro-micro component variations. These waters usually have high iron-manganese-strontium content. Bromine content is at the average value level of ocean water (Horns value – 67 mgr/liter) [4], iodine is not found.

Table 1. Chemical composition of studied waters.

Parameters	Measure-ment unit	Well № 2PR					Water from Sukhodol Bay
		Sample date					
		05.07.2014	15.02.2015	15.09.2015	17.09.2015	17.09.2015	15.09.2015
Sampling depth	m	250-300				0.5-10	
pH		7.42	6.89	7.1	7.42	7.43	6.6
Mineralization	g/l	35.0	31.8	35.5	33.9	0.62	5.9
HCO_3^-		0.015	0.016	0.017	0.0175	0.277	0.211
SO_4^{2-}		–	–	0.49	0.495	0.162	0.407
Cl^-		22.3	20.2	22.1	21.1	0.016	3.2
Ca^{2+}		9.5	9.4	9.6	9.5	0.12	0.095
Mg^{2+}		0.29	0.18	0.27	0.186	0.016	0.19
Na^+		3.0	2.02	3.0	2,6	0.026	1.8
K^+		0.007	0.005	0.007	0.003	0.005	0.063
Si		–	0,0026	0.003	0.004	0.015	–
Fe	ppm	1.07	0.95	1.35	1.24	0.08	0.8
Mn		5.37	4.93	52	7.37	0.0047	0.02
Br		–	–	64.2	58.5		9.11
Sr		204.8	171.5	206.8	248,1	0.732	–
Ba		0.37	0.4	0.5	0.59	0.09	0.0004
Li		1.5	3.5	3.5	1.93	0.011	0.002
		Relationship between components					
Cl/Br		–	–	344	360	–	347
rNa/rCl		0.21	0.15	0.21	0.19	7.4	0.87
Ca/Cl		0.42	0.46	0.43	0.45	2.4	0.03
Formula of ion-salt composition		Cl/Na-Ca	Cl/Ca	Cl/Na-Ca	Cl/Ca	SO_4-HCO_3/Ca	Cl/Na

The marine water analysis from Sukhodol Bay revealed the fact that this water is of sodium chloride composition, having minimum pH value and total salt content, which, in its turn, is associated with the fresh water ingression from Sukhodol and Petrovks rivers into the bay (Figure 1), as well as the low bay depth.

The prevalence of this or that chemical element in the solution is not accidental and is governed by characteristic origin aspects and succeeding groundwater conversion in each specific case. For example, to determine the genesis of high-salinity water the Cl/Br, rNa/rCl, Ca/Cl coefficient ratio was applied [5].

Calculated genetic coefficient values of investigated water is listed in table 1. These calculations showed that high-salinity water as a consequence of low sodium content exhibits minimum coefficient value of rNa/rCl, while this ratio for marine water exhibits 0.87. The extent of water metamorphization can be inferred by the distinct increase of calcite content after brine burial. According to this ratio value these waters could be correlated to intensively metamorphosed ones.

The composition and salinity of water depends on numerous factors and processes, including the following major ones: composition of host rocks, geomorphological and landscape features, hydrodynamic conditions and others. Water exchange intensity is the main influence factor for high-salinity groundwater. Hydrogeologically, the formation of calcium chloride, sodium-calcium chloride high-salinity water is governed by the impeded water exchange, which, in its turn, reveals relatively low reservoir properties of Mesozoic Primorye rocks. This is also confirmed by previous data showing that the content value of tritium in water is less than 0.3 TE. This clearly demonstrates the aquifer integrity including brines, i.e. hindering its mixing with recent waters and water exchange.

To determine the formation conditions of high salinity groundwater chemical composition in the studied area equilibrium state of water to leading minerals in host rocks was calculated (Figures 2, 3).

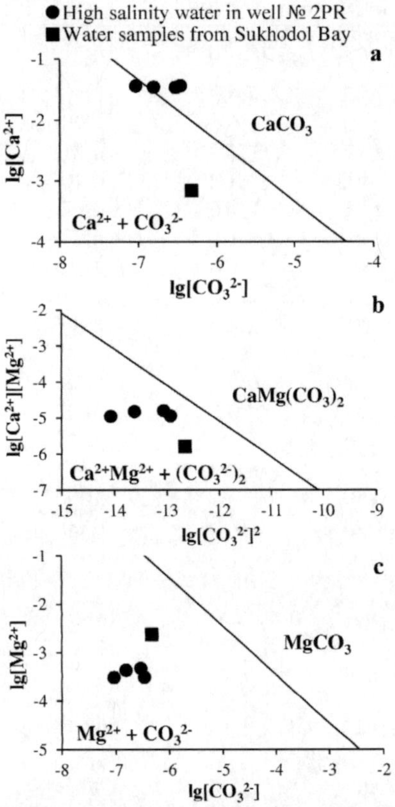

Figure 2. Water saturation intensity relative to (a) calcite, (b) dolomite, (c) magnesium

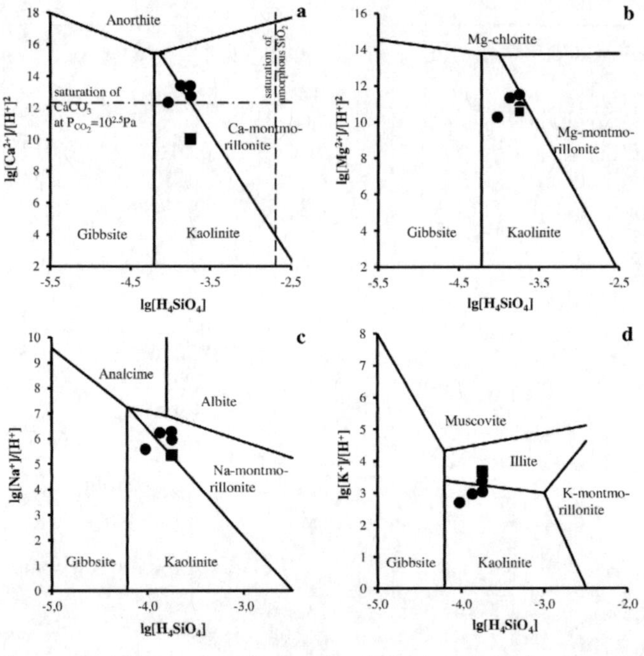

Figure 3. Diagrams of aluminosilicate mineral saturation intensity in water under standard conditions: (a) HCl-H_2O-Al_2O_3-CO_2-CaO-SiO_2system; (b) HCl-H_2O-Al_2O_3-CO_2-MgO-SiO_2 system; (c) HCl-H_2O-Al_2O_3-CO_2-Na_2O-SiO_2 system and (d) HCl-H_2O-Al_2O_3-CO_2-K_2O-SiO_2 system. (Legend – Figure 2).

Component activity in aqueous-based solution is calculated via HydroGeo program [6]. The selection criterion identified the possible calculations of activity by applying Pitzer algorithm, designed high salinity water and brines.

In model matching the most common minerals from the studied cross-section were imported into the system (aluminosilicate minerals, calcite, dolomite, magnesium and others). Calculation of component activity was performed in standard (laboratory) conditions.

Thermodynamic calculations in determining the water to carbonate minerals equilibrium (Figure 2) revealed the fact that a portion of water is saturated with calcite, while other waters are undersaturated relative to dolomite and magnesium.

Despite the continuous interaction of investigated high salinity water with rocks, the equilibrium to primary aluminosilicate minerals was not observed. Simultaneously, there exists the water to kaolinite, illite, Ca- and Na montmorillonite equilibrium (Figure 3).

Mineralization, Ca, Na, Mg, K, Si content and pH value govern the equilibrium degree of high salinity water to aluminosilicate minerals. Secondary mineral formation (carbonates, clays, etc.) is the barrier in establishing the equilibrium with primary aluminosilicate minerals.

5. Conclusion

Based on hydrogeological and hydrogeochemical data the formation conditions of unique high salinity water in Primorye territory, located in the coastal area of Sukhodol Bay, were determined. It was established that hydrogeochemical water behavior is governed by sedimentation water and sedimentary rock interaction within slow water exchange zone. Analysis results showed that the high salinity groundwater -rock system preserves its equilibrium- non-equilibrium state, which, in its turn, is determined by the non-consistent interaction of its components, and, respectively, consistent evolution of groundwater composition.

Acknowledgments

The research was supported by Russian Foundation for Basic Research funding, project No. 15-35-50438

References

[1] Chelnokov G, Kharitonova N, Bragin I and Chudaev O 2015 Geochemistry of mineral water and gases of the Razdolnoe Spa (Primorye, Far East of Russia) *Appl. Chem.* **59** 147-154

[2] Chelnokov G, Kharitonova N and Vasileva M 2012 Geohimija i genezis podzemnyh rassolov severo-vostochnoj chasti Ussurijskogo zaliva (Primorskij kraj) *Geoecology, Engineering Geology, Hydrogeology, Geocryology* **4** 310-319

[3] Martynov Y and Khanchuk A 2013 Cenozoic volcanism of the eastern Sikhote Alin: Petrological studies and outlooks *Petrology* **21** (1) 85-99

[4] Horn R A1969 *Marine Chemistry* (The structure of water and the chemistry of the hydrosphere) (Willey-Interscience) p 565

[5] Krainov S, Rizhenko B and Shvest V 2012 *Geohimija podzemnyh vod. Teoreticheskie, prikladnye i jekologicheskie aspekty* (CentralLitNeftGas) p 672

[6] Bukat M 2005 *Reklamno-tehnicheskoe opisanie programmnogo kompleksa HydroGeo. Gosudarstvennaja registracija algoritmov i programm vo Vserossijskom nauchno-tehnicheskom informacionnom centre* № 50200500605 (Moscow) p 7

Distribution and Geochemistry of Rare-Earth Elements in Rivers of Southern and Eastern Primorye (Far East of Russia)

O V Chudaev [1], I V Bragin [2], N A Kharitonova [3,5] and G A Chelnokov [4]

[1, 2, 4, 5] Far East Geological Institute of Far Eastern Branch of Russian Academy of Sciences. 159, prospect 100-letya Vladivostoku, Vladivostok, Russia, 690022
[3]Lomonosov Moscow State University, GSP-1, Vorobjevy gory, Moscow, Russia

E-mail: [1]chudaev@fegi.ru, [2]bragin_ivan@mail.ru, [3]tchenat@mail.com, [4]geowater@mail.ru

Abstract. The distribution and geochemistry of rare earth elements (REE) in anthropogenic, technogenic and natural surface waters of southern and eastern Primorye, Far East of Russia, are presented in this study. The obtained results indicated that most of REE (up to 70%) were transported as suspended matter, ratio between dissolved and suspended forms varing from the source to the mouth of rivers. It is shown that all REE (except Ce) in the source of the rivers are predominantly presented in dissolved form, however, the content of light and heavy REE is different. Short-term enrichment of light rare earth elements (LREE) caused by REE-rich runoff from waste dumps and mining is neutralized by the increase in river flow rate. Rivers in urban areas are characterized by high content of LREE in dissolved form and very low in suspended one.

1. Introduction

During the last decades, REE in various natural water sources have been the subject of intensive study. REE is often used as sensitive indicators of different water types and the degree of pollution by industrial, household, and agricultural wastewaters and nuclear power plant waste products. The insignificant differences in the chemical properties between the elements and their relative inertness provide grounds to assume that REE can be used as very sensitive indicators of geochemical processes that occur in water–rock–organic matter systems. The study of REE behavior in the natural water sources of the Russian Far East was initiated over 15 years ago [1-6]. First full data set on river waters of Primorye showed that obtained REE concentrations were at the detection limit of ICP-MS (Agilent 7500) for some rivers of Eastern and Southern Sikhote-Alin. A unique opportunity to measure REE concentrations in slightly mineralized waters in the rivers of eastern and southern Sikhote-Alin using the high resolution ICP-MS method appeared in 2014.

The purpose of this study is to measure the REE concentrations and the patterns of their fractionation in the rivers of eastern and southern Sikhote Alin, including relatively pure and polluted water streams crossing mining and urbanized areas.

Eleven rivers were sampled during 2014. On the north-east of Primorye, the Rudnaya River affected by mining has been investigated. From the north-east to the south-east of Primorye the following rivers with minimum antropogenic impact have been tested: the Avvakumovka, the Margaritovka, the Milogradovka, the Partizanskaya, the Sukhodol, the Petrovka, the Shkotovka, the Artemovka. The streams with anthropogenic impact are presented by the Vtoraya and the Pervaya

Rechka rivers within the territory of Vladivostok city, and a stream from purification plants of Dalnegorsk city. More details on sampling points can be found at https://www.google.com/maps/d/edit?mid=zw2CLGdOtNrE.k5G_87mVb_IM&usp=sharing. Sampling methods, sample preparation and analytical peculiarities were discussed in [3].

2. Hydrochemical characteristics of rivers

In all the eastern and southern Sikhote-Alin rivers unaffected by anthropogenic impacts, mineralization is usually low (table), which reflects the wide distribution of crystalline rocks in the drainage areas and the low intensity of their chemical denudation. Increased mineralization is recorded within Vladivostok and in the Rudnaya River which is subjected to significant impacts of the mining industry (Table 1). The pH values typically varied from 6.0 to 7.0; in the Rudnaya River and Vladivostok water streams, they substantially exceeded 7.0. The content of dissolved oxygen in the unpolluted rivers was 9 mg/l. The turbidity of the water during the sampling period ranged from a few to tens of milligrams per liter. Cations in the %-equiv form are largely represented by Ca or Ca + Mg + Na mixture. The intensively polluted water streams frequently show a notable shift toward Na. Atmospheric precipitations are characterized by the mixed Ca–Na cation composition. Among the anions, the hydrocarbonate ion is dominant. A substantial shift toward the sulfate ion is observed in the mine waters and in a creek starting from the tailing storage. In the water streams of Vladivostok and its suburbs, the anion composition frequently exhibits a shift toward notable increase in Cl^- and SO_4^{2-}.

Thus, the general hydrochemical parameters of the water in the studied rivers unaffected by anthropogenic impacts reflect insignificant erosion in the drainage areas composed mostly of acid and intermediate volcanic rocks.

3. Rare Earth Elements

Rare earth elements are transported by rivers both in solid (suspended particulate matter) and dissolved form. The content of REE transported in the form of suspended particulate matter by Primorye rivers may be as high as 70 % [6]. The proportions of REE transported in the solid and dissolved states are influenced by sorption/desorption processes and the pH of the medium [7]. Thus, to obtain a clear vision of REE fractionation, it is important to know REE concentrations in river water and suspended particulate matter.

The degree of fractionation depends on the physico-geographical conditions of the drainage area if the anthropogenic impact is negligible. The slight fractionation between light and heavy REE in the studied unpolluted rivers is explained by their hydrological features: steep profile; relatively small drainage area; low turbidity, total mineralization, and content of organic matter; and short interaction between water and suspended particulate matter. At the same time, in most large rivers with long interaction between suspended particulate matter and water, the water is enriched with heavy REE relative to LREE.

Very high concentrations of REE occur in mine water effluent and waste dumps of mining enterprises of the Rudnaya River basin. This is explained by lower pH values and higher TDS of water [6, 7]. While away from the pollution source (toward the mouth), the situation becomes similar to that in the unpolluted rivers of Sikhote Alin. Concentrations of LREE (La-Eu) in the rivers of Vladivostok city are tenfold lower. These concentrations are affected by of organic domestic wastes. LREE easily interact with organic matter and are removed from the solution. Figure 1 shows the results on total content of REEs (solid+dissolved), normalized to North-American shale composite (NASC). All lines were divided into three groups depending on presence/absence of anthropogenic or technogenic impact (A, B, C – see figure 1). The obtained data show that all lines demonstrate pronounced negative Ce-anomaly associated with the transition of Ce^{3+} to Ce^{4+} in oxygenated conditions (Eh>0), common for surface water streams. Four-valent cerium is inactive form and is derived from the solution [6, 8]. The actual fractionation between LREE and HREE doesn't occur. We can clearly observe the separation of the REE concentration levels in the upper, middle reaches and mouth of the

Rudnaya River. Upper reaches are mostly affected by runoff from the tailings and mines which have high concentration of REEs (Table).

Towards mouth of the Rudnaya River, concentration of REE is reduced by dilution and sorption processes and falls dramatically in the mouth due to seawater influence [6]. Acidity growth results in a decreased REE concentration in water and increased fractionation to the suspended phase.

Table. Geochemical characteristic of watercourses of Southern and Eastern Primorye.

Sampling point (ID number)	Chemical type of water	TDS, mg/l	pH	Eh, mV	Proportion of dissolved form* LREE	HREE	$\sum_{REE}$, ug/l
Technogenous water streams							
Streamhead of Krasnorechenskiy creek (1)	Mg–Ca-SO_4	414	6.6	234.8	0.26	0.09	67.96
Run off from the mine (2)	Na–Mg–Ca–SO_4	1006	5.59	158	0.98	0.98	34.10
Rudnaya River							
Streamhead	Na–Ca–HCO_3	70	7.08	124	0.68	0.78	2.56
After Krasnorechenskiy creek inflow	Ca–HCO_3–SO_4	103	7.13	209	0.48	0.56	3.17
After Taiga village waste water inflow	Na–Ca–SO_4–HCO_3	92.1	7.28	286	0.36	0.44	3.66
After Dalnegorsk city waste water inflow	Na–Ca–SO_4–HCO_3	143	7.33	236	0.32	0.48	1.90
Mouth	Ca–HCO_3–SO_4	59.3	6.3	290	0.23	0.37	1.66
Natural water streams (rivers)							
Avvakumovka (6)	Ca–Na–HCO_3	70	5.91	254	0.19	0.22	11.20
Margaritovka (7)	Na–Ca–HCO_3	55	6.5	258	0.50	0.60	1.84
Milogradovka (8)	Ca–Na–HCO_3	51.4	6.17	296	0.45	0.48	3.51
Partizanskaya (9)	Na–Ca–HCO_3	76.2	6.88	246	0.38	0.47	0.32
Sukhodol	Ca–Na–HCO_3	44.6	6.38	277	0.30	0.35	0.77
Petrovka (10)	Na–Ca–HCO_3	57	6.92	263	0.39	0.40	0.80
Shkotovka (11)	Na–Ca–HCO_3	53	6.99	270	0.24	0.28	1.23
Artemovka (12)	Na–Ca–HCO_3	72.6	6.89	262	0.23	0.28	5.06
Anthropogenous water streams							
Stream from purification plant of Dalnegorsk city (3)	NH_4–Na–Ca–Cl–SO_4–HCO_3	300	6.96	40	0.82	0.94	0.20
Vtoraya Rechka river (4)	Ca–Mg–Na–SO_4–HCO_3–Cl	964	7.43	183	0.20	0.76	0.37
Pervaya Rechka river (5)	Ca–Na–SO_4–HCO_3–Cl	370	7.2	196.2	0.18	0.76	0.51

*Calculated the proportion of dissolved forms of REE on the sum of dissolved and suspended phase.

With increasing pH, the adsorption of lanthanides on suspended particles decreases from LREE to HREE [9]. The same patterns are observed for the proportion of the dissolved form, i.e. with a reduction of anthropogenic impact either the total REE content in water or the proportion of the dissolved forms of occurrence of REE decreases (Table).

Even more impressive difference between light and heavy REEs is observed in the streams of Vladivostok city (Figure 1C), as well as for purification plants of Dalnegorsk city. It reflects that this type of water is slightly alkaline (pH = 6.96–7.43) contianing high concentrations of organic matter. REE Patterns for the Pervaya and the Vtoraya Rechka rivers demonstrate the same trend, but the last one has significant positive Gd anomaly, reflecting influence of heath care facilities [10]. REE fractionation between dissolved and colloid forms is also different for various types of water streams. For example, almost 98% of REE present in the dissolved form in the runoff from the mine, showing the lack of organic contaminants and impurities capable of carrying REE adsorbed on colloidal particles. At the same time, the proportion of dissolved forms of REE in rivers was the least stable and varied from 0.19/0.22 for the Avvakumovka River to 0.5/0.6 for the Margaritovka River, reflecting the dependence of the total content of rare earth elements and the physic-chemical water parameters (pH, Eh), caused by a variety of flood characteristics.

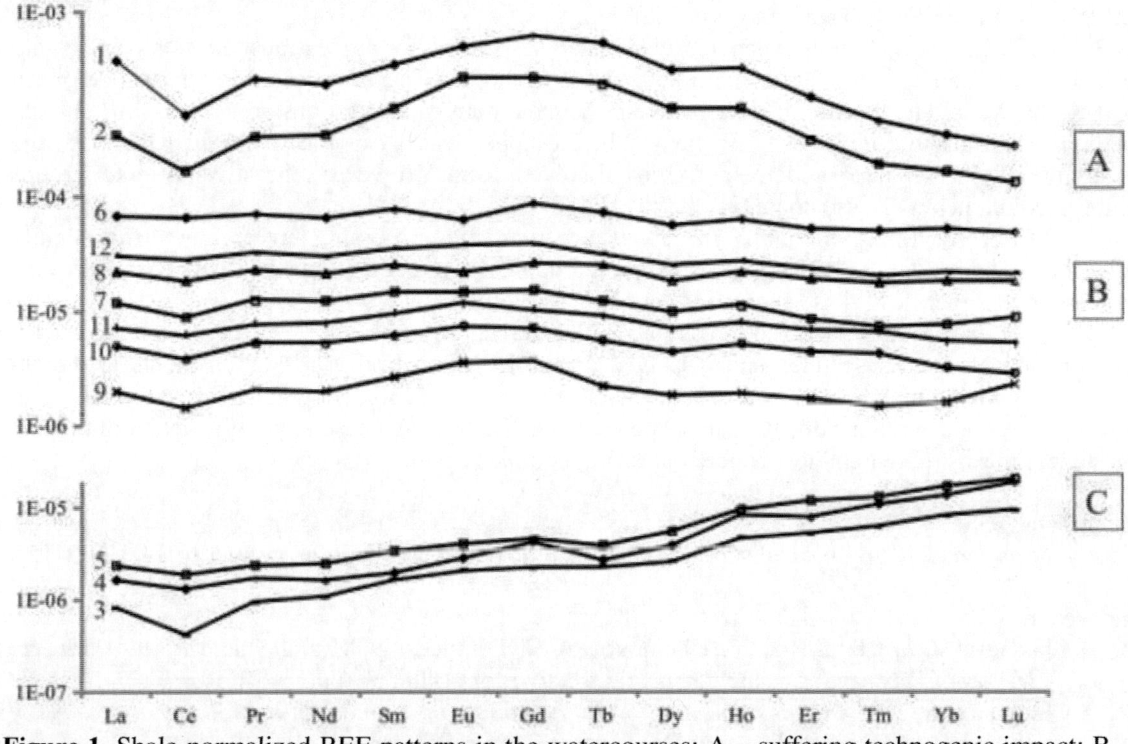

Figure 1. Shale-normalized REE patterns in the watercourses: A – suffering technogenic impact; B – located in natural conditions; B – subject to anthropogenic pressure. Ordinate - normalized to NASC REE concentrations, the scale is logarithmic.

Previously, we calculated migration forms of some REE in sampled objects using the Selector software [3]. As a whole, it should be noted that most of them are mainly represented by the ion and carbonate forms, while their migration in water streams from tailing storages is largely provided by the sulfate form. For example, in the waters of the Avvakumovka and the Milogradovka rivers, La, Ce, and Nd migrate in the ion form, while the carbonate forms are subordinate. In the Avvakumovka River, this is also true for Eu, while in the Milogradovka River, the carbonate form slightly prevails over the ion one (51 % vs. 43 %). In the case of Yb, the ion and carbonate forms play approximately equal roles in its migration in the Avvakumovka River; in the Milogradovka River, migration of this element in the carbonate form sharply increases [3].

The proportions between the migration forms of Nd, Eu, and Yb in the Rudnaya River are slightly different. The migration of these elements is mostly provided by the carbonate form. Rather different proportions between migration forms are observed in the waters from the tailing storage, where La,

Nd, Eu, and Yb migrate mostly in the sulfate forms, while for Ce the ion form is the main one. Thus, these data demonstrate that the migration forms of some REE in the surface waters of three types with different REE concentrations (unpolluted water of the Avvakumovka and the Milogradovka rivers, the Rudnaya River waters subjected to the impact of the mining industry, and water from tailing storages) are different, as well.

4. Conclusion

Studies have established the following:

1. The content of REE and their fractionation in surface waters considerably varies depending on whether they are susceptible to the effects of mining enterprises or under impact of urbanized areas or located in natural conditions.
2. Host rocks, drained by rivers and, in urban areas, anthropogenic pressures have significant influence on the REE's behavior and concentration.
3. REE fractionation between dissolved and suspended forms of transportation varies with distance from the upper reaches to the mouth. Most of the REE are transported in a suspended form (> 0.45) by the middle reaches of river, wherein the maximum concentrations are typical for light and medium REE, and the minimum – to heavy. In the upper reaches of the rivers all REE, with the exception of cerium, are predominantly in dissolved form. However, the division into groups (light+medium/heavy) is also common.
4. Effect of REE-rich runoff from waste dumps and processing enterprises drifts slightly against the general flow of the river. It is expressed in a short-term enrichment of LREE, but with the increase in the river flow, these anomalies are being leveled.
5. The rivers in urbanized areas are characterized by the most clearly expressed REE fractionation between suspended and dissolved forms. Effluents from purification plants and small rivers experiencing high human pressure are characterized by high content of LREE in dissolved form and very low concentrations of LREE in suspended one. This can be caused by high content of organic substances in wastewater and low concentrations of complex agents.

Acknowledgements
This work was made with financial support of Russian Scientific Foundation, project № 14-17-00415.

Reference

[1] Chudaeva V A, Chudaev O V, Chelnokov A N, Edmunds W M and Shand P 1999 *Mineral waters of Primorye (Chemical aspect)* (Vladivostok: Dalnauka) p 166 (in Russian)
[2] Chudaev O V 2003 *Composition and formation conditions of modern hydrothermal systems of Russian Far East* (Vladivostok: Dalnauka) p 216 (in Russian)
[3] Chudaev O V, Chelnokov G A, Bragin I V, Kharitonova N A, Blokhin M G and Aleksandrov I A 2015 Fractionation of rare earth elements in the rivers of the Eastern and Southern Sikhote-Alin under the conditions of natural and anthropogenic anomalies *Russian journal of Pacific geology* **9** 428–438
[4] Shand P, Johannesson K H , Chudaev O V, Chudaeva V A and Edmunds W M 2005 *Rare earth element contents of high pCO_2 groundwaters of Primorye, Russia: Mineral stability and complexation controls* (Netherlands: Springer) Ch7 161-186
[5] Chelnokov G A and Kharitonova N A 2008 *High-P_{CO2} mineral waters of Southern Far East* (Vladivostok: Dalnauka) p 165 (in Russian)
[6] Chudaeva V A and Chudaev O V 2011 Features of accumulation and fractionation of rare earth elements in the surface waters of the Far East in the conditions of natural and anthropogenic anomalies *Geochemistry* 3 1–27 (in Russian)
[7] Sholkovitz E R 1992 Chemical evolution of rare earth elements: fractionation between colloidal and solution phases of filtered river water *Earth Planet. Sci.Lett.* **114** 77-84
[8] Sholkovitz E R 1995 The aquatic chemistry of rare earth elements in rivers and estuaries *Aquat.*

Geochem. **1** 1-34

[9] Goldstein S J and Jacobsen S B 1998 Rare earth elements in river waters *Earth Planet. Sci.Lett.* **89** 35-47

[10] Kulaksız S and Bau M 2011 Rare earth elements in the Rhine River, Germany: First case of anthropogenic lanthanum as a dissolved microcontaminant in the hydrosphere *Environ. Int.* **37** 973-979

Ground water regimes containing country rock minerals in Southern Kuzbass (case study: Narysk-Ostashkin)

E Domrocheva[1] and O Lepokurova[2]

[1]A.A. Trofimuk Institute of Petroleum Geology and Geophysics Siberian Branch RAS, Tomsk, Russia
[2] National Research Tomsk Polytechnic University, Tomsk, Russia

E-mail: [2]LepokurovaOY@ipgg.sbras.ru

Abstract. The paper describes the calculation results revealing groundwater in equilibrium to carbonates and aluminosilicate minerals of country rocks in Narysk-Ostashkinsk area. It was proved that groundwater is in nonequilibrium to primary (endogenous) minerals in which they dissolve, however are in equilibrium to clays and carbonates which precipitate in the groundwater. The groundwater composition varies.

1. Introduction
There are numerous publications devoted to geochemistry of Kuzbass underground water [1–5]. This issue has been viewed on the basis of the concept proposed by S.L. Shvartsev: water-rock interaction [6]. According to this concept identifying the equilibrium of the groundwater to major minerals is only the beginning in understanding the formation conditions and chemical composition of groundwater. In 2012 and 2014 the investigation embraced Narysk-Ostashkinsk area, Erunakov region, Kemerov Oblast. The research target was to calculate and analyze the thermodynamic equilibrium in the water-rock system based on the data in temperature, pH, chemical composition and country rock composition.

2. Factual information and methods
From 2009 to 2012 41 well were drilled at a depth of 100-1200m within the territory of Narysk-Ostashkinsk area and one well in Chaltok field. All in all, 122 samples from 42 wells were examined. The macro / micro- component compositions of groundwater were analyzed in the Task-Solving Research Hydrogeochemical Laboratory, TPU. The software program HydroGeo, developed by M Bukat [7] was used to calculate the existing water-rock equilibrium, which, in its turn, is based on the equilibrium constant method. Based on hydrogeochemical data saturation value of groundwater relative to the common water-bearing minerals were determined.

3. Groundwater chemical composition
According to previous and new data in hydrogeology and hydrochemistry (2014) the following fact was established within the investigated area: straight vertical hydrodynamic zonality. In accordance to intensive water exchange, two zones were identified- active and moderate water exchange. Groundwater in both zones differs in chemical and gaseous composition and is governed by the straight vertical hydrodynamic zonality [2].

Groundwater in active water exchange zone can be found in the upper cross-section composed of loose Mesozoic-Cenozoic sediments (predominately Quaternary) and in the zone of intensively fractured Jurassic sediments. This fresh water (mineralization from от 0.5 to 1.0 gr/liter), which expands to a depth of approximately 150m., although this is an arbitrary level as it can expand only to 30-50m in some areas and up to 400m, in other areas (Table 1). This is governed by the intensive water exchange and permeability of the sediments. According to its chemical composition the groundwater is calcium hydrocarbonate and calcium-sodium hydrocarbonate, whereas the Na^+ fraction in the water composition increases to the bottom zones of active water exchange. Cl^- and SO_4^{2-} concentrations are not high, not more than 10 mgr/liter in average. In general, groundwater is neutral and weakly alkaline, pH is 7.0-8.6. Gaseous composition includes aerial gases, i.e. N_2 (71–80 %), O_2 (17–22 %) and CO_2 (a fraction of %). CH_4 can also be found in insignificant amounts.

Table 1. Chemical composition (allowable concentration and average value) of underground water in Narysk-Ostashkinsk area; mgr/liter (2012-2014).

Parameters	Active water exchange zone (24 samples)	Moderate water exchange zone	
		Intensively fractured rocks (73)	Fading fracturing (25)
pH	7.0–8.6 (7.7)	7.5–9.8 (8.1)	7.2–9.9 (8.3)
Ca^{2+}	32–120 (77)	3–88 (32)	0.2–61.0 (20.4)
Mg^{2+}	6.1–35.3 (16.7)	0.4–48.6 (10.0)	4.9–79.3 (22.3)
Na^+	8.5–130.0 (47.0)	44–3800 (438)	1996–7230 (3481)
Fe^{2+}	0.05–24.5 (3.9)	0.01–192.0 (10.4)	0.1–29.5 (9.2)
HCO_3^-	321–745 (433)	258–3111 (824)	1399–15494 (7375)
CO_3^{2-}	0.2–6.2 (1.6)	0.3–742.8 (31.1)	18.5–1285.7 (357.8)
Cl^-	0.7–24.0 (5.1)	1.5–182.0 (22.9)	30,0–5494,0 (919)
SO_4^{2-}	0.8–16.9 (5.6)	0.2–185.6 (18.4)	0.21–75.7 (20.3)
SiO_2	7.5–25.3 (14.6)	1.8–42.5 (22.0)	5.9–44.6 (24.6)
Total mineralization	497–1029 (615)	436–4355 (1271)	3999 25139 (12159)

Lower is the moderate water exchange zone which is confined to Mid-Upper Permian sand sediments, partially, within Triassic effusive-sedimentary sediments. This zone is characteristic of not only rather high mineralization (from 1, rarely 0.5–1.2 and 10–25 gr/liter) but also the water composition. The most common water is sodium (HCO_3-Na), typically found throughout Kuzbass with pH from 7.5 to 9.9. Due to complex hydrogeochemical conditions this zone is subdivided into two subzones: upper and lower.

Upper subzone embraces fresh and/or weakly mineralized (0.5–5 gr/liter), calcium-sodium hydrocarbonate and/or sodium-calcium hydrocarbonate water, pH 7.5–9.8. It should be noted Cl^- concentration increases up to 182 mgr/liter, and sometimes, SO_4^{2-}concentration- up to 186 mgr/liter. At the same time, chloride concentration increases to depth, while sulphate-ion decreases. The source of the latter could be sulphates (pyrite), the oxidation of which furthers the emerging sulphate ion. The pattern is reversed in the case of Ca^{2+} concentration – it decreases up to 3mgr/liter. The waster-dissolved gas includes methane (85–95%).

In the lower subzone at a depth of more than 400–500 m saline and weakly saline sodium groundwater (up to 25 gr/liter) can be found. Due to the continuous increase of HCO^-_3, Na^+, Cl^- concentration mineralization also increases. High hydrocarbonate ion concentrations (to 15 gr/liter!) is associated with the coal-bearing Permian sediments, which, in its turn, are sources of carbon. At the same time, SO_4^{2-} concentration is insignificant. In this case the water is hydrocarbonate, rarely chloride-hydrocarbonate, sodium chloride- hydrocarbonate. Methane continues to accumlate in the water dissolved gas (up to 99%).

Analyzed samples (Narysk-Ostashkinsk area, 2014) revealed that groundwater is related to the lower moderate water exchange zone, highly mineralized and increased chloride concentration from 1.0 to 2.3 gr/liter was found in some samples.

4. Equilibrium calculation results in water-rock system

Factual data of groundwater chemical composition was imported into the HydroGeo program. The most common minerals in country sediments, basically aluminosilicates (quartz, feldspar, clays, etc.), separate carbonates (calcite, magnesium, dolomite, siderite) and some salts were also selected. Calculation results are depicted in Table 2 and Figures 1 and 2. It was revealed that all water types, if highly mineralized water in the lower water exchange zone are not in equilibrium with source minerals of water-bearing rocks: feldspar, hornblende, amphibole, epidote, pyroxene and others.

Table 2. Value index of saturated groundwater (leading minerals) in (1) active water exchange zone, (2) moderate upper exchange zone and (3) lower zone.

Mineral	1	2	3	Mineral	1	2	3
Ionic strength	0.008-0.016	0.012-0.054	0.062-0,.24		Aluminosilicates		
	Oxides			Albite	-3.3-(-0.3)	-1.7-1.8	0.1-2.2
Hydargillite	1,3-1,8	1,3-2,1	1,4-2,0	Biotite	-23-(-14)	-21-(-13)	-19-(-12)
	Carbonates			Anorthite	-5.2-(-2.4)	-5.7-(-1.3)	-7.7-(-0.3)
Calcite	-0.1-**1.3**	-0.4-**1.8**	**0.7-2.7**	K-feldspar	-7.1-(-6.4)	-7.7-(-5.3)	-6.1-(-4.7)
Siderite	-1.2-**2.9**	-0.6-**5**	**0.3-4.1**	Microcline	-5.3-(-0.8)	-4.4-**1.9**	-2.3-**2.1**
Dolomite	-0,5-**2,4**	-0,8-**4,1**	**1.4-5.6**	Moscuvite	-6.4-(-2.5)	-9.2-(-1.3)	-8.3-(-0.8)
Magnesium	-1.7-(-0.3)	-3-(-2.5)	-0,5-**1,6**	Pyroxene	-12-(-9)	-10-(-7)	-9-(-6)
	Sodium minerals			Chlorite	-2.0-**0.9**	-2.4-**2.5**	-4.8-**3.1**
Tron	-120	-120-(-100)	-110-(-95)	Ca-Mt	-0.1-**4.9**	**0.1-5.8**	**1.9-9.6**
Natron (soda)	-130-(-120)	-115- (-120)	-110- (-115)	Na-Mt	-1.2-**2.3**	**0.8-5.3**	**2.7-6.1**
	Sulphates			Mg-Mt	**0.1-5.5**	**0.4-8.9**	**2.3-10.0**
Gypsum	**1.2–2.4**	-3.4-**0.6**	-5.1-(-0.7)	K-Mt	-2.8-**0.6**	-4.1-**3.1**	-4,7-**2,3**
Anhydrite	-7.3-(-6.1)	-10-(-7.9)	-11.8-(-7.5)	Illite	-0.6-**1.3**	-1.7-**3.0**	**-2,3-3,4**
	Chlorides			Kaolinite	**0.1-1.0**	**0.5-1.4**	**0.6-1.2**
	-20-(-19)	-17-(-15)	-13-(-8)	Daphite	-2.4-**2.6**	-2.0-**4.5**	-2.7-**4.5**

Groundwater in active water exchange zone is undersaturated relevant to initial aluminosilicate minerals, due to the low mineralization value and pH. Under such conditions these minerals are unstable and actively dissolve, as a result of which, Ca, Mg, Na, Fe, K, Si, Al and other elements migrate into the solution. Some of the elements migrate out (as the water is oversaturated by these minerals) as secondary formations: oxides and hydrooxides Fe, Mn and Al, clays (kaolonite and different montmorillonites, except potassium), carbonates (calcite, dolomite, often siderite). They are in bold in Table 2.

Water in moderate water exchange zone is more alkaline and mineralized, however, it is undersaturated in endogenous minerals. At the same the groundwater in the upper water exchange zone is highly saturated in secondary aluimnosilicates. (illite appears, montmorillonites, chlorites and even albite and microcline) and carbonates (siderite and magnesium).

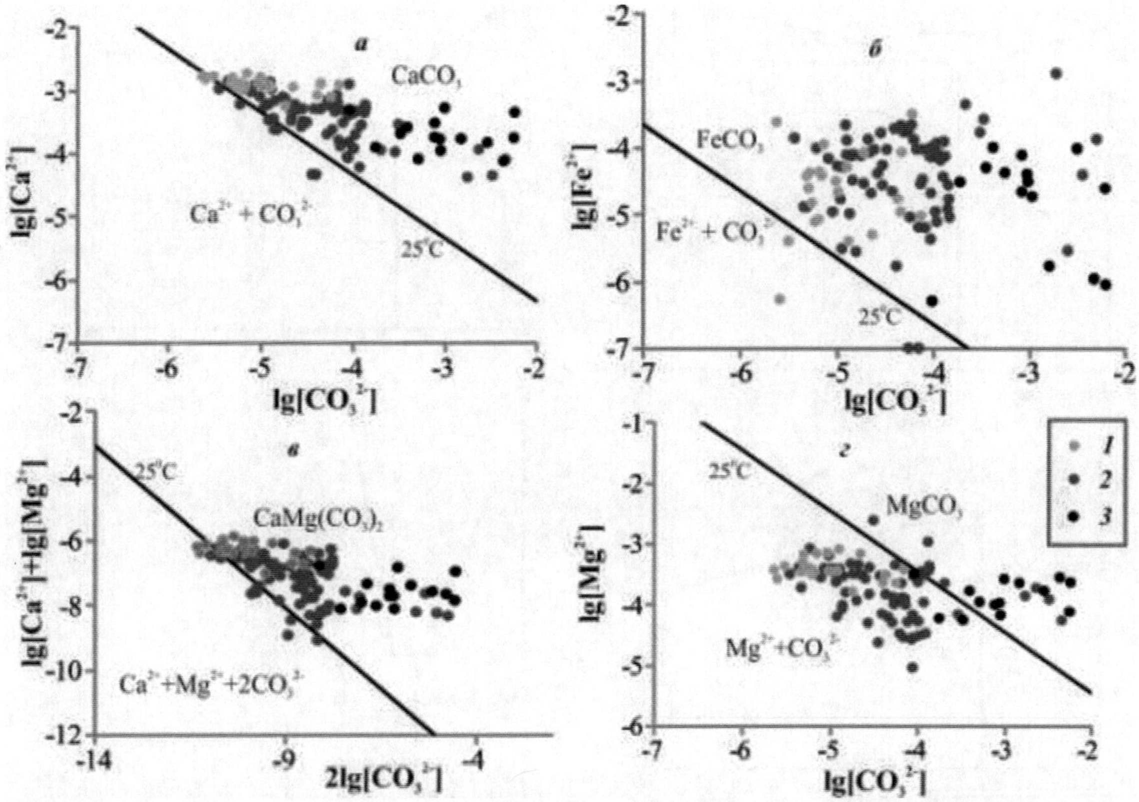

Figure 1. Groundwater equilibrium with (a) calcite, (b) siderite, (c) dolomite, (d) magnesium at standard temperature : 1- groundwater form active water exchange zone; 2- groundwater from moderate water exchange zone (upper) and (3) groundwater from moderate water exchange zone (lower).

According to Shvartsev definition this equilibrium is the so-called "equilibrium-non equilibrium": water is always in non equilibrium with primary aluminosilicates, but under specific geochemical conditions could be in equilibrium with secondary ones. In this case, water composition is determined as the difference between dissolved and precipitated minerals. Thus, at depth water composition alters from HCO_3-Ca to HCO_3-Na. Carbonates are bounded by Ca, Mg, Fe, C, clay minerals- Al, Si, Ca, Mg, Fe, and, partially by Na and K, oxides and hydrooxides- Fe. Na is less bounded to secondary sediments as water remains rather in non equilibrium to sodium minerals (table 2). In this case, Na continues accumulating in water and its composition becomes more sodium saturated. In the upper moderate water exchange zone SO_4^2 is observed in the water composition which is associated with pyrite oxidation. In the lower zone- there is a significant amount of Cl^-. As there are no salt minerals in this zone, aluminosilicate minerals (in impurities) could be the source of chloride and sodium. In time it accumulates in the solution under conditions of rather moderate water exchange. As the solution is undersaturated in Cl (table 2-halite) such conditions are favorable for its accumulation.

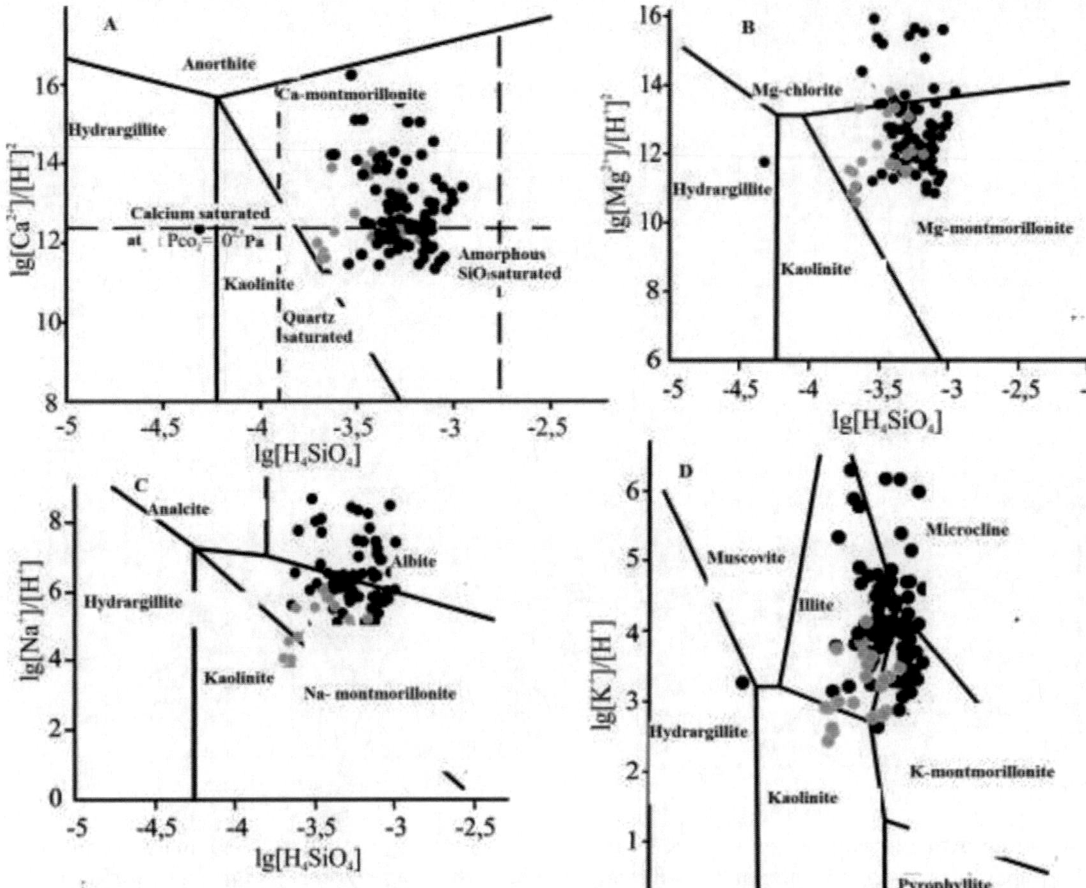

Figure 2. Equilibrium diagrams: (a) calcium, (b) sodium, (c) magnesium, (d) iron-containing, (e) calcium-sodium at lg[H_4SiO_4] = –3.25 and (f) groundwater minerals.

5.Conclusion

According to the recent data it has been revealed that in Narysk-Ostashkinsk area there is straight vertical groundwater zonality: fresh HCO_3-Ca water of active water exchange zone changes to saline (M up to 25 gr/liter) of moderate water exchange zone. Thermodynamic calculations revealed that all groundwater types are in non equilibrium to common aluminosilicate minerals which are the sources Na, Ca, Mg, K, Si, Al and even positive Cl is not an exclusion. Pyrite in the upper level of moderate water exchange zone imports sulphates into the groundwater. A portion of these elements move out of the solution into secondary sediments, where groundwater is in equilibrium with clays (montmorillonite, hydromica, chlorite), carbonates (calcite, dolomite, siderite) and even gypsum. Na is less bounded in the secondary sediments and continues accumulating in the groundwater. Water composition changes from sodium to chloride in depth, as HCO_3^- migrates with carbonates and Cl concentrates in the solution.

Acknowledgments

The authors thank professor S.L. Shvartsev for his expert guidance and sound thinking in writing this article and Dmitri Sizikov, Laboratory Head in OAO "Gas promgas" for his valuable assistance.

References

[1] Domorocheva E 2005 *Gidrogeohimicheskie osobennosti ugol'nyh rajonov Kuzbassa* Dissertation abstract 22

[2] Shvartsev S L, Khryukin V T, Domrocheva E V, Kuzevanov K I, Rasskazov N M, Popova T S, Lepokurova O E and Shvachko E V 2006 Hydrogeology of the Erunakovo region of the Kuznetsk basin in the context of coal methane formation and mining *Russian Geology and Geophysics* **47** (7) 881–891

[3] Shvartsev S L 2004 Hydrogeochemical conditions of formation of dawsonite mineralization in the Berezovoyarka area, Kuznetsk Basin *Geochem. Int.* **42** (10) 938–949

[4] Shvartsev S L 2004 Hydrogeochemistry of processes of dawsonite formation *Doklady Akademii Nauk* **394** (6) 818–821

[5] Kopylova Y G, Lepokurova O E, Tokarenko O G and Shvartsev S L 2011 Chemical composition and genesis of the carbonic-acid mineral waters of the Tersinskoe deposit (Kuzbass) *Doklady Earth Sciences* **436** (2) 284–289

[6] Shvartsev S L 2008 Geochemistry of fresh groundwater in the main landscape zones of the earth *Geochem. Int.* **46** (13) 1285–398

[7] Bukat M B 2002 Razrabotka programmnogo obespechenija dlja reshenija gidrogeologicheskih zadach *Izvestija Tomskogo politehnicheskogo universitet* **305** (6) 348–365

Chemical composition of natural waters of contaminated area: The case for the Imandra Lake catchment (the Kola Peninsula)

Z A Evtyugina[1,2], N V Guseva[3], J G Kopylova[4], D A Vorobeva[5]

[1] Apatity branch of «Murmansk State Technical University», Russia
[2] Kola branch of Geophysical Survey of RAS, Apatity, Russia
[3, 4 ,5] National Research Tomsk Polytechnic University, Tomsk, Russia

E-mail: [1]zina_evt@mail.ru, [3] guseva24@yandex.ru

Abstract. The study of the current chemical composition of natural waters in the eastern and western parts of the Imandra Lake catchment was performed using ion chromatography, potentiometry and inductively coupled plasma mass spectrometry. It was found that the content of trace elements in the surface water is considerably higher than that in the groundwater. The nickel and copper concentrations exceed the background levels over 19 and 2 times respectively in groundwater, and 175 and 61 times in the surface waters. These data show that the Severonikel influences negatively air and surface water.

1. Introduction

The Kola Peninsula is one of the regions of Russia with the most unfavorable ecological situation. A large number of mining and metallurgical enterprises is located here. The enterprises of the Kola Mining and Metallurgical Company have polluted the atmosphere with sulfur compounds, copper, nickel for many years. This led to degradation of forest landscapes until complete destruction of the forests and the formation of anthropogenic wasteland [4].

The impact of anthropogenic factors leads to a change of all landscape elements, including deterioration of natural water quality. The water bodies in Arctic zone reveal a low ability for self-purification, but they are often used as drinking water sources for the local population. Therefore, the chemical composition of subarctic lakes is studied clearly in the following aspects: anthropogenic impact, mechanisms of chemical element accumulation in water and bottom sediments, transformation of aquatic ecosystems [11, 12].

In 1992-1996 an ecogeochemical project was implemented in the territory of Russia (north of the Kola Peninsula) and neighboring areas of Finland and Norway. The project objectives were to study catchments, the regional geochemical mapping, to research the sources and migration of chemical elements [6, 7]. The parts of the project such as the description of the content and source of chemical elements (Cu, Ni, isotopes of S etc.) were published in the article [7]. In Finland, the composition of the aerosols, the solubility of elements of anthropogenic origin and their sources were studied [8].

The Kola Peninsula is not the only region in the Arctic territory of Russia where the anthropogenic influence is so significant. The Norilsk industrial complex in the Taimyr Peninsula is also mining and processing of copper-nickel ore. In this region the concentration of heavy metals (Cu, Ni, Zn, Hg, Cd and Hg) and sulfate ion in the ecosystems was studied taking into account the orography and the wind rose [10].

At the same time, soil, as a landscape element, is a biogeochemical barrier for chemical elements input to ecosystems from the polluted atmosphere. Soil degradation leads to decrease in their sorption capacity and, consequently, to groundwater contamination with heavy metals. According to the Central Kola Expedition study in 2008-2010, in the territory with the local impact of «Severonikel» plant where soils are undestroyed, the groundwater is protected from aerotechnogenic input of copper and nickel [2].

Nickel, copper and SO_2 emissions from the «Severonikel» plant decreased from 287 thousand tons in 1990 to 45.8 thousand in 1999 and to 36.6 thousand tons in 2013 [5]. This led to the reduction of anthropogenic impact on the environment, including natural water.

The aim of this research is to study the current chemical composition of natural waters in the lake Imandra catchment, exposed to significant anthropogenic impact.

2. Methods

The studied area is located in the south-western part of the Kola Peninsula in the western and eastern part of the Lake Imandra catchment (Figure 1). Aluminosilicate rocks are widely spread in the studied area. In the western part of the lake catchment geological structure is represented by acidic and basic metavolcanics of the Upper Archean and peridotites, pyroxene, gabbronorites, basalt porphyrites, diorites of the Early Proterozoic age. Vendian deposits composed of arcosic and polymictic sandstones, siltstones, argillites are locally spread. In the eastern part of the Lake Imandra catchment near Kirovsk city the Devonian-age rocks such as alkaline ultramafic, nepheline syenite, carbonatite predominate.

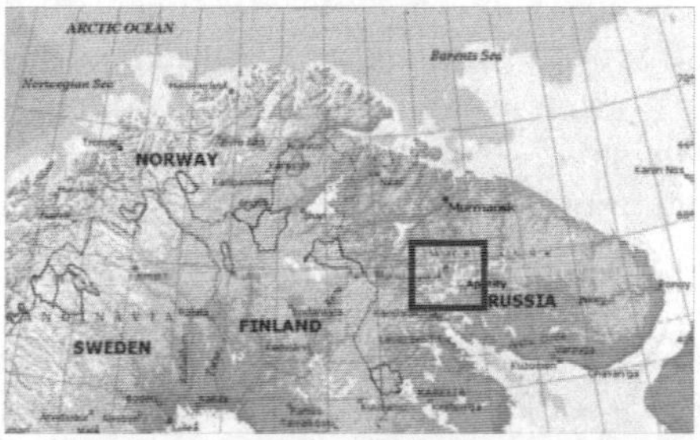

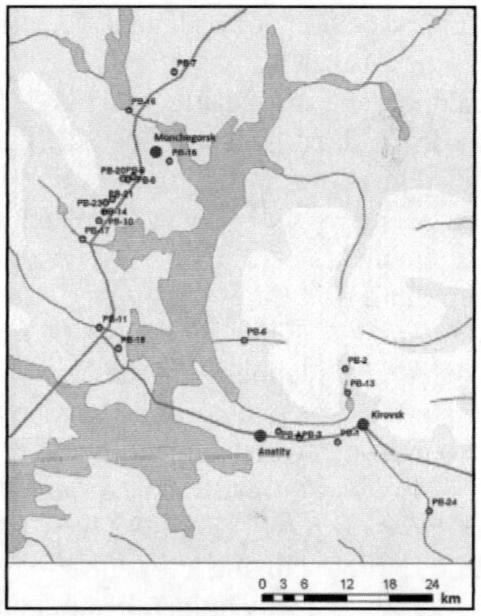

Figure 1. Map of the studied area and sampling points.

In July 2014, 24 samples of surface water and groundwater were taken (Figure 1). In the east of the lake Imandra catchment 3 springs: Prihibinskij (PB-1), Molodezhnyj (PB-3), Poddorozhnyj (PB-4), 3 rivers: Malaya Belaya (PB-6), Vudyavrjok (PB-13), Ajkuajvenjok (PB-14) and Lake Malay Vudyavr (PB-2) were sampled. In the western part of the lake Imandra catchment 5 springs : Bolotnyj (PB-7), Sportivnyj (PB-15), Kislaya Guba (PB-18), Dorozhnyj (PB-9(19)) and Gornyj (RV-10(22); 6 surface water points: the River Kurka (PB-11), the River Moncha (PB-16), the river that flows into the lake Sopchyavr (PB-21), a stream 7 kilometers from the plant (PB-14 (23)), drainage lake of the River Vita – Devichya Lambina (PB-17) and the lake near spring Dorozhnyj (PB-8(20)) were sampled.

The samples were taken in 0.5 liter plastic bottles for the major element analysis and two 50 ml vials for the trace element analysis. The temperature, pH, Eh were measured in situ. The water chemical composition was studied using ion chromatography, potentiometry and inductively coupled plasma mass spectrometry in the laboratory of the Research and Development Center «Voda», Institute of Natural Resources, TPU.

3. Results and discussion

Taking into account the peculiarities of the water chemical composition, geological structure and the degree of anthropogenic impacts, two principally different areas were identified. The first area is the Khibiny massif area (the eastern part of the lake Imandra catchment) and the second one is area exposed to anthropogenic impact of «Severonikel» plant (the western part of the lake Imandra catchment).

The groundwater of the Khibiny massif area is slightly alkaline HCO_3 and SO_4-HCO_3 Ca and Na-Ca composition. The TDS value is 111-201 mg/l. The high TDS value and SO_4-HCO_3 Na-Ca composition is in alkaline groundwater of the spring PB-4. In this point the nitrate ion concentration is 33 mg/l. Features of anthropogenic contamination are detected in the spring PB-1, where the potassium and nitrate ion concentrations are high. The high concentration of chlorine ion and the TDS value of 115 mg/l are observed in the spring PB-3.

In this groundwater the calcium concentration is higher than the concentration of sodium and silicon, which significantly exceeds the concentration of potassium and magnesium (Table 1). This is the main cationic composition peculiarity of the groundwater of the Khibiny massif area. In the studied groundwater the concentration of such trace elements as lithium, titanium, nickel, copper, strontium, yttrium, molybdenum, antimony, barium, lanthanum, cerium, rhenium and bismuth is high. This association of chemical elements reflects the geochemical peculiarities of alkaline rocks.

In surface waters of this area the sodium and potassium content exceeds the concentration of calcium and magnesium at TDS 27 mg/l and the silicon concentration is 3.5 mg/l. It is also explained by the influence of alkaline rocks rich in sodium and potassium. As compared with the background concentrations in the groundwater, in the surface water of the Khibiny massif twice or more times as many excess is observed for magnesium (7 times), for aluminum (2.4 times), for calcium (9.9 times), for manganese (3.3 times), for zinc (3 times), for gallium (19 times), for arsenic (2.8 times), for rubidium (4.2 times), for zirconium (4.6 times), for niobium (2.8 times), for molybdenum (twice), for cesium (3.2 times), for cerium (4.9 times), for hafnium (2.8 times), for tungsten (13 times), for iridium (twice), for platinum (2 times), and for lead (twice).

In the area exposed to anthropogenic impact of «Severonikel» plant, the chemical composition of groundwater varies depending on the degradation degree of soil and vegetation layer. In the area with forest cover, groundwater (springs PB-15 and PB-18) is neutral HCO_3 Mg-Ca and contains sulfate ion of 7.6-11.0 mg/l, TDS 67-87 mg/l. In the spring PB-7 (TDS 63 mg/l) pH decreases with increase of free carbon dioxide concentrations to 22 mg/l. The high concentration of free carbon dioxide is conditioned by the mineralization of organic matter.

In the area with disturbed soil and vegetation layer, groundwater (springsPB-9(19), PB-10 (22)) is slightly acidic. The TDS value is 30-39 mg/l and free carbon dioxide concentration is 16-25 mg/l. The recharge of this groundwater occurs by rainfall, which infiltrate through destroyed vegetation and soil layer in the technogenic wasteland [1]. This peculiarity of groundwater formation influences the trace element composition of water. In comparison with the groundwater of the Khibiny massif, the excess of trace elements in these groundwater is observed for boron (5 times), for manganese (7 times), for iron (2.5 times), for gallium (4.3 times), for bromine (5.9 times), for rubidium (2.6 times), for rhodium (4 times), for palladium (3 times), for hafnium (4.3 times), for tantalum (5.6 times), for tungsten (3.9 times), for iridium (4.1 times), for aurum (2.5 times), for uranium (10 times). The nickel and copper concentrations are 0.0065 mg/l and 0.001 mg/l, respectively, in the spring PB-9(19), 0.003 mg/l and 0.00013 mg/l, respectively, in the spring PB-10(22).

Table 1. The chemical composition of waters in the Lake Imandra catchment area, mg/l

Element	Background water of the Khibiny massif (eastern part of the lake Imandra catchment)		Water exposed to anthropogenic impact of «Severonikel» plant (western part of the lake Imandra catchment)		
	Surface water (N=4)	Groundwater (N=3)	Groundwater in an undisturbed landscape (N=3)	Groundwater in a disturbed landscape (N=4)	Surface water (N=5)
pH	7.09	7.46	6.67	5.84	6.4
HCO_3	17.4	77	39.4	12.2	12
SO_4	1.7	19	10.7	10.3	5.36
Cl	0.63	5.2	2.46	1.77	1.24

IOP Conf. Series: Earth and Environmental Science **33** (2016) 012004 doi:10.1088/1755-1315/33/1/012004

NO_3	0.13	12	0.65	1.42	0.1
Total hardness	0.09	1.5	0.74	0.36	0.3
Ca	1.33	27	9.8	5.31	4.69
Mg	0.23	1.7	3.1	1.17	0.86
Na	3.68	9.5	3.07	1.77	1.4
K	1.09	2.6	1.43	0.41	0.33
TDS	26.1	142	70	32.9	25.5
Li	0.000016	0.00011	0.001	0.00004	0.00012
Be	0.000015	<0.000003	<0.000003	0.0000342	<0.000003
B	0.00012	0.0062	0.0013	0.002	0.0021
Al	0.039	0.0085	0.016	0.01	0.03
Si	2.68	5.46	7.30	4.00	2.81
P	0.0081	0.0095	0.014	0.006	0.014
Sc	0.00011	0.00027	0.00026	0.00014	0.0001
Ti	0.00023	0.00019	0.00074	0.00012	0.00067
V	0.00012	0.00041	0.00061	0.0002	0.001
Cr	0.00013	0.00055	0.00054	0.0003	0.00028
Mn	0.00047	0.00102	0.00014	0.0004	0.0025
Fe	0.007	0.019	0.0077	0.0075	0.0037
Co	0.0000051	0.000024	0.000015	0.00004	0.0004
Ni	0.00012	0.0000198	0.00051	**0.0049**	**0.046**
Cu	0.00029	0.00013	0.00041	**0.00056**	**0.017**
Zn	0.00085	0.00038	0.00028	**0.0004**	**0.002**
Ga	0.00016	0.000035	0.0000082	0.000004	0.00001
Ge	0.0000053	0.0000037	0.0000039	0.000004	0.0000086
As	0.000082	0.00014	0.000029	0.00004	**0.00042**
Se	0.00032	0.0004	0.00034	0.00026	0.0004
Br	0.0058	0.08	0.014	0.007	0.011
Rb	0.0019	0.0012	0.00045	0.0009	0.0006
Sr	0.064	0.013	0.032	0.019	0.016
Y	0.000022	0.000027	0.000059	0.00004	0.000019
Zr	0.000061	0.000011	0.000013	0.0000035	0.000009
Nb	0.000011	0.0000025	0.0000039	0.0000021	0.0000025
Mo	0.0014	0.00028	0.00063	0.00003	0.000089
Ru	0.000012	0.000021	0.000015	0.0000008	0.0000056
Rh	0.000001	0.0000032	0.0000008	0.0000007	0.0000012
Pd	0.0000024	0.0000075	0.0000025	0.0000019	0.0000030
Ag	0.0000025	0.0000015	0.000002	0.000002	**0.000029**
Cd	0.0000029	0.000003	0.0000017	0.0000016	**0.000023**
Sn	0.0000041	0.000002	0.0000068	0.0000039	0.0000062
Sb	0.000015	0.000011	0.000029	0.00001	0.000034
Te	0.0000033	0.0000183	0.0000120	<0.000005	0.0000390
Cs	0.0000104	0.0000018	0.0000032	0.0000016	0.0000034
Ba	0.0112	0.0129	0.0273	0.013	0.0139
La	0.00008	0.000022	0.00008	0.00004	0.00005
Ce	0.000082	0.0000064	0.00002	0.000009	0.000052
Pr	0.000014	0.0000048	0.00019	0.000014	0.00001
Nd	0.000039	0.000018	0.000078	0.00006	0.000039
Sm	0.0000054	0.0000027	0.000014	0.000009	0.0000062
Eu	0.0000023	0.0000024	0.0000054	0.0000034	0.0000027
Gd	0.0000049	0.0000029	0.000011	0.000008	0.0000049
Tb	0.0000005	0.0000004	0.0000017	0.0000011	0.0000006
Dy	0.0000033	0.0000022	0.0000055	0.000005	0.0000021
Ho	0.0000008	0.0000005	0.0000016	0.0000010	0.0000007
Er	0.0000011	0.0000011	0.0000039	0.0000012	0.000001
Tm	0.0000003	0.0000003	0.0000008	0.0000006	0.0000004
Yb	0.0000008	0.0000012	0.0000015	0.0000022	0.0000005
Lu	0.0000002	0.0000005	0.0000008	0.0000006	0.0000003
Hf	0.0000017	0.0000026	0.0000006	0.0000004	0.0000012
Ta	0.000001	0.0000062	0.0000011	0.0000006	0.0000011
W	0.0000323	0.0000095	0.0000024	0.0000022	0.000031

Re	0.0000004	0.0000036	0.000011	0.0000057	0.0000059
Ir	0.000033	0.000067	0.000016	0.0000010	0.000014
Pt	0.0000004	0.0000003	0.0000002	0.0000004	0.0000012
Au	0.0000055	0.000013	0.0000054	0.000004	0.0000039
Hg	0.0000019	<0.00005	0.0000049	<0.00005	0.0000068
Tl	0.0000007	0.0000014	0.0000021	0.0000026	0.0000035
Pb	0.000031	0.0000091	0.000015	0.0000190	**0.00014**
Bi	0.0000008	0.0000004	0.0000045	0.000001	0.0000033
Th	0.0000025	0.0000013	0.0000019	0.0000016	0.0000032
U	0.000019	0.00013	0.000013	0.0000024	0.0000069

In the area exposed to anthropogenic impact, the river water is neutral, the TDS value is 19-28 mg/l. The pH value decreases with TDS increase up to 40 mg/l. Bicarbonate is a predominant anion in the studied area, but the content of sulfate ion is 8-27 %-eq and the content of chloride ions varies from 5 to 10 %-eq. The concentration of free carbon dioxide increases in the Lake Dorozhnoe (PB-8(20)) and the River Kurka (PB-11). The pH value increases due to organic matter mineralization. The cationic composition of the studied surface water (point PB-17, PB-11, PB-16, PB-14 (23)) is Na-Mg-Ca, the content of silicon is 2.4-2.7 mg/l. The silicon content in the stream PB-14(23) is 4.4 mg/l.

As the results of the previous studies [3], high nickel and copper concentrations were found in the stream PB-14(23). This microstream drains the destroyed forest landscape (technogenic woodlands). Over the past period, technogenic woodlands changed radically. Many conifer trees had been killed, but the ratio of deciduous trees and shrubs increased. The ground cover transformed - the area uncovered by vegetation increased and the soil was destroyed until the mineral horizon.

Reduction of anthropogenic impact on the environment has shown a positive trend in changing abundances of chemical elements in natural waters due to an increase of natural carbon compounds and cations in the water. In July 2014 in the stream PB-14(23), the content of sulfate ion (the main air pollutant sulfur compounds) was less than the minimum concentration in comparison with the sampling period 1987–1990. The hydrocarbonate ion concentration in the water increased by 1,5 times, the concentration of calcium and magnesium is close to the maximum value of the previous period. The content of copper and nickel in the stream PB-14(23) has not changed and corresponds to the previous study [3].

The highest nickel content 0.0269 and 0.0649 mg/l was found in the Dorozhnoe Lake (PB-8(20)). The copper content in this point is 0.0152 and 0.0211mg/l. The nickel concentration (0.0045 mg/l) in the waters of the river (PB-21), the River Kurka (PB-11) and the River Vita (PB-17) are nearly the same as that in the springs Gornyj and Dorozhnyj (0.0049 mg / l nickel), but higher than the nickel content in the water of the stream PB-14(23), and in the Moncha River PB-16.

Table 1 shows that, in addition to nickel and copper, in surface waters the high concentration of trace elements is observed. As compared with background concentrations of groundwater, a wide range of chemicals is observed in the surface waters of the Severonikel area. There is an excess for Mn (40 times), for Fe (6.7 times), for Co in 27 times, for Ni (89 times), for Cu (41 times), for Zn (7 times), for Ga (2.4 times), for Ge (2.2 times), for As (15 times), for Ag (14 times), for Cd (13 times), for Te (3 times), for Ce (5 times), for Yb (twice), for Hf (twice), for W (13 times), for U (twice).

4. Conclusions

The chemical composition of waters in the lake Imandra catchment is various. Taking into account the peculiarities of the water chemical composition, geological structure, and the degree of anthropogenic impacts, the two principally different areas were identified. The first area is the Khibiny massif area (the eastern part of the lake Imandra catchment) and the second one is area exposed to anthropogenic impact of «Severonikel» plant (the western part of the lake Imandra catchment). The chemical composition in both areas differs in pH and TDS values, the concentration of trace elements.

Moreover, the groundwater chemical composition differs markedly from the surface water chemical composition. Groundwater is characterized by the high concentrations of sulfate ion, calcium, magnesium and silicon. Nevertheless, the concentration of sodium, potassium, and chloride ion is the same in surface and groundwater. However, the trace element concentration in surface water is significantly higher than that in groundwater. In the surface waters, excess over background concentrations in groundwater reaches for Ni 175 times, for Cu 61 times, for Co 20 times, for Ag 15 times, for Pb 14 times, for Cd 8 times, for Pt 6 times, for As 5 times, for W 5 times, for Te 4 times, for Al 3 times, for V, Ge, Sn, Sb, Tl, Bi, Th 2 times.

These data allow us to recognize the negative impact of «Severonikel» plant firstly on the atmosphere and surface water. The degree of groundwater contamination depends on soil condition. If the soil is destroyed, the groundwater is polluted. If there is ground cover and soil, the groundwater is not significantly affected by anthropogenic activity. Thus, the main negative impact is referred to the top of biosphere that implies the development of measures for protecting groundwater from pollution.

Acknowledgements
This work was supported under the state assignment of the Ministry of Education and Science of Russia “Nauka”.

References
[1] Ananev V N 2010 *Rodniki Murmanskoy oblasti spravochnik (*Murmansk) p 88 [in Russian]
[2] Ananev V N, Karpova R V and Sycheva L B *Poiski podzemnyh vod dlya obespecheniya hozyaystvenno-pitevogo vodosnabzheniya Monchegorska Murmanskoy oblasti* Otchet za 2008-2010 gody. Gos. registracionnyy № 47-08-16b Monchegorsk, fondy CKE 2011 [in Russian]
[3] Evtyugina Z A and Nikonov V V 1990 Osobennosti migracii medi i nikelya po profilyu podzolistyh pochv v usloviyah aehrotekhnogennogo zagryazneniya // Issledovanie pochv na Evropeyskom Severe: *Cbornik materialov nauchnoy sessii posvyashchennoy 130 letiyu so dnya rozhdeniya N.M. Sibirceva (13-15 fevralya 1990) (Arhangelsk)* 107–106 [in Russian]
[4] Evtyugina Z A and Asming V EH 2013 Osobennosti formirovaniya sostava infiltracionnyh vod v usloviyah aehrotekhnogennogo zagryazneniya // *Vestnik MGTU: trudy Murmanskogo gosudarstvennogo tekhnicheskogo universiteta* **16** 73–80 [in Russian]
[5] Norilsk Nickel URL:http://www.kolagmk.ru/rus/ecology/
[6] de Caritat P, Reimann C, Ayras M and et al. 1996 Stream Water Geochemistry from Selected Catchments on the Kola Peninsula (NW Russia) and in Neighbouring Areas of Finland and Norway: 1 Element Levels and Sources *Aquati. Geochem.* **2** 149–168
[7] de Caritat P, Krouse H R and Hutcheon I 1997 Sulphur isotope composition of streamwater, moss and humus from eight arctic catchments in the Kola Peninsula region (NW Russia, N Finland, NE Norway) *Water, Air, and Soil Pollution* **94** 191–208
[8] Laing J R, Hopke P K, Hopke E F and et al. 2014 Long-term particle measurements in Finnish Arctic: Part I - Chemical composition and trace metal solubility *Atmospheric. Environ.* **88** 275–284
[9] Motuzova G V, Lukina N V, Nikonov V V and et al. 2004 Effect of Natural and Anthropogenic Factors on Soils, Soil Water, and Subsoil Water in Kola Peninsula *Water Resources* **31** 297–302
[10] Zhulidov A V, Robarts R D, Pavlov D F and et al. 2011 Long-term changes of heavy metal and sulphur concentrations in ecosystems of the Taymyr Peninsula (Russian Federation) North of the Norilsk Industrial Complex *Environ. Monit Assess* **181** 539–553
[11] Moiseenko T I, Gashkina N A, Sharov A N and et al. 2009 Anthropogenic Transformations of the Arctic Ecosystem of Lake Imandra: Tendencies for Recovery after Long Period of Pollution *Water Resources* **36** 296–309
[12] Moiseenko T I, Gashkina NA, Dinu M I and et al. 2013 Aquatic Geochemistry of Small Lakes: Effects of Environment Changes *Geochem. Int.* **51** 1031–1148

IOP Conf. Series: Earth and Environmental Science **33** (2016) 012005 doi:10.1088/1755-1315/33/1/012005

Geological-hydrogeochemical characteristics of a "silver spring" water source (the Lozovy ridge)

I S Ivanova[1,2], I V Bragin[3], G A Chelnokov[4], K Yu Bushkareva[5] and E V Shvagrukova[6]

[1]Institute of Ecological Problems of the North, Ural Branch of the Russian Academy of Sciences, Arkhangelsk, Russia
[2, 3, 4, 6]Far East Geological Institute, Far Eastern branch of the Russian Academy of Sciences, Vladivostok, Russia
[5]National Research Tomsk Polytechnic University, Tomsk, Russia

E-mail: [1] IvanovaIS_1986@mail.ru, [3]bragin_ivan@mail.ru,

Abstract. Geological and hydrogeological characteristics of the Lozovy ridge (Southern Primorye) are studied, as far as karst phenomena are widely distributed within its boundaries. Water-bearing rocks of the karst water source "Silver Spring" ("Serebryany Klyuch"), which is located near the bottom of the "Bear's fang" ("Medvezhiy klyk") cave, are investigated. It is found that karst rocks are presented by calcite (CaCO3), and an accessory mineral is barite (BaSO4). It is determined that among the trace elements forming the composition of carbonate water-bearing rocks the maximum concentrations are typical for Sr, Ba, Fe, Al, Za, Mn, Cu, and Ni. Also, the chemical composition of the waters taken from the "Silver Spring" water source is studied. These waters are fresh, hydrocarbonate, calcium, and weakly alkaline. Among the elements of the spring, such elements as Sr, Ba, Fe, Al, Zn, Mn, Cu, and Ni have the maximum concentration. The other elements have concentrations less than 1 μg/l.

1. Introduction

Karst phenomena are widely spread. Geologically, about one third of the Earth's land has the conditions for karst development. However, karsts do not have uniform distribution. Thus, limestone karst forms karst areas in the Pyrenees, the Alps, and the Carpathians. The Alps are characterized by intensively developed karst within mountain ice highlands bordered with limestone. Karst develops well in the Ordovician and Silurian limestone on the southern slope of the Baltic crystalline shield: on Aland and Gottland islands. In Central Europe karst develops in the Paleozoic rocks and Mesozoic calcareous formations embracing the territory of the Krakov Jura, numerous karst areas of Czechoslovakia, and Czech karst [3, 4]. Karst development is typical for Balkan Peninsula: Kras (Karst) plateau, Slovenian highland, Istrian Peninsula, and the Dinaric Alps. In the Carpathians carbonate karst develops, mostly, in reef limestones. In the Crimea karst phenomena are spread in mountain and steppe areas. Karst is rather wide spread in the Urals, Southern Siberia, in spots of Central and Eastern Siberia, and in the Far East. Therefore, in Eastern Primorye karstland the carbon-Permian and high-Triassic rocks are carstified, as well as reef limestones. In addition, sodded and bare karst is distributed with swallow holes, craters, rock shelters, and caves. In Southern Primorye karstland reef limestones of the late-Permian age are carstified. Also, sodded and bare karst is spread with swallow holes, craters, sink holes, and pepino hills [3].

2. Research object

The research objects are the karst waters of the Lozovy ridge and water-bearing rocks. The materials obtained by the authors as a result of their field works, having been carried out in August, 2015, are used in the paper. In addition, the published materials of the other authors are cited [1,6]. Rapidly changing parameters, such as temperature, Eh, pH, electric conductivity, and hydrocarbonate-ion content have been determined during field investigations and *in situ* hydrochemical testing. Rock and water samples have been investigated in the Prymorsky centre of local elemental and isotope analysis (the Far East Geological Institute, the Far Eastern Branch of the Russian Academy of Sciences).

The studied region is related to the Partizansky karst area of the Primorsky Territory (Figure1). The Lozovy ridge is the remnant massif formed by reef limestones, and it continues down to 7 km. The massif is stretched between the Partizanskaya and Novorudnaya rivers, northward of Nakhodka. It is bordered by low highly destroyed denudation watershed mountains. Alongside the distinguished geological features, the dominating role of the Partizanskaya river influences the territory hydrology. The massif slopes directed towards the Partizanskaya valley are flatter than the opposite stiff slopes of the ridge, which break down almost vertically.

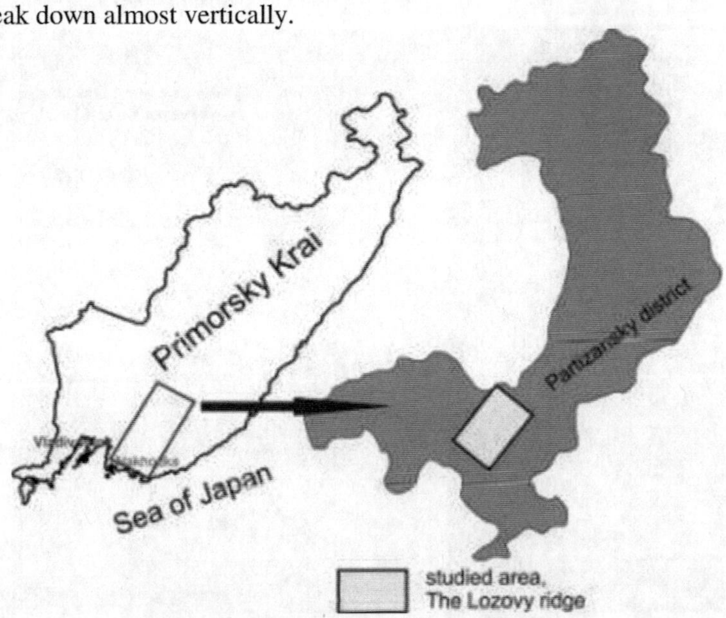

Figure1. Sketch map of the studied area.

Karst in the southern part of the massif is mostly bare, and in the northern part it is partially sodded. Karst surface is presented by swallow holes, grooves, rock remnants, rock shelters, craters, pot holes. As to subsurface karst structures, a few tens of caves are known, including the deepest cave in the Primorsky Territory – "Solyanik", as well as the caves called "Bliznets", "Dalnyaya", "Skvoznaya", "Malaya", "Mechta Speleologa", "Medvezhiy Klyk", etc.

The geology of the Lozovy ridge is studied in details [2,6,10,11]. The high-Permian sediments are widely spread within the studied area. They are presented by marine, marginal-marine and continental formations (Figure 2).

The Sicinsky suite (P2sc) is formed by terrigenous, igneous-sedimentary and igneous rocks being in the equal ratio. The thickness of the suite is 280 meters in the northern part of the area, and it increases up to 630 meters in the Nakhodka region.

The Chandalazsky suite (P2cn) is spread as a broken band, and it extends from Trudny Peninsula to the Povorotnaya river basin. It overlies the Sicinsky suite. The suite is formed by limestones, aleurolite horizons (the first scores of meters), and bands of conglomerates, sandstones and aleurolites interbedding. On the whole, the suite composition does not change in the strike. The suite thickness is 385-570 m.

Alluvial sediments (aQ) build up an accumulative part of high terraces. They are presented by inequigranular sands with occasional fine pebble and clays with interbeds of loamy and sandy soils. The thickness of sediments reaches 50 m [1, 12].

Hydrogeologically, the territory of the Partizansky karst region including the Lozovy ridge refers to the southern part of Sikhote-Alin hydrogeological area. The area is characterized by low mountain relief and arterial drainage. The zone of active water exchange with the thickness of 300-500 m. occupies the upper fractured part of horizontal section: quaternary deposits, the upper section part of the Cenozoic intermountain area. Ground waters are formed, mostly, due to atmospheric precipitation infiltration and surface waters. Deep carbonate karst rock are present. Mainly, they are connected with a perched water subzone and the zones of alternating, horizontal, and, partially, siphon circulation. Solutional caves, karts channels, coves and caverns are attributed to this group. Quaternary deposits are the most water-abundant. Water-bearing strata and complexes of a fracture-karst-stratal type are associated with carbonate and terrigenous-carbonate deposits of the late Paleozoic.

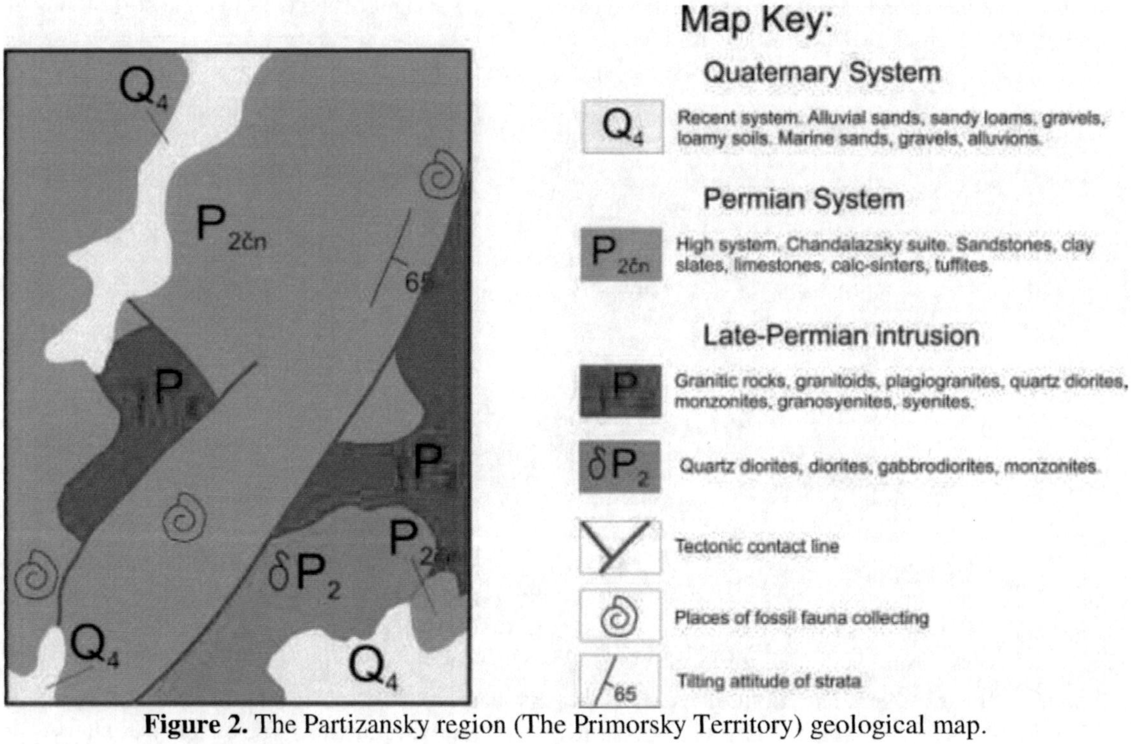

Figure 2. The Partizansky region (The Primorsky Territory) geological map.

3. Results and discussion

One of the main conditions for karst development is permeability of karsting rocks determined by their porosity, fractures, and cavernosity. The authors have investigated the adjacent rocks and waters of the "Silver Spring" source (N43°0'49", E133°1'18") coming out near the bottom of the karst massif, close to the "Bear's Fang" cave. The "Bear's Fang" cave is located at the height of 430 m. above sea level on the narrow rocky ridge. Cave rocks are divided by fracture systems of various thickness and directions. The expansion of cave cavity has finished due to mountain massif destruction and its drainage area elimination. The "Silver Spring" discharge does not exceed 50 l/sec. The chemical composition of waters is shown in Table 1.

In basic physical and chemical properties, spring waters are fresh (mineralization – 243mg/l), weakly alkaline (pH 8,1), hydrocarbonate, and calcium ones. Ions of SO_4^{2-}, Cl^-, Na^+,K^+ and Mg^{2+} have strictly subordinate value in the salts composition. The concentration of organic carbon (C_{org}) in waters reaches 1.6 mg/l, which makes only 4% of the total dissolved carbon.

Trace elements in karst waters do not determine the chemical type of water and its general mineralization; they are contained in insignificant quantities determined by the first mg, μg, and fractions of μg per 1 liter [7]. The trace elements are determined using the method of atom-emission spectrometry with inductively coupled plasma at the iCAP 6500Duo spectrometer (the analysts: Gorbach G.I., Tkalina E.A., Khurkalo N.V.). The trace elements composition of the "Silver Spring" waters is shown in Table 1. Only Ca is notable for its high concentration with regard to an average value for the waters of hypergenesis zones. It is worth saying that the "Silver Spring" karst waters are poor with trace elements, and only such elements as Sr, Ba, Fe, Al, Zn, Mn, Cu, and Ni have concentrations ranging from 6 to 46 μg/l. Such elements as Li, Pb, As, Se, Rb, V, Cr, Ag, U, Cs, Co, Y, Cd, Ce, La, Sc, Nd, Ga, Tl, Gd, Pr, Dy, Sm, Er, Be, EuTb, Ho, Yb, Th, Tm, and Lu are characterized by decreased values (the concentration is less than 1 μg/l).

Table 1. Physical and chemical parameters of the "Silver Spring" karst waters

Components	Units	Karst waters	Average value for waters of hypergenesis zones [8]	Components	Units	Karst waters	Average value for waters of hypergenesis zones [8]
T	°C	8	–	Cu		1.17	5.58
pH		8.1	6.9	Zn		6.18	41.4
Eh	mV	152	–	Ga		0.0083	0.37
TDS		243	469	As		0.40	1.46
Ca^{2+}		72.3	39.2	Se		0.321	0.72
Na^{+}		2.22	67.6	Rb		0.261	1.86
Mg^{2+}		1.23	18.2	Sr		46.0	183
K^{+}		0.54	5.15	Y		0.0281	–
Si		4.00	8.3	Ag		0.181	0.26
HCO_3^-	mg/l	156.10	187	Cd		0.0243	0.24
SO_4^{2-}		5.33	70.7	Cs		0.041	0.26
Cl^-		2.05	59.7	Ba		13.86	18.3
NO_3^-		1.34	2.4	La		0.0151	0.67
TC		40.50	–	Ce	μg/l	0.0186	–
IC		38.90	–	Pr		0.0028	–
TOC		1.60	8.29	Nd		0.0120	–
Li		0.84	13.0	Sm		0.0026	–
Be		0.0018	0.19	Eu		0.0017	–
Al		8.43	226	Gd		0.0032	–
Sc		0.0150	0.07	Dy		0.0027	–
V		0.252	1.34	Er		0.0019	–
Cr	μg/l	0.233	3.03	Yb		0.0015	–
Mn		1.40	54.5	Tl		0.0037	–
Fe		11.11	481	Pb		0.493	2.97
Co		0.0392	0.39	Th		0.0004	0.24
Ni		1.06	3.58	U		0.0799	1.31

Note: «–» – no data; T – water temperature; TDS – total dissolved salts; TC- total carbon; IC – inorganic carbon; TOC – total organic carbon.

Strontium has good migration properties, and it is easily leached out of rocks. This fact is proved by one of the highest concentration values among the trace elements in the studied waters (46 μg/l). The content of strontium in waters is determined, mostly, by the time of water interaction with rocks. Barium has lower chemical activity than strontium, therefore, its content in ground waters is much lower (14 μg/l). According to the results of ICP-MS, **t**otal distribution of heavy metals in the "Silver Spring" karst waters has the following sequence: Fe>Zn>Mn>Cu>Ni, though their concentrations are small, and they range from 11 to 1 μg/l, respectively.

To determine the sources of chemical elements in waters, the microelement composition of adjacent rocks has been studied. It is performed using the method of mass-spectrometry with inductively coupled plasma at the Agilent 7500 spectrometer (the analyst: Elovsky E.V.). The main elements of the studied rock are CaO– 54.9 wt.% и SiO_2– 0.92 wt.%. Among the elements found in the composition of carbonate rocks, the maximum concentrations are typical for Sr (374 g/t), Cu (15.9 g/t), and Ba (18.6 g/t). Cr, Ni, Zr, La, Ce, Nd, Y, Sc,V, As, and Pb are characterized by the content ranging from 1.0 to 8 g/t. The minimum concentrations are found among Sn, Cs, Tb, Tm, Lu, Hf, and Ta.

Strontium should be set apart all the other trace elements, because its concentration is the highest one. It could be met in rocks of any lythologic composition. Mostly, it is found in the composition of carbonate rocks, either as an impurity element, or as calcium isomorphous substitutions in the structure of carbonate minerals' crystalline lattice (limestone and dolomite). Strontium is fully engaged in the processes of dissolution and leaching occurring under the influence of atmospheric and ground waters. Moreover, strontium may form separate minerals – strontianite ($SrCO_3$) and celestine ($SrSO_4$), though they have not been found on the investigated territory [9]. Strontium has not been observed in the sample studied by the electron microscope, either.

Petrographic research of rocks has been performed at a JeolJSM 6490LV scanning electron microscope with an INCAEnergy 350 add-on unit (the analyst: Bushkareva K.Yu.). Based on the sample analysis, it is possible to make the following conclusion: the karst rocks of the "Bear's Fang" cave are presented by calcite ($CaCO_3$) with a small part of magnesium (Figure 3, a). The obtained results are proved by the archival materials [2]. It is stated that the main rock components are CaO (55 wt.%) and MgO (0.5 wt. %). FeO (Figure 3, b), SiO_2, TiO_2, as well as single crystals of pyrite (FeS) and Au, have been identified as microinclusions. Barite ($BaSO_4$)(Figure 3, c) is the most widespread accessory mineral. Also, the alloys of CuNi- and CuZn-types are met, which have, probably, a technogenic origin.

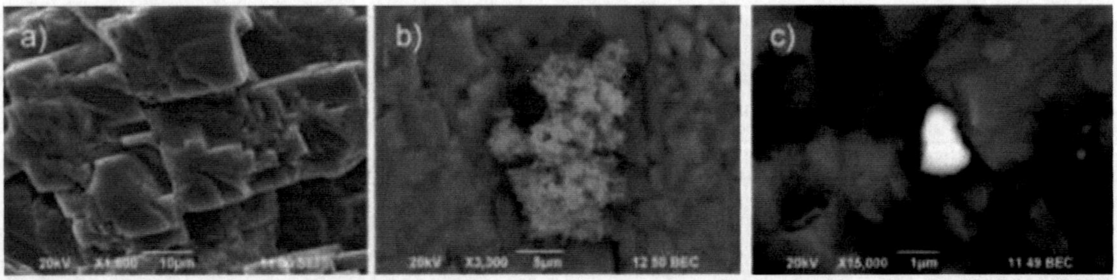

Figure 3. Photos of the Lozovy ridge carbonate rocks samples obtained using the JeolJSM 6490LV electron microscope: a) $CaCO_3$ morphology, b) FeO morphology, c) $BaSO_4$ microinclusions in carbonate.

4. Conclusion

The analysis of the obtained material allows confirming the fact that the carbonate karst type is typical for the Lozovy ridge, where karsting rocks are the limestones of the Chandalazsky suite. Carbonate rocks contain Sr, Ba, and Ni in the high amount. Their mineral forms, except Sr, have been proved by the use of electron microscopy. Ground waters held in Paleozoic terrigenous-carbonate sediments form the complexes of a fracture-karst-stratal type. The spring waters are also characterized by the maximum concentrations of such elements as Sr, Ba, Fe, Al, Zn, Mn, Cu, and Ni, the sources of which are carbonate adjacent rocks. The obtained geochemical characteristics for the karst rocks of the "Spring Silver" water source demonstrate that water aggressiveness is insignificant in relation to carbonate rocks. The vertical down-flow water communications occur only on the slopes due to the significant vertical filtration gradient.

Acknowledgements
The work is financially supported by grant № 15-35-50435, № 16-35-00002 of the Russian Foundation for Basic Research, by grant № 15-17-10009 of Russian Science Foundation and by grant of President of Russian Federation MK-4984.2016.5.

References

[1] Anohin V M, Rybalko V I, Alenicheva A A, Lelikov E P and et al. 2011 *Gosudarstvennaja geologicheskaja karta Rossijskoj Federacii.* Masshtab 1:1000000 (tret'e pokolenie). Serija Dal'nevostochnaja. List K– (52), 53 (Vladivostok. – SPb.: Kartograficheskaja fabrika VSEGEI) p 332
[2] Bersenev Ju I 1989 *Karst Dal'nego Vostoka* (M.: Nauka) p 172
[3] Gvozdeckij N A 1981 *Kars* / N.A. Gvozdeckij. (M.: Mysl') p 214
[4] Gvozdeckij N A 1972 *Problemy izuchenija karsta i praktika* (M.: Mysl') 1972 p 392
[5] *Gosudarstvennaja geologicheskaja karta Rossijskoj Federacii* 2011 Masshtab 1:1 000 000 (tret'e pokolenie). List (L-(52), 53; (K-52, 53) – oz. Hanka. Objasni tel'naja zapiska (SPb.: Kartograficheskaja fabrika VSEGEI) p 684
[6] Demin L V, Bersenev Ju I and Tatarnikov V A 1980 *Karst Primorskogo, Habarovskogo kraev i Amurskoj oblasti*//Karst Sibiri i Dal'nego Vostoka (Vladivostok) pp 5–54
[7] Krajnov S R, Ryzhenko B N and Shvec V M 2012 *Geohimija podzemnyh vod. Teoreticheskie, prikladnye i jekologicheskie aspekty*. 2-e izd., (M.: CentrLitNefteGaz) p 672
[8] Shvarcev S L 1998 *Gidrogeohimija zony gipergeneza* (M. : Nedra) p 366
[9] Judovich Ja Je, Majdl' T V and Ivanova T I 1980 *Geohimija stroncija v karbonatnyh otlozhenijah* (L.: Nauka) p 152
[10] Kotlyar G V, Belyansky G C, Burago V I, Nikitina A P ,Zakharov Yu D, Andrey V and Zhuravlev A V 2006 South Primorye, Far East Russia – A key region for globalPermian correlation *Journal of Asian Earth Sciences* **26** 280–293
[11] Kruk N N, Simanenko V P, Gvozdev V I, Golozubov V V and et al. 2014 Early Cretaceous granitoids of the Samarkaterrane (Sikhote-Alin'): geochemistry and sources of melts *Russian Geology and Geophysics* **55** 216–236
[12] Kotlyar G V, Shu-Zhong Shen, Kossovaya O L and ZhuravlevA V 2007 Middle Permian (Guadalupian) biostratigraphy in South Primorye,Russian Far East and correlation with Northeast China *Palaeoworld* **16** 173–189

IOP Conf. Series: Earth and Environmental Science **33** (2016) 012006 doi:10.1088/1755-1315/33/1/012006

Determination of water-soluble forms of oxalic and formic acids in soils by ion chromatography

E Karicheva[1], N Guseva[2] and M Kambalina[3]

[1, 2, 3] National Research Tomsk Polytechnic University, 30 Lenin Avenue, Tomsk, 634050, Russia

E-mail: [2] guseva24@yandex.ru

Abstract. Carboxylic acids (CA) play an important role in the chemical composition origin of soils and migration of elements. The content of these acids and their salts is one of the important characteristics for agrochemical, ecological, ameliorative and hygienic assessment of soils. The aim of the article is to determine water-soluble forms of same carboxylic acids – (oxalic and formic acids) in soils by ion chromatography with gradient elution. For the separation and determination of water-soluble carboxylic acids we used reagent-free gradient elution ion-exchange chromatography ICS-2000 (Dionex, USA), the model solutions of oxalate and formate ions, and leachates from soils of the Kola Peninsula.

The optimal gradient program was established for separation and detection of oxalate and formate ions in water solutions by ion chromatography. A stability indicating method was developed for the simultaneous determination of water-soluble organic acids in soils. The method has shown high detection limits such as 0.03 mg/L for oxalate ion and 0.02 mg/L for formate ion. High signal reproducibility was achieved in wide range of intensities which correspond to the following ion concentrations: from 0.04 mg/g to 10 mg/L (formate), from 0.1 mg/g to 25 mg/L (oxalate). The concentration of formate and oxalate ions in soil samples is from 0.04 to 0.9 mg/L and 0.45 to17 mg/L respectively.

1. Introduction

Carboxylic acids (CA) play an important role in the chemical composition origin of soils and migration of elements [1]. The content of these acids and their salts is one of the important characteristics for agrochemical, ecological, ameliorative and hygienic assessment of soils [2-4]. According to the modern notions oxalic acid is a harmful substance that is naturally present in various foodstuff [5-6]. It is characterized by demineralizing factors that inhibit the uptake of Ca^{+2}, Fe^{+2}, Zn^{+2} and some other elements due to the formation of sparingly soluble compounds [7-8].

Identification of individual carboxylic acids remains quite challenging due to the low concentrations of these compounds in natural water and soils and their physicochemical properties. Carboxylic acids are unstable compounds which easily undergo degradation during sample preparation, such as extraction or concentration [9-10]. Gas chromatography, mass spectrometry and ion chromatography are still the most common methods to determine the carboxylic acids. Ion chromatography (IC) as the most preferred method of analysis has been widespread in recent years due to relatively simple equipment, the most effortless sample preparation, and the use of different modes of operation, such as gradient elution [11-12].

IOP Conf. Series: Earth and Environmental Science **33** (2016) 012006 doi:10.1088/1755-1315/33/1/012006

The aim of the article is to determine water-soluble forms of same carboxylic acids – (oxalic and formic acids) in soils by ion chromatography with gradient elution.

2. Methods

To separate and determine water-soluble carboxylic acids we used reagent-free gradient elution ion-exchange chromatography ICS-2000 (Dionex, USA) with conductivity detection, equipped with vacuum degas, carbonate removal device CRD 200 (Thermo Scientific, USA), a separating column 2*50 mm IonPac AG 20 (Thermo Scientific, USA), background eluent conductivity suppressor ASRS 300 (2 mm) (Dionex, USA) and Dionex EGC III KOH Potassium Hydroxide Eluent Generator Cartridge. The eluent was delivered at a flow rate of 0.25 ml/min, injected sample volume was 1 ml Instrument control and data processing was performed using Chromeleon software.

The model solutions of oxalate and formate ions were prepared using the corresponding standards (Oxalate Standard, Fluka Analytical, USA, c=1 mg/ml and SSRS 7278-96, Ekmets, Russia) and deionized water (Simplicity, Millipore, USA). The concentration of oxalate and formate ions in the model solutions is 10 mg/L. Calibration solutions (0.01 to 10 mg/L) were prepared by serial dilution of a standard solution just before analysis.

The distribution of oxalate and formate ions was studied in soil samples collected from five sites in the study area, which was the Imandra lake surrounding area (the Kola Peninsula). Samples were taken from various depths. The scheme of soil sample preparation for analysis is shown in Figure 1. According to the scheme, one part of each soil sample was mixed with 5 parts of distilled water. After 24 h, the pH was measured and the samples filtered through a 0.45 μm cellulose nitrate filter.

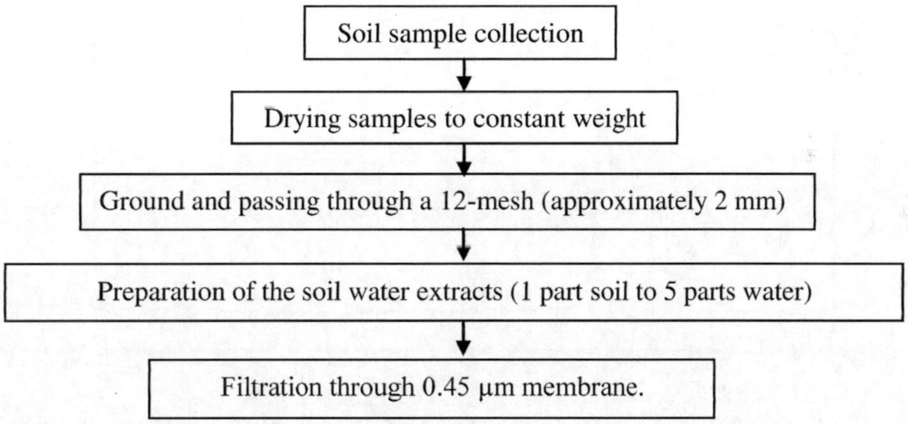

Figure 1. Scheme of soil sample preparation for analysis.

3. Results and discussion

The conditions for separation of oxalate and formate ions were optimized to obtain reproducible measurements of analytical signal. Optimized gradient program was established and presented in Figure 2.

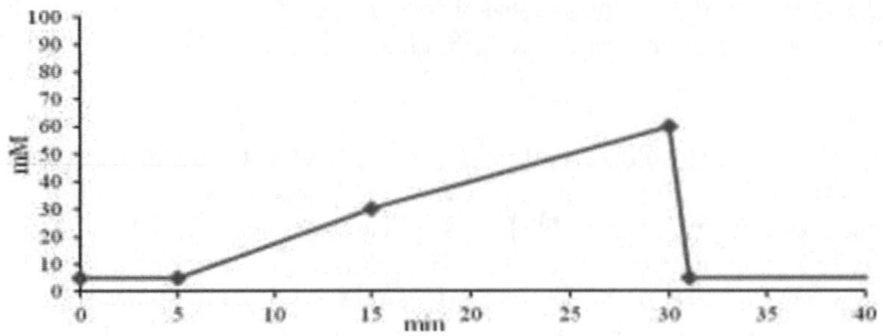

Figure 2. Multi-Step Gradient for carboxylic asid separation.

As shown in Figure 2, separation starts with a weak eluent. Either initially or as soon as the most weakly adsorbed components are eluted, ionic concentration of the eluent starts slowly increasing, and by the end of the separation reaches some maximum value, enough to elute quickly the most strongly adsorbed components. Thus, each component of mixture is fully separated at the optimum eluent composition for shorter period than in isocratic mode.

To establish the retention times for both anions, the model solutions of oxalate and formate ions were analysed. The retention time was experimentally determined that was 16.9±0.1 min and 7.6±0.1 min for oxalate and formate ions respectively (Figure 3).

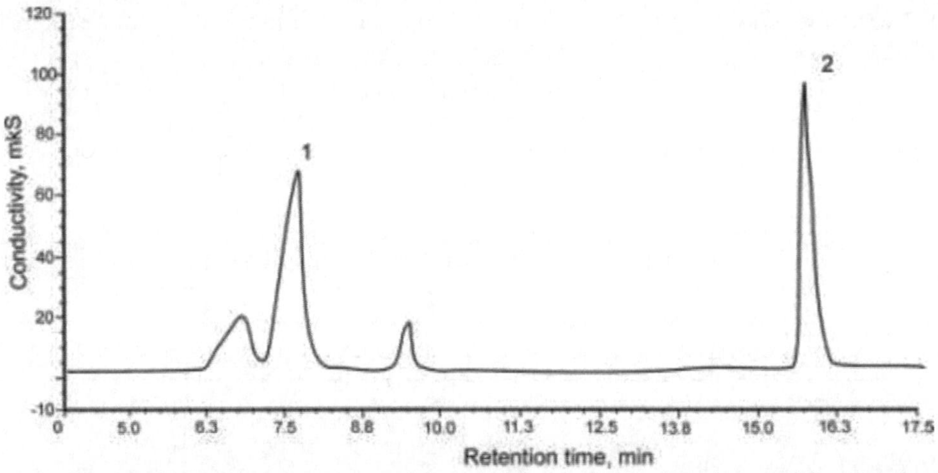

Figure 3. Ion chromatogram of the model solutions of oxalate and formate ions.

The basic metrological characteristics of oxalate and formate ion determination water solution are presented in Table 1. The concentrations of carboxylic acid ions were determined based on previously created calibration curve.

Table 1. Basic metrological characteristics of oxalate and formate ion determination by ion chromatography (n = 3; P = 0.95).

Element	Retention time, min	Detection limit, mg/L	Linear range*, mg/L
$HCOO^-$	7.6±0.1	0.02	0.04 – 10
$C_2O_4^{-2}$	16.9±0.1	0.03	0.10 – 25

*- dilution factor is unaccounted for

Based on the results of model solution research, the distribution of water soluble form of carboxylic acids in soil of the Kola Peninsula was found. The results of the determination of oxalate and formate ions are shown in Table 2.

The concentration of formate and oxalate ions in soil samples ranges from 0.04 to 0.9 mg/L and from 0.45 to 17 mg/L respectively. Thus, the content of water-soluble formic acid form in the studied soils is significantly lower than that of oxalic acid.

Table 2. Determination of formate and oxalate ions in leachates of the soils of the Kola Peninsula (n = 2; P = 0.95).

Samples	pH	Concentration, mg/L	
		$HCOO^-$	$C_2O_4^{-2}$
1	5.1	0.29±0.01	0.45±0.02
2	5.2	0.040±0.002	0.52±0.02
3	6.4	0.38±0.01	1.1±0.04
4	4.9	0.30±0.01	5.90±0.18
5	4.5	0.93±0.04	17.00±0.51

The method of standard addition was applied to confirm analytical validity of oxalate and formate ion determination in the leachates of the soils by ion chromatography. To the studied soil leachates, the addition was made in quantity equal to the initial concentration of ion in the samples. The results are shown in Table 3.

Table 3. Validation of results (n = 3; P = 0.95).

Sample	pH	Concentration, mg/L					
		$HCOO^-$			$C_2O_4^{-2}$		
		Sample	Standard	Total concentration	Sample	Standard	Total concentration
1	5.1	0.29±0.01	0.20	0.45±0.02	0.45±0.02	0.50	0.94±0.05
2	5.2	0.040±0.002	0.05	0.090±0.005	0.52±0.02	0.50	1.00±0.05
3	6.4	0.38±0.01	0.20	0.55±0.02	1.1±0.04	1.0	2.00±0.08
4	4.9	0.30±0.01	0.30	0.60±0.02	5.90±0.18	5.0	11.0±0.55
5	4.5	0.93±0.04	1.0	1.90±0.04	17.00±0.51	15	32.0±1.6

The present results show that statistically significant systematic errors are not identified while determining formate and oxalate ions by the developed technique.

4. Conclusions

1. The optimal gradient elution mode has been found to determine oxalate and formate ions in aqueous solutions.
2. A stability indicating method was developed to determine simultaneously water-soluble form of carboxylic acids in soils. The method has shown high detection limits such as 0.03 mg/L for oxalate ion and 0.02 mg/L for formate. High signal reproducibility was achieved in wide range of intensities which correspond to the following ion concentrations: from 0.04 mg/L to 10 mg/L (formate), from 0.1 mg/L to 25 mg/L (oxalate).
3. The accuracy of the obtained results is confirmed by the method of standard addition.
4. The concentration of formate and oxalate ions in soil samples ranges from 0.04 to 0.9 mg/L and from 0.45 to 17 mg/L respectively. Thus, the content of water-soluble formic acid form in the studied soils is significantly lower than that of oxalic acid.

Acknowledgements
This work was supported under the state assignment of the Ministry of Education and Science of Russia "Nauka" № 5.1931.2014/K.

References

[1] Vranova V, Rejsek K and Formanek P 2013 Aliphatic, cyclic, and aromatic organic acids, vitamins, and carbohydrates in soil: a review *J. Sci. World* 15

[2] Ström L, Owen A G, Godbold D L and Jones D L 2001 Organic acid behaviour in a calcareous soil: Sorption reactions and biodegradation rates *Soil Biol. Biochem.* **33** 2125-2133

[3] Zhi-An LI , Bi ZOU, Han-Ping XIA, Yong-Zhen DING, Wan-Neng TAN and Sheng-Lei FU 2008 Role of low-molecule-weight organic acids and their salts in regulating soil pH *Pedosphere* **18** 137–148

[4] Strobel B W 2001 Influence of vegetation on low-molecular-weight carboxylic acids in soil solution *Geoderma.* **99** 169–198

[5] Dolan L C, Matulka R A and Burdock G A 2010 Naturally occurring food toxins *Toxins (Basel)* **2** 2289–2332

[6] Agarwal S K 1988 The importance of naturally occurring toxicants in foods *Biochem. Educ.* **16** 212–213

[7] Chai W, Liebman M, Kynast-Gales S and Massey L 2004 Oxalate absorption and endogenous oxalate synthesis from ascorbate in calcium oxalate stone formers and non-stone formers Am *J. Kidney Dis.* **44** 1060–1069

[8] Reginato A J and Kurnik B 1989 Calcium oxalate and other crystals associated with kidney diseases and arthritis *Seminars in Arthritis and Rheumatism* **18** 198–224

[9] Sweetman C, Sadras V O, Hancock R D, Soole K L and Ford C M 2014 Metabolic effects of elevated temperature on organic acid degradation in ripening Vitis vinifera fruit *J. Exp. Bot.* **65** 5975–88

[10] Smith R L and Oremland R S 1983 Anaerobic Oxalate Degradation: Widespread Natural Occurrence in Aquatic Sediments *Appl. Environ. Microbiol.* **46** 106–113

[11] Harvey S 1991 Comparison of the gradient elution separations for organic and inorganic anions with chemically suppressed conductometric detection *J Chromatogr* **546** 125–137

[12] Widiastuti R and Haddad P R 1992 Approaches to gradient elution in ion-exclusion chromatography of carboxylic acids *J. Chromatogr.* **602** 43–50

Geochemical features of Kulunda plain lakes (Altay region, Russia)

M N Kolpakova[1,2], S L Shvartsev[1,3], S V Borzenko[4], V P Isupov[5] and S S Shatskaya[5]

[1]National Research Tomsk Polytechnic University, Tomsk, Russia
[2]Sobolev Institute of Geology and Mineralogy, Siberian Branch of Russian Academy of Sciences, Novosibirsk, Russia
[3]Tomsk Department of the Trofimuk Institute of Petroleum-Gas Geology and Geophysics, Siberian Branch of the Russian Academy of Sciences, Tomsk, Russia
[4]Institute of Natural Resources, Ecology and Cryology SB RAS, Chita, Russia
[5]Institute of Solid State Chemistry and Mechanochemistry, Siberian Branch of Russian Academy of Sciences, Novosibirsk, Russia

E-mail: [1]marina.kolpakova@gmail.com

Abstract. Geochemical specifics of lake water of the Kulunda Steppe territory (Altay region, Russia) are studied. The results show that in the territory mainly chloride and less soda lakes with sodium compound are developed. It is presented that calcite and soda saturation indexes (SI) of lake water increase with growth of pH, but decrease in such minerals as gypsum and barite. The opposite situation is typical for SI depending on the salinity. It is revealed that evaporation, secondary mineral formation and various biological processes have the greatest impact on accumulation of elements in solution.

1. Introduction

The studying of regularities of chemical elements distribution in natural water, features of their migration, and identification of possible sources of these elements is one of the fundamental problems of modern hydrogeochemistry. The research presented in this paper were carried out at the territory of Kulunda Steppe (Altay region, Russia) where more than 5000 saline lakes with total area about 4000 km^2 are located. In 1931 – 1933 the first extensive work of Kulunda expedition was held there under the leadership of N.S. Kurnakov (Academy of Sciences of the USSR) in order to study the main issues of formation of saline lake composition. In subsequent years, there were studies in calculation of reserves and conditions of a bedding of solid salts, also various methods of their operation were considered [1, 3]. The irreplaceable contribution to establishment of chemical types of lakes water on this territory was made by Yu.P. Nikol'skaya [9]. The research of saline lakes in the steppe area of Altai proceeds at present: the main directions of hydrological and morphological features [3, 8], flora and microbiological structure [7, 8], bottom sediments [10], etc. are being studied. For monitoring the current state of saline lakes we studied geochemical features of about 30 saline lakes and natural water which are in the territory of their reservoir within interdisciplinary expeditions of 2011-2015.

2. Materials and Methods

2.1. Study area

The studied lakes are located in forest-steppe and steppe landscapes and widespread in the southwest of Altay region, Russia (Figure 1). The lakes vary in sizes, chemical element sources, hydrogeological conditions, etc. For the territory of Kulunda steppe, which is characterized by salt accumulation and droughty climate, soil is one of the main sources of salts in lakes, their TDS can reach 0,15 % [7]. Interaction of internal and external factors [11] leads to a dynamic development of saline lake systems which can be easily affected by change of the environment conditions. Therefore, it is necessary to carry out regular comprehensive studies of water composition in such systems.

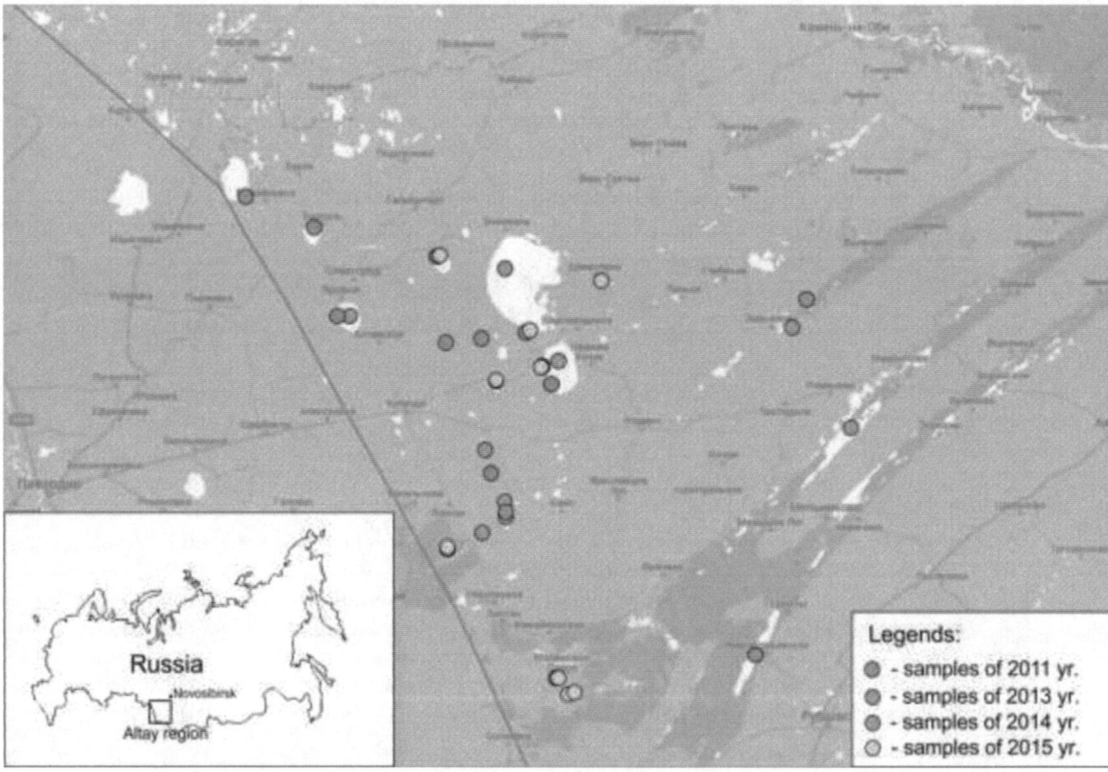

Figure 1. The location of Altay region and sampling points.

In most cases, the saline lakes of Kulunda steppe are shallow reservoirs with average depth less than 2,5m, the maximum depth of lake B. Yarovoe is 7.5m. Kuchukskoye and Kulundinskoye lakes are the largest system of connected lakes in this territory, the total area of their catchment area is about 24000 km^2. Water level in lakes fluctuates within 0.4-1m; in droughty years the area of lakes is sharply reduced [4].

2.2. Water sampling

The collected samples of water were filtered and loaded into plastic bottles (0.5 L) for analysis of major components. Before sampling all bottles were prepared to prevent pollution. Lake waters were analyzed in the field conditions (pH, Eh, electric conductivity, temperature), as well as in the Fundamental Research Laboratory of Hydrogeochemistry of Education and Research Centre "Water" (Tomsk Polytechnic University, Russia).

3. Results and Discussion

Parameters of considered lakes vary in chemical composition and the degree of their salinity. The data obtained during fieldworks (Table 1) indicate the existence of two chemical types of lakes in the

IOP Conf. Series: Earth and Environmental Science **33** (2016) 012007 doi:10.1088/1755-1315/33/1/012007

territory of research (according to Kurnakov-Valyashko's classification): chloride and carbonate (soda).

The carbonate (soda) type of lakes differs in the wide range of TDS values which vary from the brackish to strong brines: TDS reaches 107 g/L, thus its average value is only 29 g/l. The share of carbonates (CO_3^{2-} + HCO_3^-) in lakes seldom reaches 50%, the average value is 25%. The cationic component of the lakes water is always predominant by Na^+ (average – 85%) against very low concentration of calcium and magnesium (2 and 11%, respectively). pH of water changes from 8,9 to 9,9 that defines reference of these waters to carbonate (soda) type.

The chloride lake type in comparison with soda type is characterized by higher salinity (20 – 590, average is 247 g/L, tab. 1), but considerably smaller values of pH (7.2 – 9.0, average – 7.75). Sodium prevails among cations (average – 74%), the content of magnesium is slightly higher than in soda lakes and can reach 26%.

For lakes of both types the content of sulfate ions is the same and changes from 2 to 28%, average is 11-13%. The lack of accumulation of calcium and carbonate complexes in lakes is explained by formation of calcite at the early stages of chemical evolution of water [11]. The disproportional growth of sulfate relative to chloride [5] is associated with the formation of gypsum and microbiological processes in lakes [6].

Table 1. Minimum, maximum, and average values of chemical composition of Kulunda lakes, mg/L.

Chemical element	Carbonate (soda) type (number of objects – 10)			Chloride type (12)		
	Min.	Max.	Average	Min.	Max.	Average
pH	8.9	9.89	9,49	7.20	8.95	7.75
CO_2	n/d	4	1.2	n/d	255	75
CO_3^{2}	24	25800	5044	n/d	396	54
HCO_3^-	415	8845	3156	46	1610	828
SO_4^{2-}	88	18840	4011	4865	97224	37359
Cl^-	188	26306	6451	6200	190400	120025
Ca^{2+}	5	32	16	0	993	218
Mg^{2+}	7	228	85	970	81630	14936
Na^+	250	35615	9216	5580	130000	70571
K^+	11	294	72	20	700	230
Si^{4+}	0.2	42	11	0.1	42	8
Total Dissolved Solids, g/L	1.0	107	29	20	591	247
Total Organic Carbon *	25	477	124	34	599	225
Fulvic Acid**	15.7	25.7	21.7	3.9	19.7	10.7
Humic Acid**	1.1	3.0	2.1	<0.5	0.8	0.6

n/d – not detected, * - 2014 and 2015, ** - 2015 yr.

It should be noted that main anion composition of lakes in Kulunda Steppe is very similar to that of chloride lakes located in the nearby Ishimsk steppe territory (Northern Kazakhstan) but have higher values in comparison with lakes of chloride type in Western Mongolia [12]. As shown in Figure 2, the composition of considered lakes differs in smaller values of calcium (1 to 24 mg/L) in comparison with lakes of Northern Kazakhstan where Ca^{2+} content in lakes of this type reaches 3 g/L. At the same time, there are smaller concentration of Mg and SO4 in Kulunda steppe which is probably connected with the lack of additional sources of these elements.

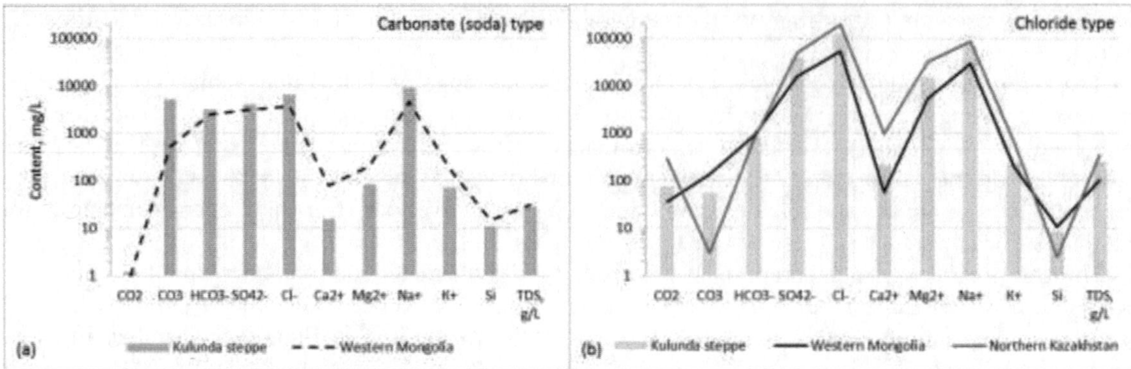

Figure 2. Comparative characteristics of the composition of the lake water of carbonate (soda) and chloride types in the territories of the Kulunda steppe (Altai Territory), the Ishim steppe (Northern Kazakhstan) and the Great Lakes basin (Western Mongolia).

The important factor of the lakes geochemistry formation is the equilibrium-nonequilibrium state of water-rock interaction which results in secondary mineral formation connected with saturation of water by mineral phases. According to the thermodynamic calculations, the carbonate type of sedimentation is characteristic for the majority of lakes, the gypsum stage of the mineral formation is not typical for the studied lakes, which is mainly caused by calcium mobilization by carbonate minerals (Figure 3a). In some cases (pH>9.5 and TDS>100 g/L), lakes reach equilibrium with respect to soda (Figure 3 a, b).

The sulphate content of saline lakes with pH values below 9.0 (chloride type) is sufficient for the formation of gypsum and barite (Figure 3c, d). It is obvious that the content of sulfates in the water can be limited only by the establishment of the equilibrium state of water with gypsum. Therefore, after this stage (gypsum precipitation) of the chemical composition transformation, the equilibrium of water with mirabilite does not reach.

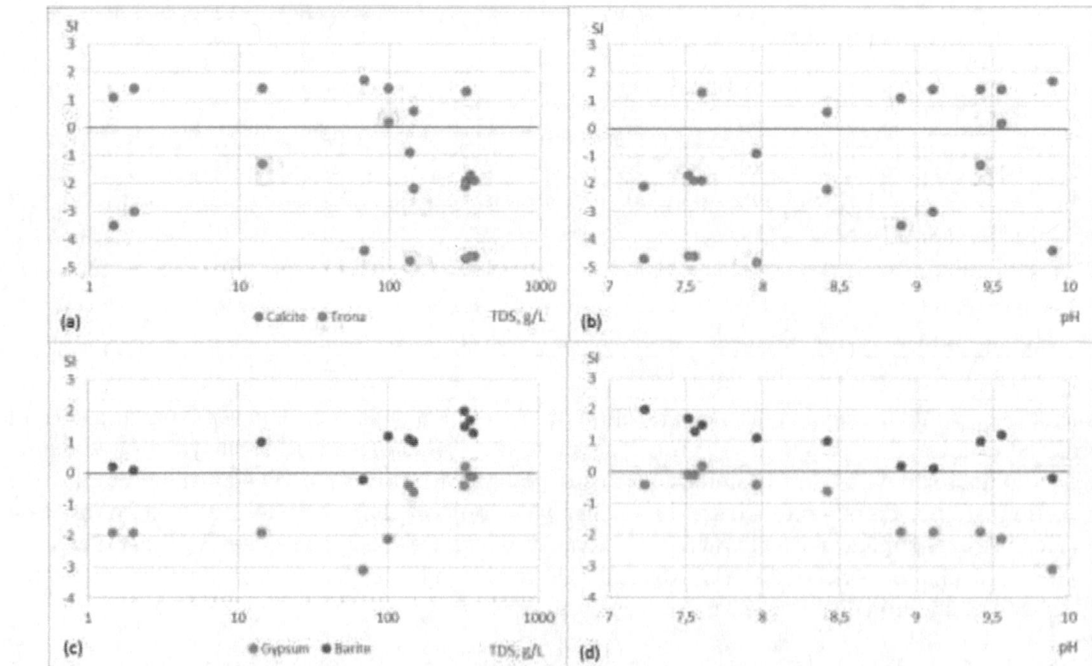

Figure 3. Saturation index of lakes water with calcite and sodium carbonate (a, b), gypsum and barite (c, d) as a function of pH and the salinity of lakes.

Our studies [6] show that the accumulation of sulfate ions in the equivalent amounts with chloride does not occur due to, firstly, their bacterial reduction into H_2S and, secondly, their removing from the solution predominantly in the hydrotroilite form. Although there is an evidence that a portion of the hydrogen sulfide is oxidized, but in this case there is a possible loss in the form of elemental sulfur and iron polysulfide [6]. Nevertheless, the precipitation of gypsum and, rarely, other sulfate minerals is the main geochemical barrier of sulfate accumulation in chloride type of lakes (Figure 3). The study in the behavior of saturation index (SI) of gypsum with increasing of TDS (Figure 3c) and pH (Figure 3d) of lake water shows an inverse dependence. Thus, the saturation of water by gypsum and barite can reach faster in high saline lakes than in brackish lakes and, vice versa, the saturation by these minerals reduces by the increasing of pH.

It is known that the evaporative concentration is the major factor of both sodium and chlorine accumulation in solution [2]. However, in fig. 4a it is visible that the growth of these ions with the increasing of salinity is not uniform and sodium concentration becomes much smaller than the content of chloride ions in highly mineralized waters. It is explained by binding of sodium ions in clay minerals. At this stage of the "lake water - rock" system development, chloride ion does not form its own minerals and continues to accumulate in the solution up to the point of saturation with respect to halite or sylvite.

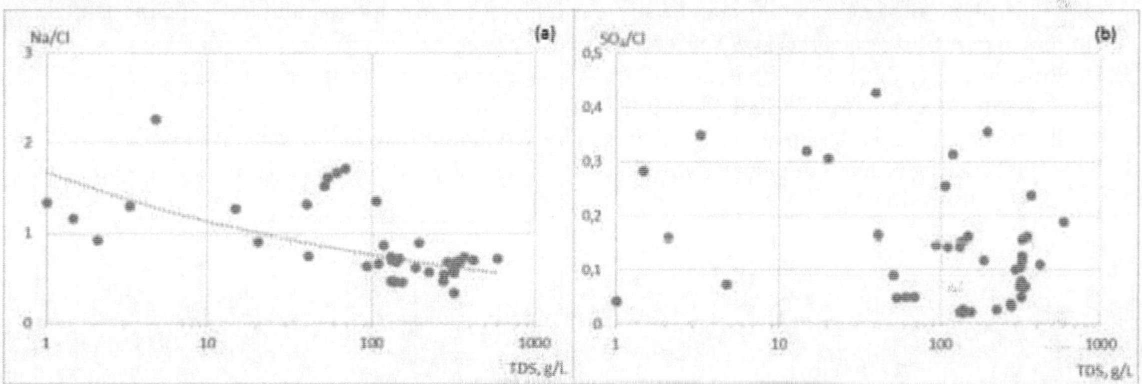

Figure 4. The relationships between the salinity of Kulunda steppe lakes and Na/Cl (a) and SO4/Cl (b).

4. Conclusion

Thus, in the territory of Kulunda Steppe mainly chloride and soda lakes with sodium cationic composition are developed. Lakes of sulfate type in the considered territory are not revealed. The TDS of lakes (in some years) reach 590 g/L, pH – 9.9. The growth of salinity of lakes water is followed by the increasing in the content of chemical components, however, their accumulation in lakes are uneven since the formation of a chemical composition of water affect the degree of evaporation, as well as geochemical and biological processes. The lack of sulfate ion accumulation in lakes (fig. 4b) is connected with sulfate reduction and precipitation of secondary minerals, in particular, gypsum.

Acknowledgements

The study of lakes in 2013-2014 supported by RFBR (project No 13-05-00556), the analytical results obtained in 2015 yr. were carried out by financial contribution of RSF (Russian Scientific Fund, project no 15-17-10003) (to K.M.N. and B.S.V.).

References

[1] Abramovich D I 1960 *Water of Kulunda steppe* (Novosibirsk: SB RAS) p 214
[2] Valyashko M G 1962 *The regularities of salt deposits formation* (Moscow: MSU) p 397

[3] Grohovsky L M 1972 *Lacustrine deposits of salts, their study and commercial evaluation.* (Moscow: Nedra) p 168

[4] Dobrovolsy G V, Sergeev E M and Gerasimova A S 1977 *The natural conditions of the central part of the West Siberian Plain* (Moscow: MSU) p 363

[5] Zamana L V and Borzenko S V 2010 Hydrochemical regime of saline lakes in the Southeastern Transbaikalia *Geography and Natural Resources* **31** 4 370 – 376

[6] Kolpakova M N, Borzenko S V, Isupov V P, Shatskaya S S and Shvartsev S L 2015. Hydrochemistry and geochemical typification of salt lakes steppes of the Altai territory *Water: chemistry and ecology* **1** 11 – 16

[7] Leonova G A, Bobrov V A, Bogush A A, Bychinskii V A and Anoshin G N 2007. Geochemical characteristics of the modern state of salt lakes in Altai Krai *Geochem. Int.***45**, 10 1025 – 1039

[8] Luzgin B N 2010. Strokes to the equilibrium carbon Kulunda system *Geography and natural resources of Siberia* **12** 97 – 110

[9] Nikol'skaya Yu P 1961 *The processes of salt formation in lakes and water Kulunda steppe*. (Novosibirsk: SB RAS) p 481

[10] Strakhovenko V D, Shcherbov B L, Malikova I N and Vosel' 2010 The regularities of distribution of radionuclides and reare-earth elements in bottom sediments of Siberian lakes. *Russian Geology and Geophysics* **51** 11 1167 – 78

[11] Shvartsev S L 2008 Geochemistry of fresh groundwater in the main landscape zones of the Earth *Geochem. Int.* **46** 13 1285 – 398

[12] Shvartsev S L, Kolpakova M N, Isupov V P, Vladimirov A G and Ariunbileg S 2014 Geochemistry and Chemical Evolution of Saline Lakes of Western Mongolia *Geochem. Int.* **52** 5 388 – 403

Sodium groundwater in SE Western Siberia

O Lepokurova

A.A. Trofimuk Institute of Petroleum Geology and Geophysics Siberian Branch RAS, Tomsk, Russia
National Research Tomsk Polytechnic University, Tomsk, Russia

E-mail: LepokurovaOY@ipgg.sbras.ru

Abstract. The paper describes the updated results in the study of the formation conditions of sodium water within SE Western Siberia. The authors identified a classification of the types of sodium water, location conditions and chemical composition.

1. Introduction

Sodium groundwater are widely found throughout the Earth upper crust. The existence and formation of these groundwater is still the issue of the day. The investigation of specific formation conditions of sodium groundwater in nature, isotope studies and calculations of fossil water- secondary aluminosilicate equilibrium system deduces the possible identification of the sodium accumulation sources and mechanisms. In this case, the most significant sodium groundwater occurrences are being described. This, in its turn, embraces the SE region of Western Siberia. This choice is based on the following factors (1) sodium water is extensive and diverse in this region and (2) influence of the complete water-rock (aluminosilicate)-gas (methane, carbon dioxide) – organic matter (coal, swamp) system on formation of water composition can be found in this region, if anywhere.

2. Investigation background

Most specialists from different countries associated the formation of sodium water with ion exchange. This recognized fact has been confirmed in numerous publications [1–5]. This proves the fact that the importance of aluminosilicate rocks in the formation of sodium water remains out of view of hydrogeologists, geochemists and geologists, which, in its turn hampers the further development of geochemistry. S.L. Shvartsev [6, 7] proposes to decipher the genesis of sodium water on the basis of the existing concept: water-rock interaction. This concept embraces the experimental data results from different European and American laboratories investigating such aspects as mineral dissolution in various geochemical environments under different temperatures; physico-chemical modeling of different hydrogeochemical processes; and calculations of water-rock system equilibrium [8–18]. However, the most significant aspect of this concept is that it involves the data results of numerous geological and hydrogeological observations. S.L. Shvartsev proved that sodium formation is the natural consequence of water-aluminosilicate interaction and is attributed to a specific evolution phase of this system if water is carbonate saturated. This interaction could be as a result of the equilibrium-nonequilibrium behavior within the system itself: water is not always in equilibrium relative to endogenous minerals which are consequently dissolved and, this provides the continuous evolution of the water composition; however, in this case, water is in equilibrium to a definite set of secondary minerals.

3. Regional distribution and specific features of chemical, gaseous and isotope composition of sodium groundwater

Distribution conditions (Figure 1) and specific composition characteristics of sodium water (Table 1) are discussed below; each region – Western Siberia artesian and Kuznetsk artesian basins are described separately.

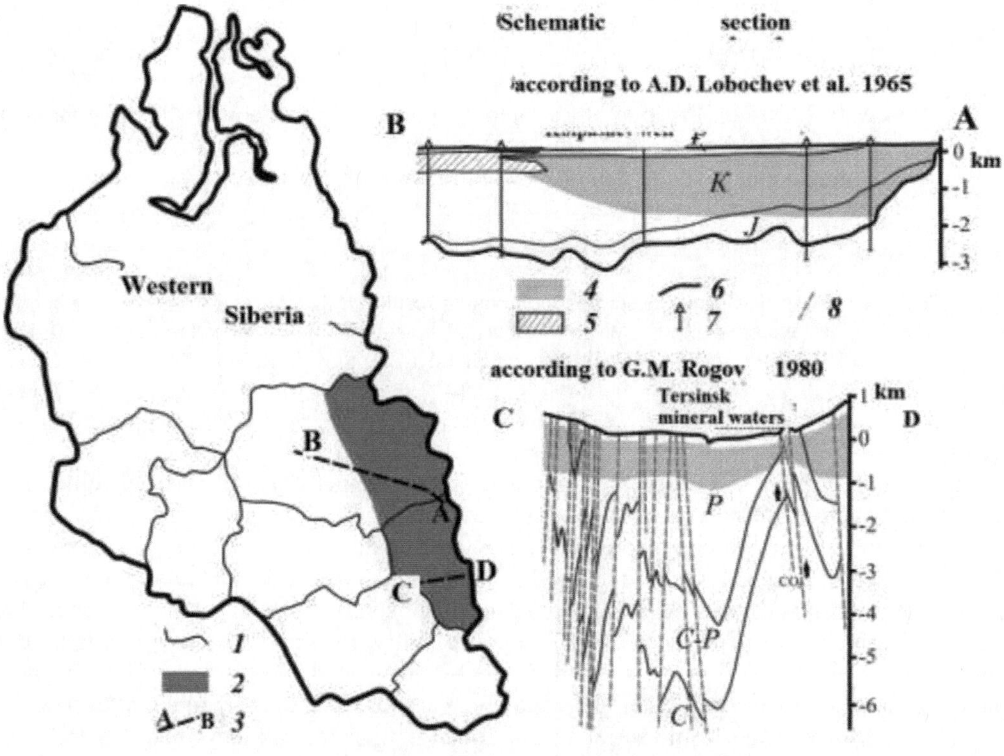

Figure 1. General map and cross-sections[19, 20]: 1 – civil division boundary; 2 – investigated area; 3 – cross section line; 4 – distribution area of natrium hydrocarbonate (sodium) water; 5 – regional confining layer; 6 – stratigraphic boundaries; *7* – wells; *8* – faults.

Hydrogeologically, S-E region of Western Siberia artesian basin is disclosed. There is no Cretaceous- Paleogene water confining layer, (figure 1, cross-sections AB) as a result the influence of filtration water was not distinctly observed. The boundary of fresh water can be found at 1.5km. Sodium water can be widely found in Cretaceous continental sedimentary rocks (sandstones and aleurolites). Groundwater intake is from Altai-Sayan Mountain bordering, where sodium water can also be found in subsurface conditions within intrusive rock fractures (mainly granites and granodiorites). In the Chulim basin sodium water can be found at a depth from several tens to hundreds of meters (up to 2 m). Maximum distribution area of above-mentioned water is confined to the artesian basin slope, gradually declining in the NW direction, i.e. in Mid Ob basin.

Unique sodium water can be found in Cretaceous sediments, Illekian suite within Chulim basin at a depth of 600-1300m. The uniqueness of sodium water is that it is rather fresh (salinity: 0.25–0.6 gr/liter), however, strongly alkaline (pH 9.1–10.3). Such alkaline water can be found in many regions of the world as it is associated with magmatic rocks. In this case, such water prevails in Lower Cretaceous continental sedimentary rocks and aleurolites (K_1il).Water produced from Chulim wells

(1266–1277 m) is therapeutic-table water with high silica content (H_4SiO_4 up to 50 mgr/liter) [21] and is labelled as "Omega" (first mineral water produced in Tomsk Oblast).

Table 1. Chemical composition of some sodium water types in SE Western Siberia (numerator- allowable concentration; denominator- average, dash- no data), mgr / liter.

Parameters	Sodium water in Kuznetsk basin; moderate water exchange [22]	Sodium water in Kusnetsk basin; moderate water exchange in lower zone area	Ter-sian acidulous water [23]	Borisovian nitrate-methane mineralsБ [24]	Fresh alkalines in Chulim basin [21	Kolpashev mineral water; well - 2^{hrs}
Host rocks	Permo-Carbonifeous carbonaceous-terrigenous sediments				Cretaceous argillo-arenaceous sediments	
Sampling depth,m.	30–1197 / 538	190–1495 / 645	370	575	1266–1277	744–760
pH	7.2–9.8 / 8.2	7.2–10.0 / 8.5	6.4–6.9	8.2	9.1–10.3	7 9
∑ ions	436–4627 / 1364	1499–25139 / 10276	4054–5809	2875	261–404	815–1111
Oxidability	0.2–7.1 / 2.5	0.5–19.2 / 5.0	–	10	1.1–1.6	–
HCO_3^-	258–3289 / 863	630–15494 / 6110	3002–3953	2013	73–209	370–568
CO_3^{2-} (CO_2)	0.3–742.8 / 33.4	8.4–1285 / 266	(1656–3060)	54	19–82	–
Cl^-	0.5–248 / 30.5	21–5494 / 779	111–192	92	1.5–17.7	106–354
SO_4^{2-}	0.2–596 / 28.5	0.2–75 / 16.5	0.5–14	1.3	2.0–20.4	0–2 5
Ca^{2+}	1.0–165 / 34	0.2–60.5 / 18	235–307	2	0.4–5	6–8
Na^+	44–1200 / 317	426–7230 / 2976	732–1054	862	85–99	236–380
Mg^{2+}	0–77.0 / 11.6	1.2–79.3 / 18	50–119	0.6	0.1–2	0–2,4
K^+	0.3–100 / 6.5	1.0–41 / 14.8	11–20	1.3	0.3–0.6	2–3 7
SiO_2	4 1–42 5 / 21,5	5 8–44 6 / 20 1	51 4–117.7	7.7	16.1–31	–
N samples	90	35	94	1	13	7

Sodium water distribution thickness in Mid-Ob basin is much less and could extend to full pinching out in regional occurrence areas of Cretaceous-Paleogene water confining layer. B In Kolpashev region in well 2 (Figure 1, cross-section AB)sodium mineral water was found in Cretaceous sediments of Pokursk suite at the depth of 744–760 m. Low-mineralized water (up to 1.1 gr/liter), weak alkaline (pH 7.9) with high chloride concentration [11]. This water is used as drinking therapeutic-table water "Siberian Resort."

Sodium water is significantly developed in Kuznetsk artesian basin under forest-steppe conditions, even at near subsurface conditions. Comparably to above-described basin, the sodium water distribution thickness is continuous throughout the cross-section (figure 1, cross-sections CD) and is 2000 m. [22]. This water is usually found at the depth of 100–200 m. and is confined to the zone of moderate water exchange in Permian carbonaceous-terrigenous sediments. Mineralization ranges from 0.4 to 5gr/liter; pH from 7.2 to 9.8, sometimes high chloride and sulphate concentration; in gas-predominately methane. Some are used as mineral water. For example, Borisovsksya therapeutic-table water includes nitrate-methane concentration, analogous to North Caucasian water "Essentuki", due to its similar medical treatment properties [24].

Unique sodium water with quite high salinity (up to 25 gr/л\liter) was discovered in Narikskо-Ostashkin area, Erunakovsk in 2012–2014. This water was assigned to the lower part of the moderate water exchange zone. Besides the significant mineralization values, this water embraces has heavy isotope composition of water-dissolved carbon (to +30.3‰), which, in its turn, что требует requires further detailed investigation (Table 2). Investigation by positive $\delta^{13}C$ in CO_2 appeared only in 2000s, i.e. only recently. Such anomaly values $\delta^{13}C$ are rare: predominately in pore water of lake and marine bottom sediments (up to 34,9‰), which, in its turn, coordinates with geochemical reduction conditions (China, Argentina, USA) [25], cold mud volcano methane waters [26] and in coal layers (up to 22%) within the production areas of shale gas and coal-bed methane (USA, Australia) [27–28].

Table 2. Carbon isotope concentration in water dissolved gases and hydrocarbonates from Nariksk-Ostashkinsk area.

№ samples	□$^{13}C_{CH4}$, ‰	□$^{13}C_{CO2}$, ‰	$\delta^{13}C_{DIC}$, ‰	($\delta^{13}C_{CO2} - \delta^{13}C_{DIC}$)
1	-67.3	13.7	25.1	11.4
2	-44.3	17.1	24.6	7.5
3	-51.2	13.3	21.5	8.2
4	-46.9	18.0	26,4	8.4
5	-45.9	-3.2	4.4	7.6
6	-43.3	6.7	14.0	7.3
7	-46.0	22.3	30.3	8.0
8	-53.1	14.6	23.7	9.1
9	-53.0	19.6	28.9	9.3
10	-49.15	12.6	21.2	8.6
11	-46.7	12.7	21.4	8.7

It should be noted that all scientists agree that heavy positive CO_2 isotope is biogenic and is always related to alkaline sodium water. Based on the above-discussed facts, the following conclusion was made- groundwater enrichment mechanism by heavy isotopes ^{13}C is associated with the sodium water formation mechanism. This could be a separate research topic.

It should be stated that in this region at conditions of significant methane gas distribution carbon dioxide gases can be found in the lower zones (Figure 1, cross-sections CD); especially, in the conjunction zone of Kuznetsk depression with Kuznetsk Altai Tersinsk carbonate mineral water fields are confined to the regional fault. Uniqueness of Tersinsk field is that it can not be attributed to recent volcanism occurrences as the youngest igneous rocks in Kuznetsk are only Triassic ones. This sodium carbonate water is cold (T at head = 12–13 °C), salinity plus total mineralization 2.0–5.5 gr/liter, slightly acidic (pH 6.2 – 7,0). Comparative analysis of Tersinsk carbonate water composition to Kuznetsk sodium water in the zone of moderate water exchange (which is well-developed throughout the whole area) shows that genetically they are close in proximity. Moreover, obtained value $\delta^{13}CHCO_3^-$ of carbonated water differs from sodium water in this region. In the, isotope-light carbon dominates which explains the biogenic genesis of CO_2 in HCO_3 composition. In Tersinsk field mineral water- "heavy" value $\delta^{13}C$ (HCO_3^-) indicates metamorphic genesis, whereas source of CO_2 are carbonates buried deep in the earth crust. Thus, carbonated water is sodium water (one and the same thing). In this case, the composition significantly changes due to CO_2 ingression from the deepest zones of the earth crust through tectonic dislocations [23].

4. Conclusion

In conclusion, sodium water is widely distributed in SE Western Siberia: from ultradispersed (0.2 gr/liter) to saline (25 gr/liter); from weakly carbonate (pH 6.2) to strongly alkaline (10.3),with carbon dissolution anomaly (+30.3‰) and ultralight (-30.3‰), high and low organic matter concentrations. The gaseous composition is also rather diverse.; nitrate, carbon dioxide, methane which are widely distributed in Mesozoic sediments on artesian water slope and Paleozoic sediments in Kuznetsk intermountain basin at a depth ranging from 100-200 to 2000 m.

Acknowledgments
The research was supported by Russian Foundation for Basic Research funding, project No. 14-05-31095

References

[1] Matthess G, Frimmel F H, Hursch P, Schulz H D and Usdowski E 1992 *Progress in hydrogeochemistry* (Berlin: Springer-Verlag)
[2] Kimura K 1992 *Ground Water Hydr* **32** (1) 5–16
[3] Appelo C A J and Postma D 1994 *Geochemistry, groundwater and pollution* (Rotterdam: A.A. Balkema)
[4] Parkhurst D L, Christenson S and Breit G N 1996 *U.S. Geol. Surv. Water-Supply Pap* 101
[5] Kim K, Yun S-T 2005 *Geochemical Journal* **39** 273–284
[6] Shvartsev S L 2008 *Geochemistry International* **46** (13) 1285–398
[7] Shvartsev S L, Ryzhenko B N, Alekseev V A, Dutova E M, Kondrat'eva I and Lepokurova O E 2007 *Geological Evolution and Self-Organization of the Water-Rock System* **2**
[8] Lasaga A C 1984 *J. Geophys. Res* **89** (B6) 4009–25
[9] Helgeson H C, Murhpy W M, and Aagaard P 1984 *Geochim. Cosmochim. Acta* **48** 2405–32
[10] Murphy W M and Helgeson H C 1987 *Geochim. Cosmochim. Acta* **51** 3137–53
[11] Murphy W M and Helgeson H C 1989 *Am. J. Sci.* **289** 17–101
[12] Casey W H & Ludwig C 1995 *Mineralogical Society of America* (Washington: A. F. White and S. L. Brantley) pp 87–118
[13] Brantley S L and Chen Y 1995 *Mineralogical Society of America* (Washington: A. F. White and S. L. Brantley) pp 119–172
[14] Nagy K. L. 1995 *Mineralogical Society of America* (Washington: A. F. White and S. L. Brantley) pp 173–233
[15] Dove P M and Czank C A 1995 *Geochim. Cosmochim. Acta* **59** 1907–15
[16] Rubie D C, Thompson A B 1985 *Advances in Physical Geoghemistry* **4** (New York: Springer-Verlag) 26–79
[17] Kerrick D M, Lasaga A C, Raeburn S P 1991 *Mineralogical Society of America* (Washington: A. F. White and S. L. Brantley) pp 583–671
[18] Sverdrup H 1990 The Kinetics of Base Cation Release Due to Chemical Weathering (Lund, Sweden: Lund University Press)
[19] *Hydrogeology of the USSR* 1972 **18** 479 [in Russian]
[20] Rogov G M and Popov V K1985 *Gidrogeologija i katagenez porod Kuzbassa* (Tomsk: Publishing house of Tomsk State University) p 191
[21] Shvartsev S L and Lepokurova O E 2014 Doklady Earth ScienceGidrogeologija i katagenez porod Kuzbassas **459** (1) 1464–1469
[22] Shvartsev S L, Khryukin V T, Domrocheva E V, Kuzevanov K I, Rasskazov N M, Popova T S, Lepokurova O E and Shvachko E V 2006 *Russian Geology and Geophysics* **47** 7 881–891
[23] Kopylova Y G, Lepokurova O E, Tokarenko O G and Shvartsev S L 2011 *Doklady Earth*

IOP Conf. Series: Earth and Environmental Science **33** (2016) 012008 doi:10.1088/1755-1315/33/1/012008

Sciences **436** (2) 284–289
[24] Lepokurova O E, Tokarento O G and Kopilova Yu G 2007 Geohimija Borisovskih mineral'nyh vod *Materialy nauchnoj konferencii Sovremennye problemy geohimii* 103–105
[25] Zhu Z, Chen J and Zeng Y 2013 *Quaternary Int.* **286** 85–93
[26] Lavrushin V Y, Polyak B G and Pokrovskii B G 2009 Lithology and Mineral Resources **2** 171
[27] Colding S, Boreham C J and Esterle J S 2013 *Int. J. of Coal Geology* **120** 24–40
[28] Sharma S and Frost C D 2008 *Ground water* **46** (2) 329–334

Features of water chemical composition of oligotrophic and eutrophic bogs in the South of the Tomsk region

O Naymushina

Vernadsky Institute of Geochemistry and Analytical Chemistry of Russian Academy of Sciences, Kosygina Str. 19, Moscow, 119991, Russia
National Research Tomsk Polytechnic University, Lenin Ave. 30, Tomsk, 634050, Russia

E-mail: olgnaim@mail.ru

Abstract. On the basis of the actual material the analysis of chemical composition of bog waters in the territory of the South of the Tomsk region is carried out. The data on average concentration of macro and trace components, organic matter, pH of bog waters are obtained. Significant distinctions in a chemical composition of surface water for different types of bogs are revealed. The composition and macrostructure of humic acids by the example of eutrophic bogs is studied.

1. Introduction

Western Siberia represents one of the largest accumulative low plains on Earth, which considerable part is boggy. Currently, wetlands range from 37 to 50% [1], and process of peat formation progresses. In regions with humid climate the content of dissolved organic matter in bog waters exceeds several times the content of ash elements [2]. Such waters are organogenic [3] and of keen interest for scientists of the world that defines the relevance of study in bog systems of the territory, including detection of the main features of chemical composition of bog waters, and study in organic matter structure.

The aim of this work is to study bog water for determination of the quantitative contents of macro- and trace components and characteristics of the elemental composition of organic matter.

2. Objects and methods

Field works were performed in the territory of right-bank part of the valley of the lower reach of the Tom River in 20–25 km from its confluence with the Ob River within the second upland fringe. The climate of the area is sharply continental, the average annual air temperature makes -0.6 °C. The amount of precipitation is of 400-600 mm/year. The evaporation capacity makes 450-470 mm/year. Precipitation evaporates incompletely because of the low humidity deficit, low air temperatures and the properties of bog soil and peat to detain moisture. The actual evaporation makes about 350 mm/year, keeping excess moistening of the area [4]. In the upper part of the section Neogene-Quaternary, Paleogene and Upper Cretaceous aquifers developed. The composition of water up to the depth of 500 m is HCO_3 Ca and becomes Na deeper [5]. The landscape is presented by generally ridge pattern.

Investigations included sampling of bog water, the subsequent determination of their chemical and organic composition, and statistical analysis of the received materials. Parameters of the rapidly changing components were measured in situ: temperature, conductivity, ORP (oxidation-reduction potential), pH – by means of the WATER TEST Hanna instruments; NO_2^-, NO_3^-, NH_4^+, HCO_3^-, SO_4^{2-}, Ca, Mg, Fe – by photometric and colorimetric methods. The water analysis was carried out in the Accredited Problem research laboratory of hydrogeochemistry of the scientific and educational center "Voda" of Tomsk Polytechnic University.

Elemental analysis of the humic and fulvic acids was performed in the Baikal Nature Management Institute of the Siberian Branch of the Russian Academy of Sciences according to [6, 7] by automatic elemental analyzer «CHN-1106» of Carlo Erba firm, the oxygen content was calculated by the difference. The functional groups were approved by the A.F. Dragunova and T.A. Kukharenko methods.

3. Results and discussion

3.1. Bog water chemistry

The results of the summarized data on chemical composition of bog waters are presented in Table 1. Investigation of the chemical composition of oligotrophic bog water shows that it is mostly acid, slightly acid and neutral (pH varies from 3.1 to 7.2, averaging 4.4); as for the value of TDS, it is ultra-fresh (average value makes 18 mg/L). According to ion composition, the water of oligotrophic bogs is hydrocarbonate calcium-magnesium, sulfate calcium-magnesium, and calcium-sodium.

The water of eutrophic bogs considerably differs under the terms of inflow and chemical composition. Eutrophic peat underlies on the unconsolidated sediments and receives its primary recharge from surface and underground water [8]. In terms of pH, natural waters of these bogs are characterized as slightly acid, more rarely acid and neutral (pH varies from 4.0 to 7.6, on average 5.5); as for the value of TDS, it is ultra-fresh (on average 38 mg/L). According to ion composition, the water of eutrophic bogs is hydrocarbonate calcium-magnesium, sulfate calcium-magnesium being not typical.

There are close synergies between the values of pH of bogs water and its chemistry, especially the concentration of HCO_3. In acid water of oligotrophic bogs this ion often is not found at all, and the carbon is contained therein as dissolved dioxide and organic compounds [9]. The latter is formed in the maximum concentration in water of oligotrophic bogs, reducing the pH to 3.1, and considerably exceeding the total content of the dissolved salts. The general content of organic substances in eutrophic bog water in terms of C_{org} makes from 7 to 110 mg/L that is much lower than their content in waters of oligotrophic peat sediments, where C_{org} varies from 20 to 190 mg/L. As compared to eutrophic bogs the average concentrations of organic substances in oligotrophic bogs water are almost twice as higher. Among organic compounds, humic and the fulvic acids sharply dominate which are formed at decomposition of biological material and make the most share of gross organic substance [10, 11]. The average value of fulvic acids content in water of oligotrophic sites is much higher than in eutrophic ones, and can reach 160 mg/L, accounting for 85% of the total organic content in the water. The concentrations of humic acids are lower, 13 on average in the eutrophic bogs water, up to 25 mg/L in the oligotrophic ones.

The high content of organic substances in the studied bodies contributes to concentration of ammonium ions significantly. The average concentration of NH_4^+ in oligotrophic bog water is 4.1 mg/L, in the eutrophic one - 2.3 mg/L, which exceeds twice the Clarke value [2].

Among trace components, in bog water heavy metals prevail. The maximum content of Zn, Pb, Cu, Mn and Al can testify the possibility of their forming complex compounds with organic substances which are found in large amount in the studied water. Among the metals, the content of Fe is the highest; concentration in 3-9 mg/L prevails. Especially, they are abnormal in water of eutrophic bogs, where they can reach 46 mg/L. This is due to the large amount of organic matter in the water, since iron acts as the main element in the competition forming organometallic complexes that promotes its retention and accumulation in water [12].

IOP Conf. Series: Earth and Environmental Science **33** (2016) 012009 doi:10.1088/1755-1315/33/1/012009

Table 1. Average, minimum, and maximum concentrations of the dissolved substances and value of physical and chemical parameters of bog water.

Parameter	Unit	**Oligotrophic bogs**			**Eutrophic bogs**		
		Average	Min.	Max.	Average	Min.	Max.
Conductivity	uS/cm	55.9	16.0	408.0	63.7	25.0	382.0
pH		4.4	3.1	7.2	5.5	4.0	7.6
CO_2	mg/L	72.1	2.6	141.0	57.2	3.5	154.0
HCO_3^-		5.0	N/D	159.0	17.9	2.6	238.0
SO_4^{2-}		4.3	N/D	11.3	2.5	1.5	16.5
Cl-		2.0	0.4	5.8	3.4	0.8	20.6
Ca^{2+}		4.3	0.5	13.8	6.4	1.6	46.0
Mg^{2+}		0.9	0.2	8.4	3.0	0.5	8.5
Na^+		1.4	0.2	11.5	2.5	0.3	8.5
K^+		0.9	0.2	4.0	2.1	0.1	5.2
TDS		17.8	3.4	214.0	37.6	5.5	414.0
O_2		2.6	0.6	8.4	3.8	0.5	7.7
C_{org}		79.6	19.9	185.6	46.9	7.5	108.0
Fulvic acids		69.9	15.3	158.6	38.7	3.0	96.0
Humic acids		24.2	1.1	61.5	13.2	0.3	46.1
NH_4^+		4.1	0.2	87.0	2.3	0.4	10.5
NO_2^-		0.03	0.01	0.28	0.02	0.01	0.73
NO_3^-		0.8	0.05	2.7	1.2	0.5	4.3
PO_4^{3-}		0.2	0.01	58.8	0.9	0.1	3.4
Fe_{gen}		2.8	0.4	18.3	9.3	0.4	46.2
Mn		0.03	0.01	0.7	0.2	0.01	5.9
Si		4.2	0.3	11.6	5.0	0.9	10.2
F		0.1	0.02	0.3	0.1	0.1	0.2
Al		0.2	0.04	1.2	0.5	0.3	0.8
Total petroleum hydrocarbons		2.9	0.10	16.10	0.05	0.01	0.40
Permanganate value	mgO_2/L	99.5	14.0	176.8	43.1	3.6	264.0
Zn		49.5	8.0	250.0	24.4	0.03	267.0
Cd		0.2	0.2	0.2	0.2	0.2	0.7
Pb		1.8	0.2	265.0	2.1	0.6	19.0
Cu	µg /L	2.5	0.7	25.0	3.6	0.6	706.0
As		1.8	0.3	5.0	5.7	5.0	54.0
Li		7.1	1.0	30.0	2.7	0.01	46.0
Sr		6.4	3.0	124.0	1.3	1.0	18.6
Number of samples		38			32		

3.2. Elemental composition of humic substances

To understand the mechanisms of natural organometallic complexes formation and assess the chemical properties of humic substances, first of all, it is necessary to analyze their elemental composition. In this regard three tests of water samples of eutrophic bog are analyzed.

The results of the humic substances composition are presented in Tables 2 and 3. The elemental composition of HA and FA (Table 2) corresponds to the average content of the humic acids elements released from natural waters [7, 13-15]. The amount of carbon in the HA and FA varies between 42.52 - 49.54 %. The hydrogen content of humic acids ranges from 3.82 to 4.13 %, the minimum value is characteristic for FA_1, maximum - for HA_2. The ratio of H:C reaches its highest value in the samples of HA_1 and FA_3, decreasing to 1.01 (HA_2 and FA_1). The observed fluctuation of the H:C ratio results mainly from different proportions of carbon in the aromatic lattice and carbon in the peripheral aliphatic chains.

Table 2. Elemental composition of humic and fulvic acids of bog water

	Weight percentage				Atomic percentage				Atomic ratio		
	C	H	N	O	C	H	N	O	H:C	O:C	C:N
HA_1	45.31	4.07	2.06	48.46	34.33	36.77	1.37	27.59	1.12	0.80	25.06
FA_1	42.52	3.61	1.72	52.15	33.78	33.97	1.15	31.11	1.01	0.92	29.37
HA_2	49.54	4.13	2.18	41.82	44.15	37.02	1.43	25.10	1.01	0.68	25.89
FA_2	44.74	3.82	2.11	49.33	34.73	35.11	1.40	28.76	1.02	0.83	24.81
HA_3	48.32	4.19	2.21	45.28	36.02	37.19	1.42	25.38	1.03	0.70	25.37
FA_3	43.21	4.11	1.93	50.75	32.79	36.99	1.27	28.96	1.13	0.88	25.82

The aromatic part of molecules (the degree of benzenoids) is more expressed in HA_2 than in HA_1. The degree of benzenoids of FA in all samples (except sample 1) is lower than that of HA.

The relations of O:C, which are caused by the content of oxygen in the carboxylic, alcoholic, phenolic and methoxy groups, reach the maximum values in FA_1, decreasing among FA_3 – FA_2 – HA_1 – HA_3 – HA_2. The amount of oxygen in humic substances is inversely proportional to the amount of carbon. C:N ratios for humic acids change in a narrow range (25.06-25.89), for fulvic acids – over a wider range (25.82-29.37).

Table 3. Characteristics of humic and fulvic acids of bog waters

	Oxidation level	Functional groups, mmol/g of specimen			Benzenoids
		Sum of functional groups	Carboxyl	Phenolic hydroxyl	
HA_1	0.53	10.6	6.8	3.8	0.24
FA_1	0.35	10.2	6.7	3.5	0.29
HA_2	0.38	11.2	6.3	4.9	0.42
FA_2	0.69	11.3	7.5	3.8	0.38
HA_3	0.64	11.4	6.5	4.9	0.39
FA_3	0.64	11.6	7.2	4.4	0.22

4. Summary and Conclusions

Thus, bog water of the studied area is ultra-fresh, with extremely high content of dissolved organic compounds, P, Fe, NH_4 and some trace components (Zn, Pb, Cu, Mn, Al). The tendency of increasing in concentration from the oligotrophic bogs to the eutrophic is typical almost for all elements of macrostructure; the inverse pattern identified for organic substances is presented mostly by fulvic and humic acids. In the composition of humic acids there are about 45 % carbon, 45 to 52 % oxygen, 4 % hydrogen and up to 2 % – nitrogen.

Acknowledgments
The reported study was funded by RFBR, according to the research project No. 16-35-60084 and by the State Assignment "Nauka" № 5.1931.2014/K.

References

[1] Savichev O G 2015 Geochemical Parameters of Bog Waters in the Taiga Zone of the Western Siberia *Izvestiya Rossiiskoi akademii nauk. Seriya geograficheskaya* **4** 47-57 [in Russian]
[2] Shvartsev S L 2008 Geochemistry of fresh groundwater in the main landscape zones of the Earth *Geochem. Int.* **46**(13) 1285-1398
[3] Vernadsky V I 2003 *History of Natural Waters* (Moscow: Science) p 751
[4] Evseeva N S 2001 *Geography of the Tomsk region. Environment and resources* (Tomsk: Publishing house of TGU) p 223
[5] Naymushina O 2014 Drinking water supply of tomsk (Western Siberia, Russia): Groundwater resources and quality. Paper presented at *the Int. Multidiscip. Sci. GeoConf. SGEM* **1**(3) 215-222
[6] Belchikova N P 1972 Methods of definition of functional groups of humic substances *Organic substance of the virgin and developed soils* (Moscow: Science) 239-241[in Russian]
[7] Orlov D S 1990 *Humic acids of soils and general theory of hymification* (Moscow: Publishing house of MSU) p 325 [in Russian]
[8] Rasskazov N M 2005 Major features of bog water chemistry: Case study of southeastern part of the Western Siberia *Izv. TPU* **4** 55-58 [in Russian]
[9] Naymushina O S, Shvartsev S L, Zdvizhkov M A and El-Shinawi A 2010 Chemical characteristics of swamp waters: A case study in the Tom river basin, Russia. Paper presented at the *Water-Rock Interaction - Proceedings of the 13th International Conference on Water-Rock Interaction* ***WRI-13*** 955-958
[10] Inisheva L I 2009 *Bolotovedenie* (Tomsk: Publishing house of TGPU) p 210 [in Russian]
[11] Kharanzhevskaya Y A, Voistinova E S and Ivanova E S 2011 Chemical composition and quality of bog water in the Basin of Chaya River *Contemporary Problems of Ecology* **1** 137- 145
[12] Krainov S R, Ryzhenko B N and Shvets V M 2012 *Groundwater Geochemistry Theoretical, Applied, and Ecological Aspects* (Moscow: CenterLitNefteGaz) p 670 [in Russian]
[13] Lavrik N L and Mulloev N U 2006 Effect of the Concentration of Humic Acids in Aqueous Solutions on the Structure of Their Macromolecules *Khimiya v interesakh ustoichivogo razvitiya* **14** (4) 379-390 [in Russian]
[14] Buffle J 1988 *Complexation reactions in aquatic systems: An analytical approach* (New York: Ellis Horwood Limited) p 692
[15] Schnitzer M 1978 *Soil Organic Matter* (Amsterdam: Elsevie) p 286

Thermodynamic simulation of the interaction of the system "water - mineral sediment - organic matter" at the diagenesis stage

E S Sidkina[1,2], B NRizhenko[1] and E VCherkasova[1]

[1]Vernadsky Institute of Geochemistry and Analytical Chemistry, Russian Academy of Sciences, Moscow, Russia
[2]Faculty of Geology, Moscow State University, Moscow, Russia

E-mail: [1]SidkinaES@yandex.ru

Abstract. The current study examines thermodynamic calculations of interaction of the "water - mineral sediment - organic matter" system at temperatures and pressures of diagenesis. In the course of reactions mature kerogen forms in the system, mainly $C_{292}H_{288}O_{12}$, rarely $C_{128}H_{68}O_7$ including accompanying substances (hydrocarbons, nitrogen compounds). The removal of CO_2 (g) and N_2 (g) from the system promotes the reaction. It is found that the aqueous phase during the formation of kerogen does not change significantly. In general, there occur desalination and changes of pH and Eh, as well as the increase in dissolved CO_2 content in water composition.

1. Introduction

Diagenesis is termed as a process, where the system is close to equilibrium in condition of a shallow burial and which leads to consolidation of the sediment. At the stage of subaqueous diagenesis, first of all, biogenic residues and rocks that differ in mineralogical composition accumulate at the bottom of ponds.

The analysis of materials showed that the first stages of transgressions and regressions [12] provide the most favorable conditions for the accumulation of organic matter (OM). They usually require a number of conditions: shallow water, accumulation of fine-grained material, active inflow of autochthonous and allochthonous organic matter, alternation of sea and freshwater facies. The continental slopes are regarded as another type of environment that is favorable for organic matter accumulation. They are characterized by normal sea conditions, i.e. high primary productivity of the coastal waters, as well as the inflow of materials formed on land (Figure 1). Here, as a rule, the optimum ratio between the rate of sedimentation and the dynamics of the water masses is maintained, which, in its turn, stipulates precipitation with a high content of OM. According to Tissot and Welte [7], water salinity has no direct effect on organic matter accumulation, but it largely depends on the bioproductivity of water. For high levels of biological productivity, the presence of nutrients (phosphate and nitrate), light, temperature typical for continental margins is of great importance. For preservation of organic matter on the bottom of the water body, clayey fine-grained sediment characterized by limited access of oxygen is the most favorable. The adsorption of the dissolved OM on the surface of the mineral particles increases its stability, since such an OM is more protected from the attack of bacteria and it quickly dips through the water column [7]. However, the accumulation of dissolved OM is of secondary importance from the viewpoint of petroleum formation, in comparison with the accumulation of detrital material, because the water-insoluble compounds are the main material for the formation of hydrocarbons (HC).

During silt deposition, system enters the stage of diagenesis. Freshly deposited sediments have high moisture and high porosity. Their moisture content may reach 86-96% according to [8]. When sediments go deeper, water column and organic matter deposited on the subsequent stages of sedimentation exert pressure on the buried sediments. With increasing pressure, porosity reduces and water is squeezed out from clays. In the process of diagenesis, some components of sediments dissolve, and authigenic minerals form. At the final stage of diagenesis loose sediments become a sedimentary rock.

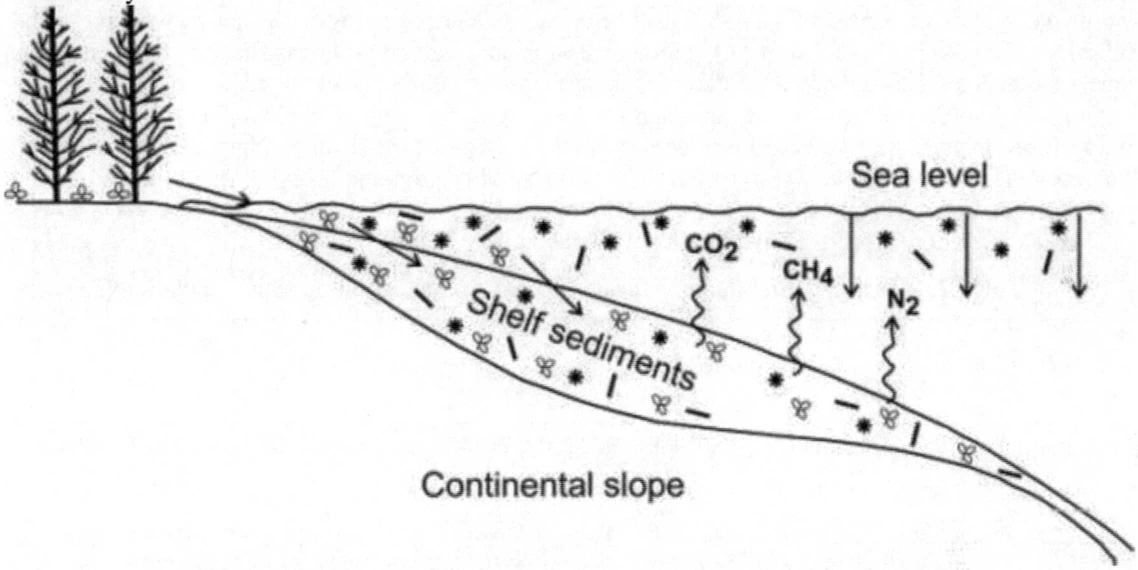

Figure 1. Schematic diagram showing environments of deposition.

2. Simulation scheme

Based on the geological conditions, the task is to model the interaction of the system components "mineral sediment - organic matter - pore water" under the condition of changing temperature and pressure. The chemical composition of three types of living matter, i.e. algae, zooplankton, and green plants, the organic matter are analyzed in the current study (Table. 1).

Table 1. The elemental composition of living matter and biosphere used in simulation.

Elements	Algae, planktonic algae			Bacteria			Plants			Biosphere
	1	*2*	*3*	*1*	*2*	*3*	*1*	*2*	*3*	*4*
C	39.1	34.5	48.2	56.4	54.0	52.4	45.2	45.4	50.0	18.0
H	4.6	4.1	7.7	4.8	7.4	7.9	6.2	5.5	6.1	10.5
O	53.2	47.0	36.9	27.4	23.0	28.0	46.7	41.0	42.5	70.0
N	1.7	1.5	7.2	11.4	9.6	11.7	1.9	3.0	0.2	0.3
S	–	1.2	–	–	0.53	–	–	0.34	–	0.05
The atomic ratio in living matter										
H/C		1.42			1.63			1.44		
O/C		1.02			0.32			0.77		

Note: 1 - authors' estimates; 2 - Ermakov, 2008; 3 - Kholodov, 2006; 4 - Vinogradov, 1954 [2]

The elemental composition of living matter is taken from [8] and [3]. For comparison, it was calculated on the basis of the chemical composition and the ratio of the mass of organic matter of the diagenesis main products. In our opinion, the data provided by V.V. Ermakov are more justified. According to the research literature, mineral sediments are mainly composed of clay (75%) and carbonate (25%) material. The chemical composition of the mineral component is taken from [5]. The chemical composition of the water is taken as a standard composition of seawater SW0 [4].

Based on thermodynamic simulation programs [1, 9], it is possible to simulate chemical reactions if the Gibbs free energy of the participants of the reactions and their masses are known. One of the solids that forms during the process is kerogen. Its free energy that can be of five types is shown in Table 2. These five types represented by Helgeson are reference models of kerogen, which correspond to the natural samples. Kerogen of $C_{515}H_{596}O_{72}$ composition is an organic matter that has not yet passed the stage of diagenesis. Other types contain fewer amounts of simple and mixed ethers, they do not contain alcohol groups which are abundantly present in the immature kerogen of $C_{515}H_{596}O_{72}$ composition. Diagenetic process causes deoxygenation of immature kerogen characterized by lower O/C ratios. The maturation is accompanied by the formation of aromatic groups. Among the discussed types, kerogen of $C_{128}H_{68}O_7$ composition is considered mature.

Thermodynamic properties of hydrocarbons are taken from [11].

Table 2. Standard thermodynamic properties at 25°C and 1 bar used in simulation [13].

H/C	O/C	Formula	ΔfG°	ΔfH°	S°	V°	$C°_p$	$C°_p=a+bT+cT^{-2}$		
								A	B($\times10^3$)	c($\times10^{-5}$)
1.68	0.053	$C_{415}H_{698}O_{22}$	80405	-2865436	2128.66	5654.1	2079.97	-2007.70	12178.7	405.88
1.30	0.047	$C_{406}H_{528}O_{19}$	402917	-1829146	1781.98	5086.4	1768.93	-944.01	8344.4	200.06
1.16	0.14	$C_{515}H_{596}O_{72}$	-1248076	-4123990	2133.34	6989.3	2211.50	-3577.45	16695.7	721.04
0.99	0.041	$C_{292}H_{288}O_{12}$	528693	-684086	1124.69	3398.2	1096.89	-107.37	3930.6	28.76
0.53	0.055	$C_{128}H_{68}O_7$	358000	52282	383.77	1320.7	390.60	-83.82	1539.9	13.60

Note: ΔfG° and ΔfH° cal mol^{-1}, S° and $C°_p$ cal mol^{-1} K^{-1}, V° cm^3 mol^{-1}

At the first stage of simulation, the system is added by a mixture of moisturized loose deposits composed of organic and mineral sediments. According to the research literature and after conversion to the unit of weight, the system represents a following mixture: 30 kg of "living" matter +14 kg carbonate mineral material + 56 kg of clay mineral material + 4258 kg of sea water. The amount of sea water is calculated based on the original porosity of moisturized silt, which can reach 90% or more. The mass of initial materials contributes to a mass of reaction products, which can be considered in the capacity of 100 kg of bottom silt. At each stage of simulation, the system is exposed to increasing temperature (from 10 to 60 ° C, step 1 °) and pressure (from 10 to 135 bar, step 2.5 bar), which corresponds to the temperature and baric conditions of diagenesis. Water is removed from the system in accordance with the data presented in [7, 8]. Thus, the system is exposed to the impact of diagenetic transformations and as a result of calculations the ultimate state of the system is reached. It is due to the fact that calculations are based on an equilibrium approach.

3. Simulation results.

As the result of the calculations, the following mineral products of reaction were obtained: quartz, kaolinite, K,Li-mica, Mg,Ca-dolomite, ankerite, siderite, pyrite, Sr, Ca-calcite. Along with CO_2, CH_4 and homologs, kerogen $C_{292}H_{288}O_{12}$ or kerogen $C_{128}H_{68}O_7$ formed; in some cases, both types of kerogen form. In addition to kerogen, the reaction mixture also yields carbon dioxide from hydrocarbons, predominantly methane: as the temperature increases, the concentration of CH_4 (aq) rises (experiments with algae and green plants), or insignificantly varies (zooplankton). At the same time, the proportion of alkanes in hydrocarbons decreases, while the concentration of CO_2 (aq) increases with increasing temperature. [6]. It is known that during sedimentation process sediments emit CO_2 (oxidation product of organic matters) and N_2 (the product of oxidation of ammonium nitrogen). In corresponding geological literature, there are no quantitative characteristics of the degassing, therefore, there was no possibility to reconstruct cleavage reaction of the living matter. Thus, the removal of gases from the system was made arbitrarily, depending on their amount. The available data on modeling is not enough to make a numerically reasoned judgment, but it is clear that degassing involves acidification of the aqueous phase: removal of CO_2 (g) reduces the concentration of bicarbonate and carbonic acid, and removal of N_2 (g) reduces the concentration of ammonium nitrogen.

In the course of simulation, sea water changes its pH and Eh values (Figure 2). In addition, the desalination effect is also registered, which corresponds to the data provided in [10]. The chemical composition of water is Cl-HCO_3 Na with a very low SO_4 content. During the simulation, ionic strength of water reduces from 4 to 2.3 in all systems, regardless of the living matter type, therefore, salinity decreases. In all systems, a high content of ammonium nitrogen is registered in water.

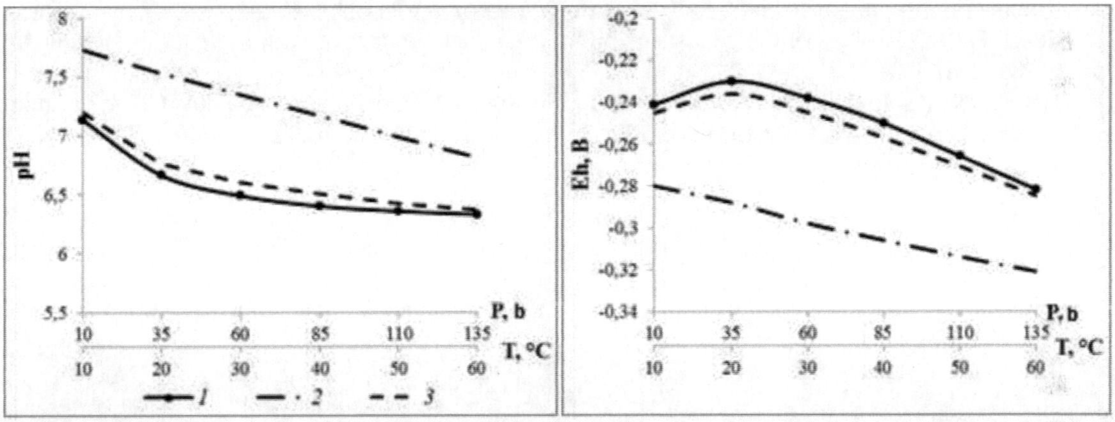

Figure 2. Changes in aqueous solution pH and Eh during the simulation process in the system with living matter: 1 - algae, 2 - zooplankton 3 - green plants.

4. Conclusions

Thermodynamic simulation of the interacting system "living matter - mineral material - standard sea water" was performed at the parameters of diagenesis. In the course of reactions, kerogen and accompanying substances (hydrocarbons, nitrogen compounds) formed in this system. It was shown that the removal of CO_2 (g) and N_2 (g) from the system stipulates the reaction. In addition, it was revealed that the aqueous phase during the formation of kerogen does not change significantly. In general, desalination, changes of pH and Eh, and increase in dissolved CO_2 content in water composition were registered.

Acknowledgments

The reported study was funded by RFBR, according to the research projects № 14-05-00266, 16-35-60047 mol_a_dk, thermodynamic calculations were performed with the support of RSF № 15-17-00010.

References

[1] Borisov M V and Shvarov Yu V 1992 *Thermodynamics of Geochemical Processes* (Moscow: Moscow State University Press) p 256

[2] Voitkevich G V, Miroshnikov A E, Povarennykh A S and Prokhorov V G 1970 *Brief guide of geochemistry* p 280

[3] Ermakov V V 2008 *Problems of Biogeochemistry and Geochemical Ecology* **3 (7)** 19–35
[4] Zaitsev I K 1986 *Hydrogeochemistry of the USSR* (Leningrad: Nedra) p 239
[5] Krainov S R, Ryzhenko B N, and Shvets V M 2012 *Groundwater Geochemistry. Theoretical, Applied, and Ecological Aspects* (Moscow: Tsentr LitNefteGaz) p 672
[6] Ryzhenko B N, Sidkina E S and Cherkasova E V 2015 Thermodynamic Modeling of Water–Rock Systems to Evaluate Their Generative Potential for Hydrocarbons *Geochem. Int.* **53 (9)** 825–837
[7] Tissot B P and Welte D H 1978 *Petroleum formation and occurrence* p 504
[8] Kholodov V N 2008 *Geochemistry of Sedimentary Process* (Moscow: GEOS) p 608
[9] Shvarov Y V 2008 HCh: new potentialities for the thermodynamic simulation of geochemical systems offered by windows *Geochem. Int.* **46 (8)** 834–839
[10] Shishkina O V 1972 *Geochemistry of Marine and Oceanic Sludge Water* (Moscow: Nauka) p 228
[11] Amend J P and Helgeson H C 1997 Group additivity equations of state for calculating the standard molal thermodynamic properties of aqueous organic species at elevated temperatures and pressures *Geochim. Cosmochim. Acta* **61 (1)** 11–46
[12] Bitterli P 1963 Aspects of the genesis of bituminous rocks sequences *Geologie on mijnbouw* **42** 183–201
[13] Helgeson H C, Richard L, VcKenzie W F, Norton D L and Schmitt A 2009 Chemical and thermodynamic model of oil generation in hydrocarbon source rocks *Geochim. Cosmochim. Acta* **73** 594–695

Isotopic composition ($\delta^{18}O$ and δD) of the shallow groundwater in the Poyang Lake basin

E A Soldatova[1,3], Z Sun[2] and N V Guseva[1,4]

[1] National Research Tomsk Polytechnic University, 30 Lenin Avenue, Tomsk, 634050, Russia
[2] East-China University of Technology, 418 Guanglan Avenue, Nanchang, 330013, China

E-mail: [3]2a61@mail.ru, [4]guseva24@yandex.ru

Abstract. The article is focused on the identification of evaporation and other natural processes which affect the isotopic composition of shallow groundwater in the Poyang Lake basin, such as water–rock interaction and mixing of the shallow groundwater and surface water. For this purpose the dual isotope approach (δD–H_2O and $\delta^{18}O$–H_2O) was used. The samples were collected from domestic wells around the Poyang Lake. The value of δD obtained for the shallow groundwater ranges from -21.5 to -42.6‰. The $\delta^{18}O$ value varies from -3.5 to -7.1‰. It was found that the shallow groundwater of the Poyang Lake catchment is of meteoric origin. The influence of evaporation on the isotopic composition of shallow groundwater is negligible and observed mainly during the dry season. The deviation from the local meteoric water line, especially during the rainy season, may be explained by the processes in the water–rock system, but this issue is required further research.

1. Introduction

The Poyang Lake basin is a unique ecosystem, which is affected by both natural and anthropogenic factors: water–rock interaction, evaporation, agricultural activity, dense population, etc. With such variety of factors, it is often difficult to identify the main processes which have an effect on groundwater origin. One of the modern techniques to identify these processes is the use of stable isotope tracers, such as ^2H and ^{18}O, which allows tracing various natural processes occurred in groundwater [1].

It is worth to note that the shallow groundwater of the study area is used by people in rural areas for drinking and other domestic purposes. Meanwhile, there is often lack of centralized water supply. In this case, special attention should be paid to identification the processes affecting the origin of groundwater chemical composition. However, these issues are not sufficiently studied in the region. Besides ecological implication, these studies are crucial for fundamental research of groundwater origin, because the dual isotope approach (δD–H_2O and $\delta^{18}O$–H_2O) enables to track processes in the water–rock system [2].

The article is focused on the identification of evaporation and other natural processes which affect the isotopic composition of shallow groundwater in the Poyang Lake basin. The potential interaction between the shallow groundwater and surface water is also considered.

2. Materials and Methods

2.1. Study area

The Poyang Lake basin is located in Jiangxi Province, the southeast of China (figure 1). The study area belongs to subtropical humid climate with the average annual temperature 17.5 °C [3]. The annual precipitation ranges from 1400 to 2400 mm [4]. Rainfall distribution during the year is irregular because it is controlled by the East Asian monsoon. The rainy season lasts from March to June, and then rainfall decreases from July to September with evaporation reaching its maximum point [5]. The dry season begins after September and lasts till February. The topography of the Poyang Lake basin varies from mountain (maximum elevation of about 2200 m above sea-level) to alluvial plains [3]. However, in the study area the relief is predominantly flat with alluvial plains and hills.

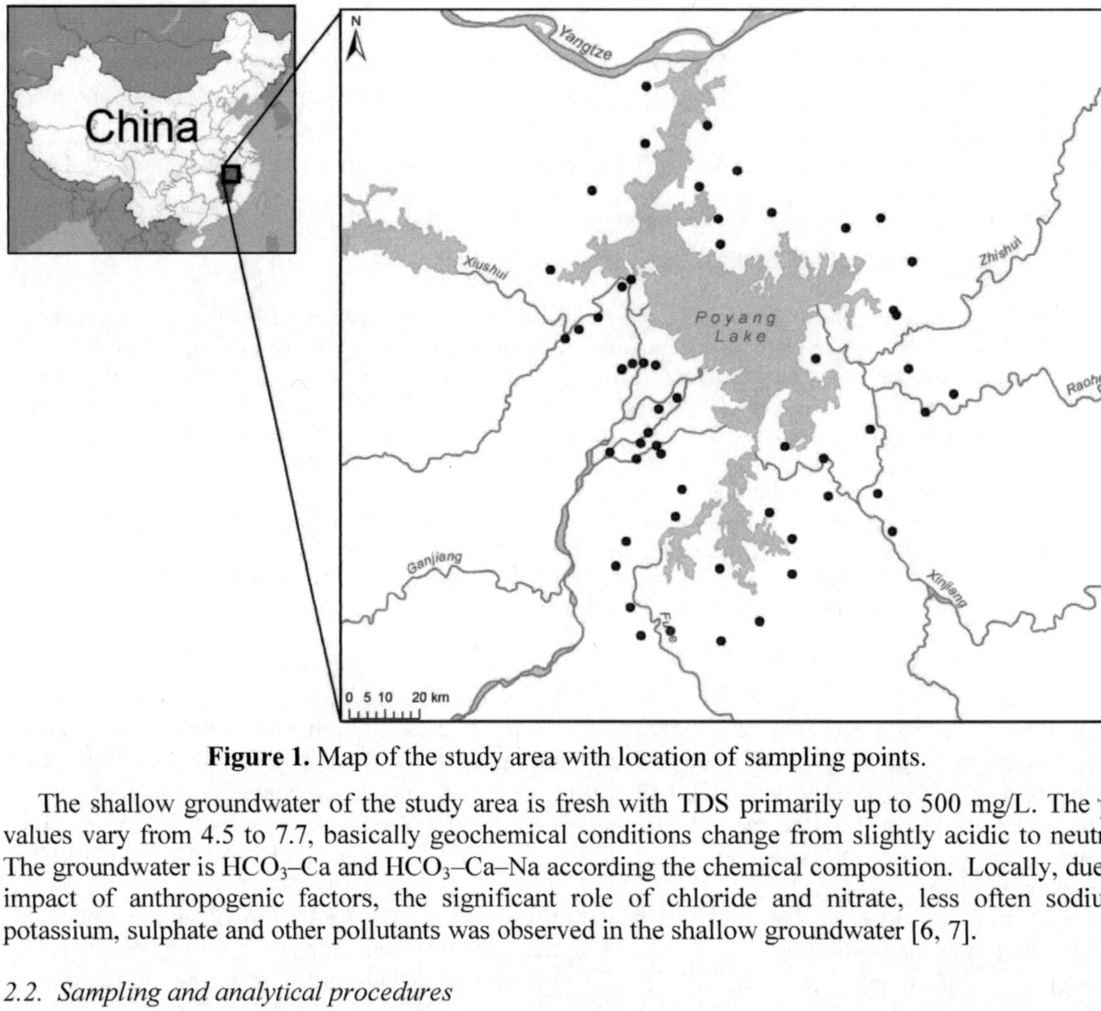

Figure 1. Map of the study area with location of sampling points.

The shallow groundwater of the study area is fresh with TDS primarily up to 500 mg/L. The pH values vary from 4.5 to 7.7, basically geochemical conditions change from slightly acidic to neutral. The groundwater is HCO_3–Ca and HCO_3–Ca–Na according the chemical composition. Locally, due to impact of anthropogenic factors, the significant role of chloride and nitrate, less often sodium, potassium, sulphate and other pollutants was observed in the shallow groundwater [6, 7].

2.2. Sampling and analytical procedures

Hydrogeochemical sampling was carried out in October 2013 and March 2014. The shallow groundwater from wells used by local people for drinking and domestic purposes was sampled in sterile 100 ml opaque polyethylene vials. Isotopic composition (δD, $\delta^{18}O$) was studied in the laboratory of the East China University of Technology (Nanchang, China) using isotope ratio mass spectrometer with element analyzer TC/EA-IRMS (Finnigan MAT 253). The isotopic ratio of H/D was determined by the sample combustion at the temperature of 1350 °C. The determination of the isotope ratios of $^{18}O/^{16}O$ was carried out at the temperature of 25 °C after addition of 0.1% CO_2 solution and keeping within 20 hours until an equilibrium state. The method error is ≤2‰ for δD and

≤0.2‰ for $\delta^{18}O$, respectively. Data on isotopic composition are given in accordance with the V-SMOW.

3. Results and Discussions

The value of δD ranges from -21.5 to -42.6°/₀₀ (Table). The $\delta^{18}O$ value in the shallow groundwater varies from -3.5 to -7.1°/₀₀. It slightly differs from the value of $\delta^{18}O$ in a marine atmosphere in the tropical and subtropical regions (0 – -5°/₀₀) presented by G. Craig and L. Gordon [8].

The study of δD and $\delta^{18}O$ enable to determine genesis of natural water and track the contribution of various factors to origin of its chemical composition. The ratio of the δD to $\delta^{18}O$ values in atmospheric precipitation was described for the first time by G. Craig (Global meteoric water line) [9]:

$$\delta D=8\delta^{18}O+10°/_{oo}, (1)$$

where 10°/₀₀ – excess parameter *d*, defining the deuterium excess in precipitation compared with its quantity in the equilibrium process, where *d*=0 [1]. However, the slope of meteoric water line and excess parameter *d* vary depending on the distance between the places of water vapour formation and atmospheric precipitation, temperature, and other factors. Since equation (1) is correct only for the world averaged annual data on the isotopic composition (δD, $\delta^{18}O$) of precipitation, it is often necessary to introduce regional coefficients [1].

Table. Isotopic composition (δD, $\delta^{18}O$) of the shallow groundwater in the Poyang Lake basin, °/₀₀.

Sample	$\delta D_{V\text{-}SMOW}$	$\delta^{18}O_{V\text{-}SMOW}$	Sample	$\delta D_{V\text{-}SMOW}$	$\delta^{18}O_{V\text{-}SMOW}$
P2	-29.1	-3.5	**P37**	-42.1	-6.3
P7	-37.2	-6.1	**P38**	-41.2	-5.3
P8	-37.4	-6.1	**P39**	-34.4	-6.6
P9	-29.1	-5.1	**P42**	-39.4	-6.7
P10	-38.2	-6.2	**P43**	-39.0	-6.8
P11	-37.5	-6.2	**P44**	-32.8	-5.2
P12	-29.6	-5.8	**P45**	-26.8	-4.5
P13	-38.0	-6.5	**P46**	-28.7	-5.0
P14	-25.5	-4.6	**P47**	-40.0	-6.9
P15	-21.5	-3.7	**P48**	-35.3	-6.3
P16	-29.5	-4.6	**P49**	-37.1	-6.4
P17	-30.4	-4.3	**P50**	-37.2	-6.5
P18	-34.6	-5.5	**P51**	-42.6	-7.1
P19	-32.4	-5.3	**P52**	-39.7	-6.8
P20	-29.6	-4.9	**P53**	-38.9	-6.7
P24	-39.2	-6.4	**P54**	-30.4	-5.1
P25	-30.2	-5.8	**P56**	-35.4	-6.3
P27	-32.2	-5.9	**P57**	-25.0	-4.7
P28	-28.9	-5.8	**P59**	-35.3	-6.2
P29	-34.9	-6.8	**P60**	-34.0	-6.0
P30	-31.3	-5.6	**P61**	-31.4	-6.3
P32	-32.2	-6.2	**P63**	-28.2	-5.6
P33	-30.3	-5.9	**P64**	-31.8	-5.3
P34	-32.3	-6.0	**P65**	-29.6	-4.6
P35	-36.4	-6.3	**P66**	-30.3	-5.4
P36	-35.0	-6.2	**P67**	-32.7	-5.7

The ratio of δD to $\delta^{18}O$ in the studied groundwater has been estimated using the global meteoric water line modified for region with average annual temperature less than 20°C. The distance between the study area and the ocean as the main source of water vapor formation has also been took into account, based on data, given in [1]. Equation of the modified meteoric water line is the following:

$$\delta D=8\delta^{18}O+9.83°/_{oo}, (2)$$

The local meteoric water line obtained by Zhou and Sun et al. for Lushan hot springs area [10–12], which situated the northwest of the Poyang Lake, has also been considered. This line is described by the following equation:

$$\delta D=7.16\delta^{18}O+8.88‰, (R^2=0.98), (3)$$

The points corresponding to the values of δD and $\delta^{18}O$ in the studied shallow groundwater are located close to the modified meteoric water line, but the local meteoric water line slightly better approximate them (figure 2). In addition, they are close to the average values calculated by G. Craig and L. Gordon [8] for continental precipitation but they are depleted in heavy isotopes.

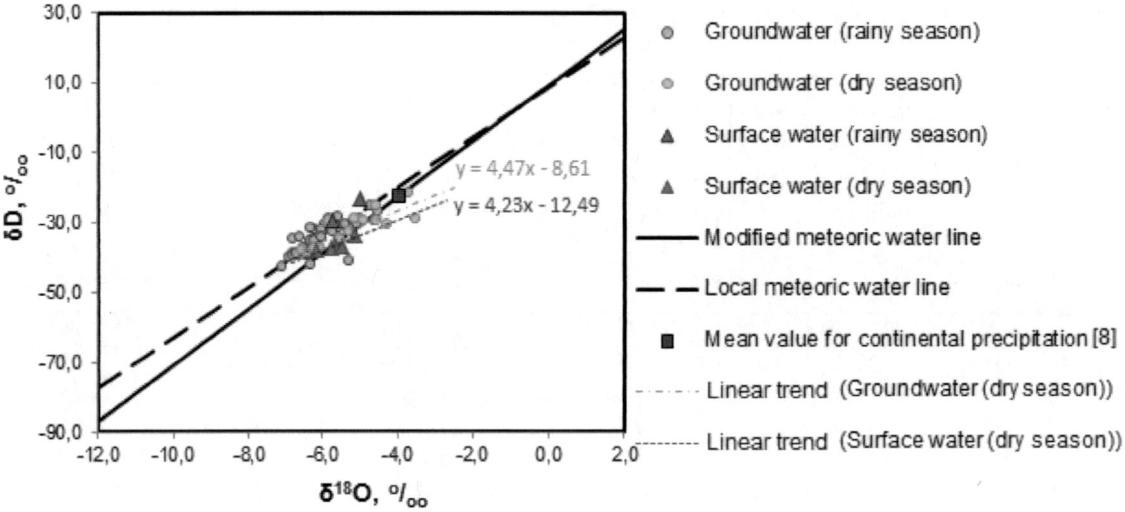

Figure 2. Values of $\delta^{18}O$ vs. δD in natural water of the Poyang Lake basin.

It should be noted that the samples collected during the dry season are slightly enriched in heavy isotopes of oxygen and hydrogen than the water taken during the rainy season (figure 3). The difference between the average values of δD and $\delta^{18}O$ in the rainy and dry seasons is statistically significant. This fact can be explained by the climate patterns. In the course of the dry season evaporation usually exceeds precipitation. As a result of evaporation, water is enriched in heavy isotopes, mainly ^{18}O, during their infiltration through the unsaturated zone. The slope coefficient of the linear trend line is close to 5 for the groundwater samples collected during the dry season (figure 2). It corresponds to evaporation process [1, 2, 13]. In this regard, in the course of the dry season, the deviation from the meteoric water line is likely connected with the influence of evaporation on groundwater. At the same time this linear trend line is very similar to the regression line for the surface water, which was also collected during the dry season. Surface water is normally subjected to more intensive evaporation than shallow groundwater and, as a result, more enriched with heavy isotopes ^{18}O and ^{2}H [13]. Therefore, we can suggest that mixing of the shallow groundwater and surface water may occur where conditions are appropriate (low-lying alluvial plains, flooded fields, etc.). The surface water within the study area may be presented not only by natural water bodies, but also by artificial ponds for farming and water for rice field flooding, which can infiltrate in the shallow aquifer.

In the early rainy season, the amount of precipitation increases sharply, which leads to the depletion of heavy isotopes of hydrogen and oxygen in the shallow groundwater. However, the samples collected during the rainy season also demonstrate the considerable deviations from the local meteoric water line. The most significant deviations are connected with the isotope ^{18}O enrichment. According to some authors [1, 2], such deviation may be explained by the processes in the water–rock system, which can lead to fractionation of oxygen and hydrogen isotopes.

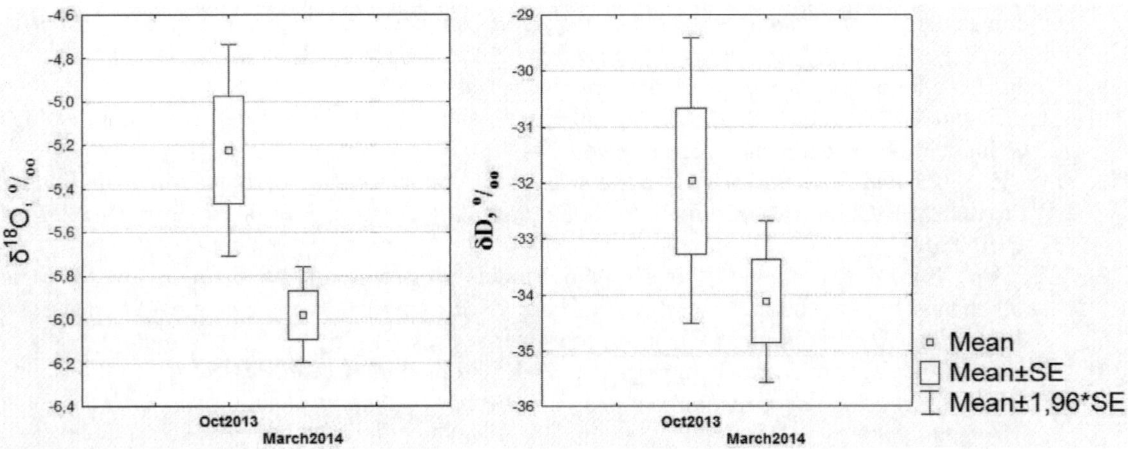

Figure 3. Box-and-whiskers plot for isotopic composition ($\delta^{18}O$, δD) of the shallow groundwater in different seasons.

4. Conclusion

We found that the shallow groundwater of the Poyang Lake catchment is of meteoric origin according to its isotopic composition. The influence of evaporation on the groundwater chemical composition is negligible and observed mainly during the dry season. This factor may affect the groundwater chemical composition indirectly due to local mixing of the shallow groundwater and surface water, subjected to more intensive evaporation. In addition to the above mentioned fact, the deviations from the local meteoric water line, especially in the rainy season, may be explained by the processes in the water–rock system, but further studies are required.

Acknowledgements

The research was supported by The Russian Foundation for Basic Research (Project No 14-05-31267) and State Assignment "Nauka" № 5.1931.2014/K. We thank Prof. S.L. Shvartsev and Dr. L.V. Zamana for valuable advice and constructive comments.

References

[1] Ferronsky V I and Polyakov V A 2012 *Isotopes of the Earth's Hydrosphere* (Springer Netherlands)

[2] Zamana L V 2012 Isotopes of hydrogen and oxygen in nitrogen hot springs of Baikal Rift Zone in terms of interaction in the water–rock system *Doklady Earth Sciences* **442** 1 81–85 DOI: 10.1134/S1028334X12010114

[3] Ye X, Zhang Q, Liu J, Li X and Xu C Y 2013 Distinguishing the relative impacts of climate change and human activities on variation of streamflow in the Poyang Lake catchment, China *J. of Hydrology* **494** 83–95

[4] Wang Q, Riemann D, Vogt S and Glaser R 2014 Impacts of land cover changes on climate trends in Jiangxi province China *Int. J. Biometeorol.* **58** 645–660

[5] Li X and Zhang Q 2011 Estimating the potential evapotranspiration of Poyang Lake basin using remote sense data and Shuttleworth-Wallace model *Proc. Environ.Sci.* **10** 1575–82

[6] Soldatova E A, Guseva N V, Sun Z and Mazurova I S 2015 Size fractionation of trace elements in the surface water and groundwater of the Ganjiang and Xiushui River basin, China *IOP Conf. Ser.: Earth Environ. Sci.* **27** 012037 doi:10.1088/1755-1315/27/1/012037

[7] Soldatova E, Guseva N and Wang G 2014 Characteristic features of groundwater pollution in the Poyang Lake catchment *IOP Conf. Ser.: Earth Environ. Sci.* **21** 012023 doi:10.1088/1755-1315/21/1/012023

[8] Craig H and Gordon L 1965 Deuterium and oxygen-18 variations in the ocean and the marine

atmosphere *Stable Isotopes in Oceanographic Studies and Paleotemperatures (Spoleto Conferences in Nuclear Geology. Proceedings)* ed E Tongiorgi (Spoleto) pp 9–130

[9] Craig H 1961 Isotopic variations in meteoric water *Science* **133** 1702–1703

[10] Sun Z and Li X 2001 Studies of geothermal waters in Jiangxi Province using isotope techniques *Science in China (Series E)* **44** 144–150

[11] Sun Z, Liu J and Gao B 2010 Hydrogeochemistry and direct use of hot springs in Jiangxi Province, SE-China *Proceedings World Geothermal Congress*; *Bali, Indonesia, 25–29 April 2010* 2807

[12] Zhou W, Sun Z, Li X and Shi W 1996 Studies of geothermal background and isotopic geochemistry of thermal waters in Jiangxi Province *China Nuclear Science and Technology Report* (Beijing: China Atomic Energy Press) 1–25

[13] Gonfiantini R, Gratzini S and Tongiorgi E 1974 Environmental isotope hydrology in the Bodna region, Algeria *Isotope Techniques in Groundwater Hydrology. Proc. Symp. IAEA* **1** 293–316

Geochemical peculiarities of nitric thermal waters in Jiangxi Province (SE-China)

Z Sun[1], S L Shvartsev[2] , O G Tokarenko[3], E V Zippa[4] and B Gao[5]

[1, 5] State Key Laboratory Breeding Base of Nuclear Resources and Environment, East China University of Technology, Nanchang, 330013, Jiangxi Province, China
[2] Tomsk Branch of Trofimuk Institute of Petroleum Geology and Geophysics, Siberian Division, Russian Academy of Sciences, 4 Akademicheskii pr., Tomsk, 634021, Russia
[3,4] Tomsk Polytechnic University, 30 Lenin Avenue, Tomsk, 634050, Russia

E-mail: [1] zhxsun@ecit.edu.cn, [2] tomsk@ipgg.sbras.ru, [3] tog@tpu.ru, [4]zev-92@mail.ru

Abstract. The chemical composition of nitric thermal waters in Jiangxi Province, SE-China, is considered. It is shown that the studied thermal waters are characterized by low TDS (293-412 mg/l), but they always have alkaline or highly alkaline pH values. It should be noted that the content of Na^+, Si, F^-, HCO_3^-, CO_3^{2-}, sometimes SO_4^{2-} is predominant, on the other hand, the content of Ca^{2+}, Mg^{2+}, Cl^- is low. It has been established that the main factors responsible for low concentrations of some chemical elements and high concentrations of others are continuous dissolution of aluminosilicate minerals and simultaneous precipitation of the secondary minerals.

1. Introduction

Jiangxi Province is located in the Southeastern part of China and is one of the provinces where nitrogen-rich thermal waters are distributed most widely in the country. A systematic compilation of Province's hot springs was done by Li Xueli in 1979 [4], later by Sun Zhanxue in 1988 [10], and it is continuing in the present time. In this Province, about 96 hot springs of different gas and chemical composition have been found. They are located in the following geothermal areas: the Ganbei (the northern part of Province), the Ganzhong (a central Province), the Gangxinan (the south-western Province) and the Gandongnan geothermal area (the southern Province). The nitric thermal waters are localized, mainly, in the Ganbei and the Gangxinan geothermal areas.

2. Geological background

Nitric thermal waters are widely spread, and they occur in the rocks of different types and ages all over Jiangxi. The majority of hot springs is located along the faults of different trends (fig.1). Nitric thermal waters are distributed all over geothermal areas, but, mainly, in the Ganbei and the Gangxinan.

The Ganbei geothermal area is situated in the north of Jiangxi. Thermal waters of the area are confined to the north-eastern trending Ganjiang deep fault, which is registered to be still tectonically active. Geologically, the studied area is composed of the Precambrian rocks represented by granites of migmatitic and biotite genesis [13].

The Ganxinan geothermal area is located in the south-western Jiangxi Province, where NNE-trending faults are widely distributed. The rocks are characterized by sandstone, conglomerates and granite of the Cretaceous and the Jurassic age [3].

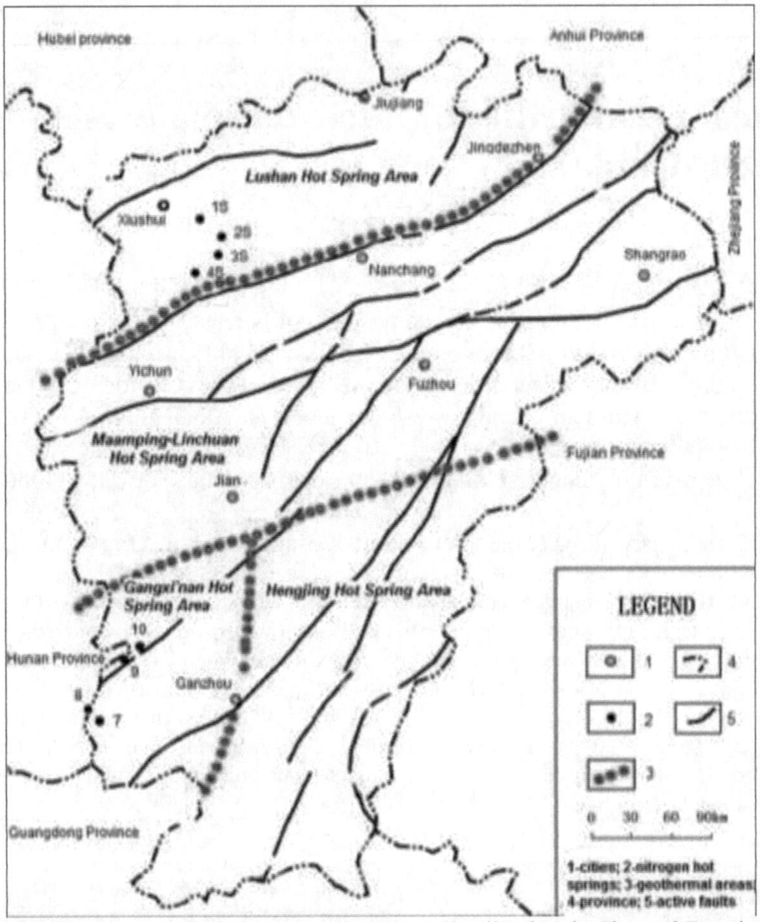

Figure 1. Scheme of nitric thermal waters distribution in Jiangxi Province.

3. Materials and methods

The factual data on chemical composition of nitric thermal waters are obtained due to the field works carried on in October, 2015. The macro component water composition has been analyzed by an ion exchange chromatography method using the ICS-000 «Dionex» device (the USA) in the accredited laboratory of «Water» Research Educational Center (Natural Research Tomsk Polytechnic University). Moreover, the earlier data of nitric thermal water chemical analysis (obtained by Sun Z. [11]) were used.

4. Results and Discussion

Experimental data on the chemical composition, temperature, pH and salinity of nitric thermal waters are given in Table.1. Firstly, it should be noted that the total mineralization dissolved solids (TDS) of thermal waters is extremely low, at the same time, pH is high. That is one of the peculiarities of nitric thermal waters all over the world [1], particularly, in Baikal Region [6]. The TDS values vary from 293 to 412 mg/l, the average value is 357 mg/l. The pH values are between 8.6 and 9.27, and the average value is 8.73.

The pH-TDS dependence in nitric thermal waters is difficult to see. However, relationship between pH and salinity of waters is characterized by a slight decline in their pH values with increasing TDS (Figure 2a). This type of correlation is not typical for the most fresh and saline natural waters [9].

According to chemical composition, nitric thermal waters are mainly sodium hydrocarbonate, excluding a single case, hot spring 8, where waters become sodium hydrocarbonate-sulfate.

Table 1. Chemical composition of nitric thermal waters in Jiangxi Province, mg/l.

№	T, C	TDS	pH	HCO_3^-	CO_3^{2-}	Cl^-	SO_4^{2-}	Na^+	K^+	Ca^{2+}	Mg^{2+}	SiO_2	F^-	Chemical type
7	36,8	293	8,70	110	12,2	5,3	14,8	71	1,3	3,8	0,6	56	16	HCO_3-Na
8	83	343	8,66	86	18,3	3,8	44,0	59	3,1	6,4	0,1	113	10	HCO_3-SO_4-Na
9	82,3	339	8,60	128	6,1	3,8	12,5	71	2,8	3,9	0,05	97	15	HCO_3-Na
10	41,1	306	9,27	67	30,5	5,3	23,0	66	2,2	2,0	0,01	96	15	HCO_3-Na
1S*	71	412	8,60	177	8,6	5,5	14,4	110	-	1,7	0	80	15	HCO_3-Na
2S*	65	396	8,78	163	11,0	5,1	13,0	96	-	2,7	0,7	90	15	HCO_3-Na
3S*	65	399	8,62	177	8,3	5,4	11,4	110	-	1,9	0	70	15	HCO_3-Na
5S*	69	367	8,61	166	8,0	5,0	13,0	87	1,7	2,1	0,2	70	14	HCO_3-Na
Average	70,6	357	8,73	134	12,9	4,9	18,3	84	2,2	3,1	0,2	84	14	HCO_3-Na

*Data from Sun [10]

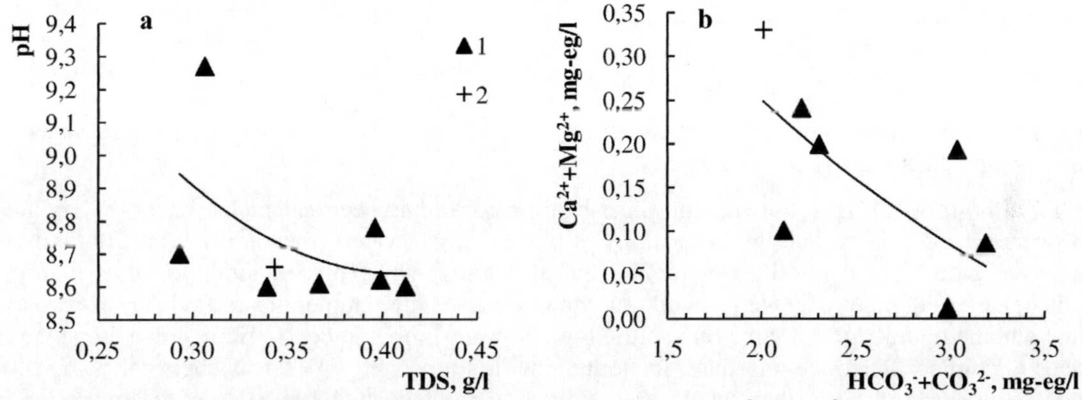

Figure 2. Dependence of TDS values on pH (a) and relation of Ca^{2+} и Mg^{2+} content to amount of carbonate ions (b). Water chemical type: 1 – HCO_3-Na; 2 – HCO_3-SO_4-Na.

As to cation content of nitric thermal waters, the dominance of sodium ion in calcium, potassium and magnesium is noteworthy. This is another peculiarity of nitric thermal waters chemical composition. It can be explained by inverse relations of Ca^{2+} and Mg^{2+} content to the amount of carbonate ions (Figure 2b), which confirms the equilibrium of nitric thermal waters with carbonate minerals, and, consequently, precipitation of these elements from the solution.

Dependence of studied waters salinity on their temperature is negligibly small. As temperature grows up, the TDS values increase (Figure 3). This tendency is related to falling of infiltration waters in the deep faults. Thus, interaction of water and water-bearing rocks takes more time, what leads to the salinity increase. But this evidence is concerned only with carbonate part of ions (Figure 4a), while sulfate ions behave in a different way: with the temperature increase the SO_4^{2-}content decreases (Figure 4b). According to this fact, it can be concluded that enrichment of nitrogen thermal waters by carbonate and sulfate ions comes from different sources. If carbonate salts are the result of aluminosilicates hydrolysis, sulfate salts are formed with sulphides and free O_2, which appears in water in the recharged areas.

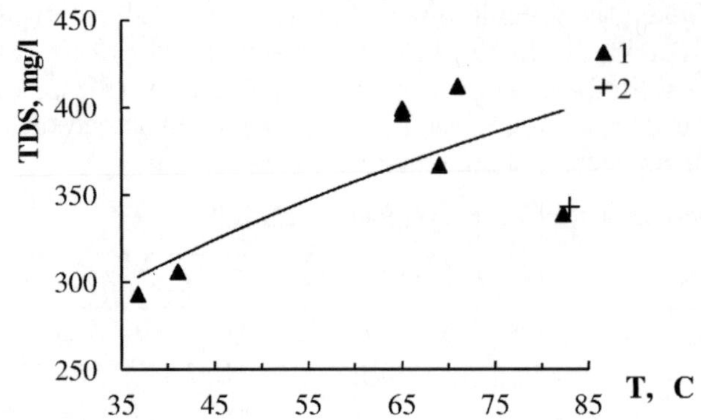

Figure 3. Dependence of nitric thermal waters salinity on temperature.

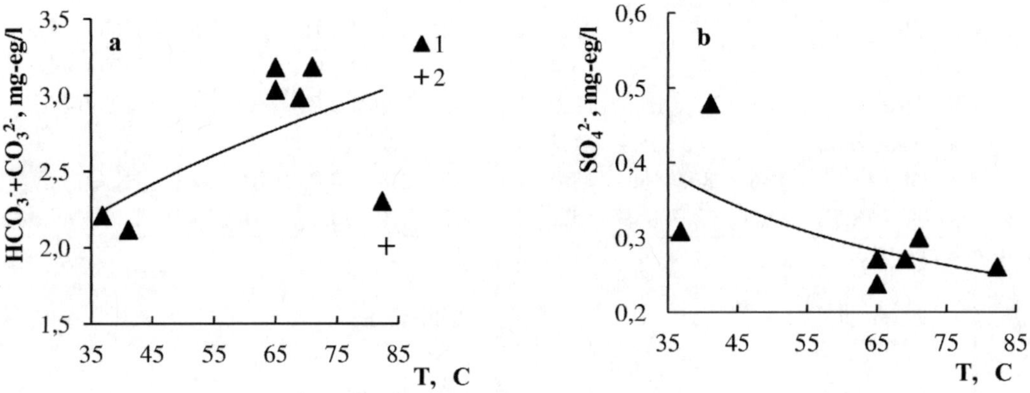

Figure 4. Dependence of content of carbonate (a) and sulfate (b) ions in nitric thermal waters on their temperature.

Equilibrium with main minerals of water-bearing rocks has been calculated using the HydroGeo software system [2]. Despite the low salinity of nitric thermal waters, they are saturated not only with calcite (Figure 5a) and fluorite (Figure 5b), but also with such exotic minerals as albit (Figure 6a), glaucophane (Figure 6b), as well as with microcline, muscovite, laumontite and others, which are not shown in the figures. At the same time, nitric thermal waters are non-equilibrium with endogenous Ca, Mg, Fe aluminosilicates of anorthite, forsterite, cheilitis types, etc. As far as chemical compositions and temperature of nitric thermal waters of Jiangxi province and Baikal Region are similar, the equilibrium character with minerals of rocks is identical, as well [6, 8].

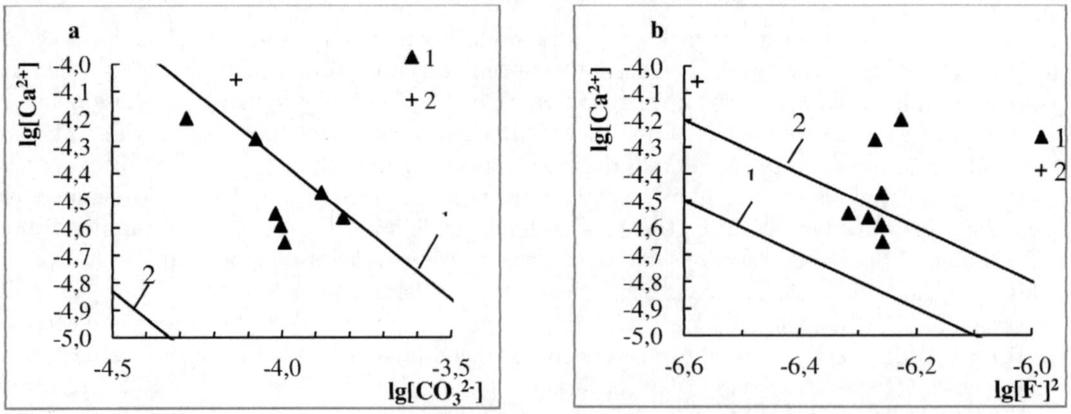

Figure 5. Equilibrium of nitrogen thermal waters with calcite (a) and fluorite (b) at 25 °C (line 1) and 100 °C (line 2).

Therefore, it can be confirmed that a water-rock system is equilibrium-non-equilibrium: water dissolves some minerals and forms the other ones. In the studied nitric thermal waters the major portions of the elements passing into solution are precipitated into newly formed minerals: Ca - into calcite, montmorillonite, zeolites, and other aluminosilicates; Mg - into glaucophane, talc, chlorite, and zeolites; Na - into albite and zeolites; K - into illite, muscovite, microcline, biotite, phlogopite, etc.; Fe - into siderite and Fealuminosilicates; and F - into fluorite and, in part, micas. This allows the system reaching both chemical and dynamic equilibrium and establishing a balance between the amounts of elements released into and precipitated from the solution. Such a balance leads to the low salinity of hot spring waters, which tends to remain constant or increase slightly with time [8].

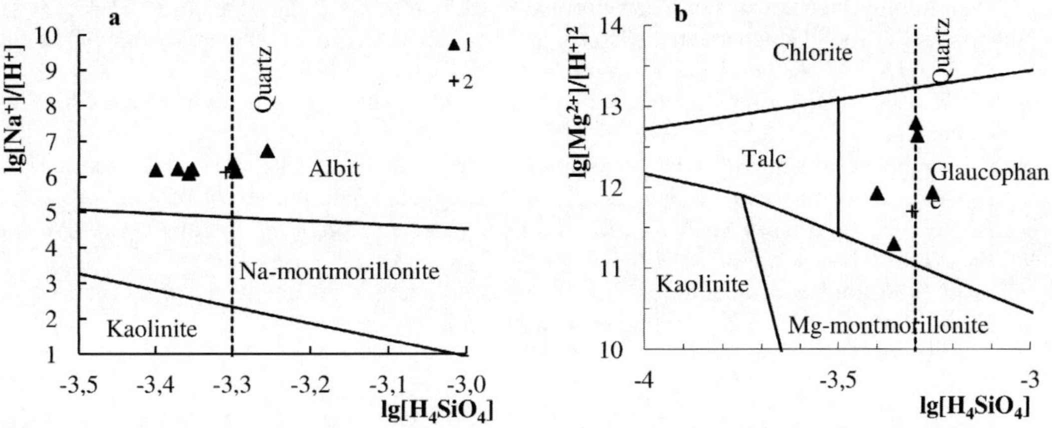

Figure 6 Equilibrium of nitrogen thermal waters with alominosilicate minerals: (a) – system SiO_2-Al_2O_3-K_2O-CO_2-H_2O at 100 °C; (b) – system HCl-H_2O-Al_2O_3-MgO-SiO_2 at 100 °C.

5. Conclusions

In Jiangxi Province, nitric thermal waters are widely distributed, and their chemical composition, salinity and pH values are similar with nitric thermal waters of other regions all over the world. These thermal waters and crustal bed rocks form a unique equilibrium–non-equilibrium system. This system evolves under conditions, when the majority of elements released from the host minerals tend to be continuously incorporated into newly formed secondary minerals. This leads to a dynamic equilibrium between the amount of elements liberated to and precipitated from the solution; as a result, the composition of the solution is stabilized, and the salinity does not longer increase, though, the interaction of water with the surrounding host rocks continues.

Acknowledgments

The study was financially supported by grant 15-55-53122 of the Russian Foundation for Basic Research «Nitric Thermal Waters Geochemistry in Transbaikal (Russia) and Jiangxi Province (SE-China)» and National Natural Science Foundation of China under the Sino-Russia Cooperation Project No.41511130031.

References

[1] Baskov E A and Surikov S N 1989 *Hydrotherm of the Earth* (Nedra) p 243 [in Russian]
[2] Bukaty M B 2002 Development of software for solving hydrogeological problems *Izv. Tomsk. Politekhn. Inst.* **305** (6) 348–365 [in Russian]
[3] Chen G, Wang G, Sun Z and Liu J 2010 The Isotopic and Chemical Characteristics of Geothermal Fluids from Two Selected Hot Spring Areas in Jiangxi Province, SE-China *Proc. World Geothermal Congress* (Bali, Indonesia) 6
[4] Li X 1979 The relationship between distribution of thermal waters and uranium mineralization in Jiangxi *J. of East China Geolog. Inst.* 21-29

[5] Lomonosov I S 1974 *Geochemistry and genesis of modern thermal waters in Baikal Rift Zone* (Novosibirsk: Science) p 168 [in Russian]

[6] Plyusnin A M, Zamana L V, Shvartsev S L, Tokarenko O G and Chernyavskii M K 2013 Hydrogeochemical peculiarities of the composition of nitric thermal waters in the Baikal Rift Zone *Russ. Geol. Geophys.* **54** (5) 647–664

[7] Shvartsev S L, Tokarenko O G, Sun Z and Zippa E V 2015 *Geochemistry of fluorine in nitric thermal water of Zabaikalie* 4 [in Russian]

[8] Shvartsev S L, Zamana L V, Plyusnin A M and Tokarenko O G 2015 Equilibrium of Nitrogen Rich Spring Waters of the Baikal Rift Zone with Host Rock Minerals as a Basis for Determining Mechanisms of Their Formation ISSN 0016 7029 *Geochem. Int.* **53** 8 713–725

[9] Shvartsev S L 2008 Geochemistry of fresh groundwater in the main landscape zone of the Earth *ISSN 0016 7029, Geochemistry International* **46** 13 1285–1398.

[10] Sun Z X 1988 The formation conditions of hot springs in Jiangxi Province, SE-China *East China Geolog. Inst.* 50

[11] Sun Z X, Liu J and Gao B 2010 Hydrogeochemistry and Direct Use of Hot Springs in Jiangxi Province, SE-China *Proc. World Geothermal Congress* (Bali, Indonesia.) 5

[12] Zamana L V and Askarov Sh A *Fluorine in nitric therms of Bauntuntovskaya group (Northern Zabaikalie)*

[13] Zhou W 1996 Studies of geothermal background and isotopic geochemistry of thermal water in Jiangxi Province *China nuclear science and technology report* 29

IOP Conf. Series: Earth and Environmental Science **33** (2016) 012013 doi:10.1088/1755-1315/33/1/012013

Groundwater composition fluctuation within technogenic zones (case study: flooded coal mines in Primorsky Krai, Russia)

I A Tarasenko[1], A V Zinkov[2], E A Vakh[3], A V Vetoshkina[4] and A B Strelnikova[5]

[1,3,4] Far East Geological Institute, 159 Prospekt 100-letiya Vladivostoka, Vladivostok, 690022, Russia
[1,2,3] Far Eastern Federal University, 8 Sukhanova St., Vladivostok, 690950, Russia
[5] National Research Tomsk Polytechnic University, 30 Lenina Ave., Tomsk, 634050, Russia

E-mail: [1]tarasenko_irina@mail.ru, [2]zinkov_a@mail.ru, [3]adasea@mail.ru, [4]Alena.Vetoshkina@mechel.com

Abstract. The paper considers groundwater composition fluctuation within technogenic zones based on evidence from the flooded coal mines of Primorye. The authors have determined the regularities of hydrogeochemical processes, specified the groundwater composition fluctuation within the technogenic complexes located in the liquidated mine areas, and identified the equilibrium phases between the studied waters and specific secondary minerals. It has been proved that water within natural-technogenic complexes in the liquidated mine areas are saturated with silicates, carbonates, sulfates, oxides, and hydroxides, which should be taken into account when designing technologies for groundwater treatment.

1. Introduction

Today, coal mining is practically interrupted in Primorsky Krai, with 15 mines being flooded. This resulted in the change of the hydrogeological properties of Partizansky, Razdolnensky, and Uglovsky sedimentary basins. Moreover, this increased the permeability in the layers above coal seams, influenced groundwater natural regime and circulation, developed hydraulically bound technogenic aquifer systems, and caused the transformation of groundwater composition and formation of new water type, termed as "technogenic". This term implies the diversity of exogenous and technogenic processes occurring within newly-formed natural-technogenic complexes.

It is important to study the chemical composition and regularities of groundwater formation within natural-technogenic complexes located in the areas of liquidated coal mines. This will allow developing hydrochemical prognosis principles, as well as measures to mitigate environmental impact.

2. Materials and methods

The present paper deals with the chemical groundwater composition changes within technogenic zones, where the 200–800 m water-bearing strata were disturbed by mining works. We have examined the following liquidated mine areas in Primorsk Territory, RF (Figure 1): Partizansky (Glubokaya, Nagornaya, Avangard, Severnaya, and Uglekamenskaya mines), Razdolnensky (Ilyichevskaya and №

4 mines), and the Uglovsky (Artem) coal basins, as well as Podgorodnensky and Khasanovsky deposits (Podgorodnenskaya and Khasanovskaya mines, respectively).

The ground water samples were collected from wells screened of a particular aquifer or connected to filter-equipped mines in undermining sections. The samples were also collected from surface water seepage. To test cations and sulphate ions, hydrochemical water was sampled through cellulose filter (0.45 μm) in situ so as to remove suspended solid particles and then was acidified with nitric acid. For testing anions, the samples were also filtered and collected into polyethylene containers without acidification. Unstable parameters were determined at the sampling point as well. Mineralogical sampling was conducted single point method in sealed glass weighing bottles.

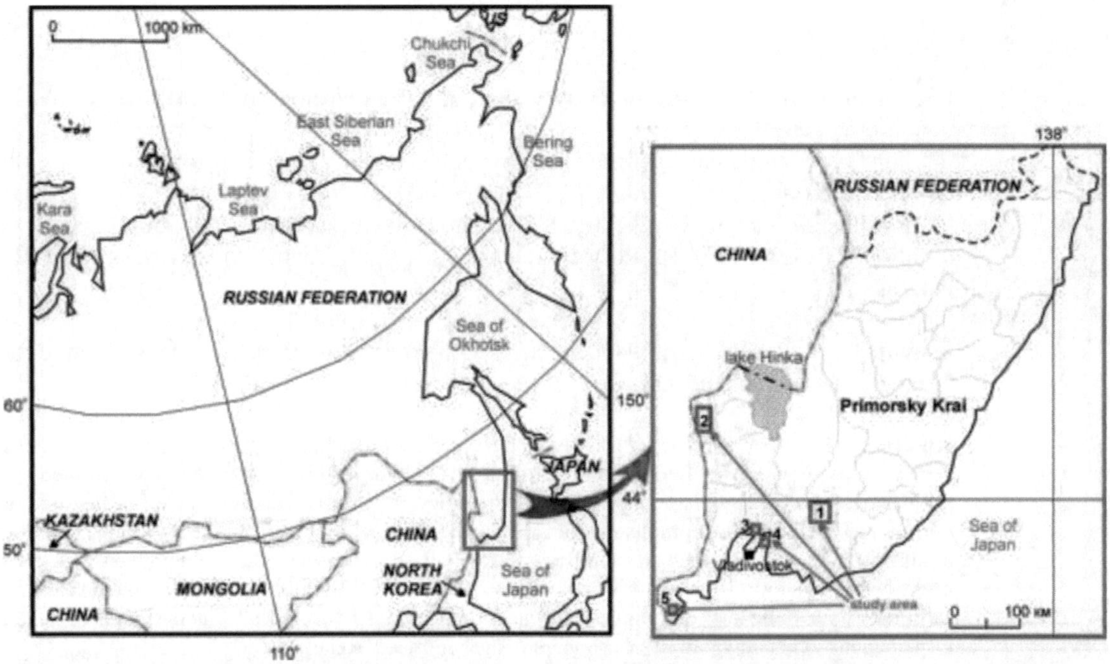

Figure 1. Study areas: 1 – Partizansky coal basin; 2 – Razdolnensky coal basin; 3 – Shkotovsky deposit; 4 – Podgorodnensky deposit; 5 – and the Khasanovsky deposit.

The water samples were examined via standard analytical methods ICP-MS; ICP-OES, Furye-spectrophotometer IR FTIR-84008. To study the composition of neocrystallizations, chemical, spectral, and roentgen-phase (diffractometer D8 Discover with CuKα; Dron-3 with FeKα) analyses were used, as well as IR spectroscopy method (IR Furye-spectrometer NICOLET, Termo).

To store, integrate, and process geochemical data, we applied the mathematical methods of statistics and physicochemical modeling – equation solving (with thermodynamic equilibrium constants – program Aqua-Chem V. 5.1 [1]) and Gibbs free energy minimization method (program Selector-Windows [2–4]).

3. Results and Discussion

The analysis data showed that within Primorye liquidated coal mine areas, the groundwater in porous Quaternary and Neogene sediments, as well as in fractured aquifer systems of Cretaceous sediments and fracture-veined aquifer systems in zones of tectonic dislocations and intrusive contacts in Cretaceous sediments are fresh (mineralization 0.3–0.4 g/dm^3), low hardness, pH-neutral, and bicarbonate. Cationic water varies from magnesium-sodium-calcium to calcium-magnesium. Chemical water composition variation within natural-technogenic complexes is attributable to mineralization (from 0.3 to 8.0 g/dm^3). Sodium, magnesium, bicarbonate, and sulfate ion accumulations are involved in the increasing water mineralization. Within newly-formed natural-technogenic complexes, the following brackish water types [5–7] exist: 1)sulphate bicarbonate, bicarbonate-sulphate, and sulphate magnesium; 2) sulphate bicarbonate, bicarbonate-sulphate, and sulphate sodium (Figure 2). The ion-exchange transformation is restricted to pH values from 5.8 to 9.4. Na^+ or Mg^{+2} accumulation in the above series is governed by the coal-overlaying formation rock. This fact is confirmed by the results of physicochemical "water-rock" system modeling based on minimization of isobar-isothermal potential (program complex "Selector-Windows")

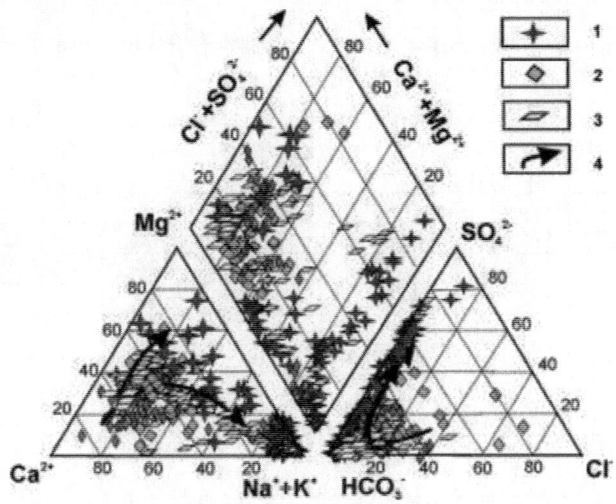

Figure 2. Piper diagram of water samples from areas of liquidated coal mines in Primorye: 1 – water within natural-technogenic complex; 2 –water within alluvial horizon; 3-water in fractured and fracture-veined complexes; 4 – water composition variation.

Modeling methodology included the calculations of "water-rock" system equilibrium. Chemically-natural water-bearing rocks were introduced in relevant proportions. Modeling involved the incongruent dissolution of rock minerals under the following conditions: 1) closed CO_2 and atmospheric gases (CO_2 – 0.01 kg/1 kg H_2O, atmospheric gases – 0.3 kg/1 kg H_2O) systems; 2) opened CO_2 system (CO_2 increased from 0.01 to 1.28 kg/1kg H_2O).

According to B. N. Ryzhenko and S. R. Krainova [8, 9], rock mass (R) and water mass (W) ratio in aluminosilicate rocks involves interaction time (conventionally, water exchange time) between rock and water solution. In this regard, the following R and W ratios were used in modeling the systems: 0.001 (step 1); 0.002 (step 2); 0.004 (step 3); 0.008 (step 4); 0.010 (step 5); 0.016 (step 6) 0.032 (step 7); 0.064 (step 8); 0.080 (step 9). Initially, step 1 of computational experiment, partial pressure of carbon dioxide is 10^{-2}bar, which corresponds to the mean groundwater within the regional fractured zone. Model adequacy to natural objects was estimated as the calculated correlation of chemical element composition in water and pH to the mineral composition of newly-formed solids of relevant natural parameters.

Geochemical properties of groundwater within technogenic complexes located in the liquidated coal mine areas generate in the catchment area itself and transform towards the filtration flow accordingly: HCO_3–Ca → HCO_3–Na(Mg) → HCO_3–SO_4–Na(Mg) → SO_4–Na(Mg) (table 1). Increased magnesium concentration in the groundwater is governed by significant mafic tuffogenic accumulations (40 %) in the stratigraphic sequence [7].

It has been revealed that closed CO_2 system (PCO_2 = $10^{-2.0}$) itself hinders HCO_3^- + CO_3^{2} concentration increase in the water phase and furthers increasing SO_4^{-2} concentration (if possible). Sulphate ions in the groundwater of study basins is governed by the sulfides (pyrite) in overlaying coal and coal thicknesses. This process is governed by the CO_2 partial pressure and the presence of O_2. Higher CO_2 partial pressure equilibrium ($> 10^{-1.0}$) and lower O_2 concentration equilibrium result in the expansion of HCO_3–Na water areas within the coal basins.

Chemical element concentration in water is governed by secondary minerals. Physico-chemical calculations identified that components migrate in natural waters as uncomplexed ions. As the mineralization process continues, the number of complexed migrating components increases, which, in its turn, results in the formation of mineral phases. There exists a distinct alteration of groundwater composition during their circulation in the natural-technogenic complexes and composition in the equilibrium secondary mineral phase (Table 1).

Table 1. Modeling data within "rock-water" system, closed CO_2 and atmospheric gases (CO_2 – 0.01 kg/1 kg H_2O, atmospheric gases – 0.3 kg/1 kg H_2O) systems.

Step number	1 (1g rock / 1 kg H_2O)	2 (2 g rock / 1 kg H_2O)	3 (4 g rock / 1 kg H_2O)	4 (8 g rock / 1 kg H_2O)	5 (10 g rock / 1 kg H_2O)	6 (16 g rock / 1 kg H_2O)	7 (32 g rock / 1 kg H_2O)	8 (64 g rock / 1 kg H_2O)	9 (80 g rock / 1 kg H_2O)
Water solution, mg/kg H_2O									
TDS	273.63	488.07	749.71	1104.24	1292.85	1898.61	3605.25	6666.75	7945.08
pH	6.69	7.03	7.21	7.33	7.38	7.48	7.56	7.63	7.61
Eh. B	0.821	0.80	0.79	0.78	0.78	0.78	0.77	0.77	0.77
Ca^{+2}	2.40e+01	4.52e+01	4.96e+01	2.94e+01	2.39e+01	1.56e+01	1.08e+01	7.80e+00	8.35e+00
Na^{+}	9.47e+00	1.88e+01	3.70e+01	7.02e+01	8.35e+001	1.02e+02	5.18e+01	5.26e+00	1.75e+00
Mg^{+2}	6.06e+00	1.15e+01	2.20e+01	4.20e+01	5.14e+01	7.84e+01	1.49e+02	2.36e+02	2.53e+02
HCO_3^-	1.03e+02	2.25e+02	3.45e+02	4.61e+02	5.18e+02	6.57e+02	8.18e+02	1.01e+03	9.99e+02
SO_4^{2-}	2.36e+01	4.49e+01	8.88e+01	1.83e+02	2.30e+02	3.69e+02	7.29e+02	1.39e+03	1.70e+03
Cl^-	1.50e+01	1.50e+01	1.50e+01	1.50e+01	1.50e+01	1.50e+01	1.50e+01	1.49e+01	1.49e+01
Si	6.42e+00	6.43e+00	6.43e+00	6.44e+00	6.44e+00	6.45e+00	6.44e+00	6.43e+00	6.41e+00
Gas parameters									
PCO_2	2.16e-02	2.16e-02	2.16e-02	2.17e-02	2.18e-02	2.20e-02	2.23e-02	2.34e-02	2.40e-02
PO_2	2.06e-01	2.05e-01	2.04e-01	2.03e-01	2.02e-01	1.99e-01	1.92e-01	1.77e-01	1.69e-01
PN_2	7.73e-01	7.73e-01	7.74e-01	7.76e-01	7.77e-01	7.79e-01	7.86e-01	8.00e-01	8.07e-01
Mineral phase. %	goethite – 3.35; montmorillonite – 29.17; quartz – 67.48	goethite – 3.32; montmorillonite – 28.98; quartz – 67.70	goethite – 3.23; montmorillonite – 28.17; quartz – 66.14; calcite – 2.46	goethite – 3.15; montmorillonite – 27.45; quartz – 64.60; calcite – 4.71; rhodochrosite – 0.10	goethite – 3.13; montmorillonite – 27.32; quartz – 64.34; calcite – 5.05; rhodochrosite – 0.15	goethite – 3.12; montmorillonite – 27.18; quartz – 64.05; calcite – 5.44; rhodochrosite – 0.21	goethite – 3.11; montmorillonite – 27.11; quartz – 63.92; calcite – 5.62; rhodochrosite – 0.24	goethite – 3.10; montmorillonite – 26.99; quartz – 63.67; calcite – 5.30; rhodochrosite – 0.24; dolomite – 0.69	goethite – 3.09; montmorillonite – 26.92; quartz – 63.51; calcite – 5.00; rhodochrosite – 0.24; dolomite – 1.23

AquaChem program V. 5.1 showed that underground water with mineralization > 0.6 g/dm^3 are saturated with calcium and magnesium carbonates, which corresponds to the findings by S. L. Shvartsev [10, 11] and other scientists [12, 13] concerning cation behavior in the groundwater within different landscape zones. If mineralization is >1.5 g/dm^3, equilibrium with magnesium sulphate – epsomite (Razdolnensky coal basin) could be found. As for surface water seepage in technogenic complexes, a steady equilibrium between above-discussed water types and sodium mineral series thermonatrite ($Na_2CO_3{\cdot}H_2O$), tenardite (Na_2SO_4), etc. is established. All waters have been identified to be in equilibrium with kaolinite. Further water-rock interaction results in increasing silicon, magnesium, calcium, sodium concentration in the solution and formation of montmorillonites and chlorites. When K^+ ions that are closely connected with the weak acid anionites (HCO_3^-) accumulate, illite and sericite are formed. If mineralization is > 4 g/dm^3, equilibrium with analcime, a typical mineral in alkaline sodium water, could be found. According to R.M. Garrelse and C. L. Kraist [14], this mineral could indicate the formation of amorphous silica.

Figure 3. Reagent-free water treatment facilities with filtration dike (Uglekamenskaya mine): A – coagulation of clay minerals at the outlet of the distributive tray; B – mineral formation on the surface of the pond-storage; C – fine-dispersed mineral fraction.

As there exists an equilibrium between the studied water and secondary minerals series on the earth surface, the intensive coagulation of newly-formed minerals and flaky fine-dispersed masses is observed (Figure 3). Phase (mineral) composition of fine-dispersed fraction was determined via IR Furye-spectrometry; and the following four mineral series were identified: carbonate, quartz, and montmorillonite; montmorillonite and quartz with insignificant iron hydroxide impurities; montmorillonite, quartz, and carbonate with significant iron hydroxide impurities; iron hydroxides.

Thus, it has been established that the water within natural-technogenic complexes in the liquidated mine areas are saturated with different silicates, carbonates, sulfates, oxides, and hydroxides. This fact should be taken into account when designing groundwater treatment facilities and developing measures to mitigate the environmental impact caused by existing coal mines.

It is noteworthy that the design and construction of the reagent-free water treatment facilities with filtration dikes are not always appropriate for mine water treatment, as there is an equilibrium between the studied waters and secondary mineral series on the earth surface, intensive coagulation of newly formed minerals and flaky fine-dispersed masses. This, in its turn, causes intensive silting of the dike filters, water level increase in the storage pond, overflow, and destruction of the hydro-technical construction.

4. Conclusion

Based on the research results, it is possible to conclude that:

- coal mine flooding in Primorsky Krai causes changes in ground water circulation and leads to formation of hydraulically bound technogenic aquifer systems;
- geochemical properties of groundwater within technogenic complexes located in the liquidated coal mine areas generate in the catchment area itself and transform towards the filtration flow;
- increased magnesium concentration in the groundwater is governed by significant mafic tuffogenic accumulations (40 %) in the stratigraphic sequence;
- closed CO_2 system ($PCO_2 = 10^{-2,0}$) itself hinders $HCO_3^- + CO_3^{2}$ concentration increase in the water phase and furthers increasing SO_4^{-2} concentration (if possible); higher CO_2 partial pressure

equilibrium ($> 10^{-1.0}$) and lower O_2 concentration equilibrium results in the expansion of HCO_3–Na water areas within the coal basins.

- existing steady equilibrium between the studied waters and secondary mineral series (silicates, carbonates, sulfates, oxides, and hydroxides) should be considered when designing measures to mitigate the environmental impact caused by the waters formed within the natural-technogenic complexes.

References

[1] *AquaChem V. 5.1. User's Manual. Water Quality Data Analysis, Plotting and Modeling* 2006 (Canada, Waterloo Hydrogeol., Inc.) p 365

[2] Karpov I K, Chudnenko K V and Kulik D A 1997 Modeling chemical mass transfer in geochemical processes: Thermodynamic relations, conditions of equilibria, and numerical algorithms *Amer. J. Sci.* **297** 767–806

[3] Karpov I K, Chudnenko K V, Kulik D A and Bychinskii V A 2002 The convex programming minimization of five thermodynamic potential other than Gibbs energy in geo-chemical modeling *Amer. J. Sci.* **302** 281–311

[4] Chudnenko K V and Karpov I K 2003 *Selector-Windows: the program facility for calculation of chemical equilibriums of the thermodynamic potential minimizations* (in Russian) (Irkutsk) p 90

[5] Tarasenko I A 2010 About the state of the natural environment in the districts of the liquidated coal mines (on the example of Partizansky district of Primorsky Krai) *Bulletin of the FEB of RAS* **3** 113–118

[6] Tarasenko I A and Zinkov A V 2013 Formation conditions and features of the ground water chemical composition of the coal-bearing districts of Sakhalin Island *Russian Journal of Pacific Geology* **32** 4 100–110

[7] Tarasenko I A and Zinkov A V 2014 Formation conditions and features of the ground water chemical composition through the "wet" liquidation of coal mines in Primorye *Water: chemistry and ecology* **7** 92–98

[8] Ryzhenko B N and Krainov S R 2001 Interaction in a thermodynamically closed system "rock–water" is the key to the formation of chloride of underground water *Geochemistry* **10** 1052–1082

[9] Ryzhenko B N and Krainov S R 2002 Physico-chemical factors in the formation of chemical composition of waters of hypergenesis zone *Geochemistry* **8** 1–28

[10] Shvartsev S L 1991 The Interaction of water with aluminosilicate rocks. Review *Russian Geology and Geophysics* **12** 16–50

[11] Shvartsev S L 1998 *Hydrogeochemistry of the hypergenesis zone* (Moscow, Nedra, 2-nd edition) p 366(in Russian)

[12] Nordstrom D K, Ball J W, Djnahoe R J and Whittemore D 1989 Groundwater chemistry and water-rock interaction at Stripa *Geochim. et Cosmochim. Acta.* **53** 8 1727–1740

[13] Morse J W, Arvidson R S and Luttge A 2007 Calcium carbonate formation and dissolution *Chemical Reviews* **107, 2** 342–381

[14] Garrels R M and Christ C L 1965 *Solutions, minerals and equilibria* (New York, Harper and Row) p 450

Thermodynamic model of Ak-Tuz deposit surface water formation

V M Alekhina[1], I V Tokaver[2], B N Ryzhenko[3] and E V Cherkasova[3]

[1]Institute of Physics of NAC of the Kyrgyz Republic, Bishkek, Kyrgyzstan
[2]Research Center "Geomodel". Research Park of Saint-Peterburg State University, Saint-Peterburg. Petergoff., Russia
[3]Vernadsky Institute of Geochemistry and Analytical Chemistry of the Russian Academy of Sciences, Moscow, Russia

E-mail: [1] alekhina@jet.kg, [2] tokarevigor@gmail.com

Abstract. In Ak-Tuz deposit surface water macro and micro components concentrations are measured. Thermodynamic model is developed for aqueous composition prognosis at variation of water exchange. The concentration of n×10-8 mg Th / kg H2O and more testifies about Th containing colloid species in aqueous solution.

1. Introduction
Ak-Tuz ore field (Kirgiz Republic) is located in the west part of Munkumo-Naratskogo median massif of the Zailisky Alatau Mountains and it is composed of amphibolite shists and Archean-Proterozoic gneises. Deposits of Y, REE, Th, Zr, Mo, Be, Pb and other metals are related to Early Permian biotite granite and alaskite granophyre of Keminsky massif. The exposed part of the granite massif, 24 km in length and 1-4 km in width, is like fracture intrusion, subducted (16o) under rock mass of amphibolite shists and gneises. The basic body of Keminsky massif is formed by meso and leucobiotite granites, their ore-bearing cupolas are leuco and alaskite granites [8, 12, 9].

Up to 90-th years of the last century Ak-Tuz deposit was the main source of rear earth elements and yttrium in the former USSR. The yttrium ore-bearings themselves are confined to upper horizons of deposit low zone with Y mineralization and lanthanide of yttrium group. From 1960 to 1995 years the deposit was exploited for rear earth elements extraction by open-cut quarry. Rear earth, yttrium oxide, plumb, tin, copper, niobium, tantalum, thorium and iron are the main products of mining. Rear ores are high radioactive due to content of thorium minerals (monazite, thorite, zircone and others). Total volume of tailing ponds and storages is 4.17 million m3, their square is 51 Ga. Beside ponds and storages, there are three large heaps of rock, their volume are 50 million m3.

It is necessary to evaluate and forecast possible environmental disasters. The area under consideration is influenced by various factors. It means that beside traditional methods of investigation (chemical analysis, statistics etc) computer simulation of water-rock-gas system should be used [2, 3].

2. Materials and methods
Ak-Tuz surface waters chemical composition (Ca, Mg, Na, K, Cl, SO_4, HCO_3. NO_3, NO_2, NH_4 and microelements) has been determined in 40 samples taken across the area. To determine

macrocomponent content a balance between cations and anions was checked as shown in (Figure1) as discord of their equivalents and the simple statistics was carried out (table 1).

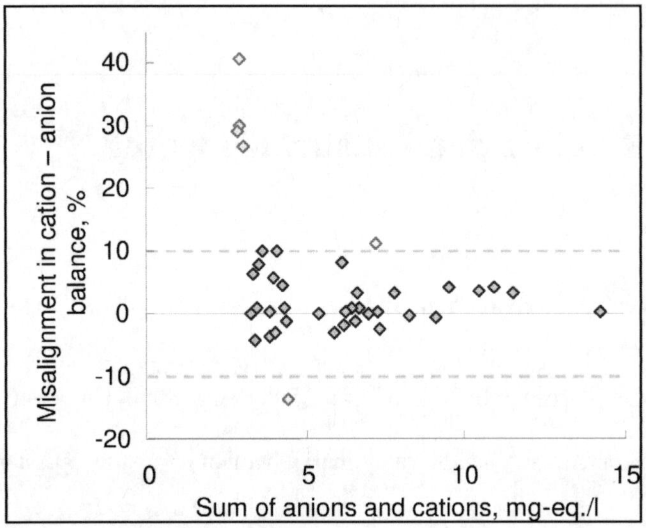

Figure 1. Discord of balance between cation and anion equivalents according to samples collected in 2014 (dotted line is ± 10 %).

It is seen the discord increases with salinity decrease. Taken 10% balance and less as boundary the aqueous analytical determinations of 100 mg/L and more (sum of cations and anions < 3 mg-eq/L) may be considered as good enough.

Table 1. Statistics of chemical analyses of surface waters (2014-th sampling).

Parameter	pH	Ca^{2+}	Mg^{2+}	Na^{+}	Cl^{-}	SO_4^{2-}	HCO_3^{-}	NO_3^{-}	Salinity. mg/L	Sum of cations and anions. mg-eq/L
		mg/L								
Max.	8.9	88	29	26	13	189	299	9.8	502	14.2
Min.	7.3	24	4.0	0.8	4.0	3.7	67	0.7	120	3.2
Average	8.1	41	11	7.0	6.1	24.4	145	2.5	234	6.2

Factor analysis is also used to identify geochemical processes responsible for surface water composition formation. The values of pH, Ca, Mg, Cl, SO_4, HCO_3 determined for all samples are used. The results of factor analysis show four significant processes responsible (97.6%) for the composition formed (table 2). Please pay attention that Factor 2 depends on chloride content. It is not obvious due to low accuracy of chloride determination.

Table 2. Factor analysis of samples collected in 2014.

	Factor 1	Factor 2	Factor 3	Factor 4
pH	-0.178	0.087	***-0.961***	-0.193
Ca	0.595	0.007	0.172	***0.746***
Mg	***0.744***	-0.242	0.193	0.533
Cl	0.025	***-0.978***	0.082	0.172
SO_4	***0.987***	0.043	0.132	-0.003
HCO_3	-0.002	-0.243	0.178	***0.949***

Surface water sample analysis (macro and micro components) was considered in terms of mineral equilibriums. The sample aqueous solutions were saturated with respect to $BaSO_4$ (barite), $CaCO_3$ (calcite), $CaMgC_2O_6$ (dolomite), $HFeO_2$ (goethite), TiO_2 (rutile) and $SrCO_3$ (strotianite). The solution samples are unsaturated with respect to ore minerals of Ag, Zn, Pb, Ni, Sb, Cr. Aqueous species are regular: they are simple cations for alkaline and alkaline-earth elements, they are hydroxides, sulfates and carbonates for two charged cations of color metals, for Sb and Fe – hydrospecies, for chromium – CrO_4^-, for nitrogen mostly NO_3^- and probably NO_2^-. Such aqueous species are explained by low concentration of the elements (mkg/L): Ag 0.n, Ni 0.n, Pb 0.n, Cr n, Cu n, Zn n×10, Ti n×10, Ba и Sr 0.n; concentration of Sb, which equals to n×100, is probably wrong. The results described above allow us to consider environmental state as satisfactory with respect to all studied elements but Sb (pmc 5 mkg/L).

Thermodynamic model of water-rock-gas system shows equilibrium concentrations of the elements in aqueous phase which contacts the rock (the rock of basic composition in this study) As [3] proves, equilibrium concentrations depend on four main factors: (a) rock composition of the sites of water-rock contact and interaction, (b) water exchange (rain, dry weather, etc.) and inflax and outflax, (c) temperature, (d) partial pressure of atmosphere gases.

For simulation of water-rock-gas system the model containing 26 chemical elements (Al-Ag-As-Ba-C-Ca-Cl-Cr-Cu-Fe-H-K-La-Mg-Mn-Mo-N-Na-Ni-O-Pb-S-Sb-Si-Sr-Th-Ti-U-Y-Zn), 110 solids (minerals and phases) and 120 simple and complex aqueous species was developed. Sorption of thorium which is of special interest of study was considered as ideal solid solution of sorbent (iron hydroxide, humic substance) and thorium adsorbed aqueous species ($Th(OH)_4^0$) using experimental value of sorption constant. Comparison of the simulation results with chemical composition of surface water samples was done.

Ak-Tuz site temperature was taken as summer average (10°C). Total pressure was taken according site position at 2 km. Chemical and mineral compositions are taken as basic rock features according to Ak-Tuz ore field description [4, 10]. Gibbs Free energy values of solids and gases, aqueous species are taken from [6, 13] and from improved base UNITHERM [2]. In some cases, BR model was used to calculate constants of unstudied steps of aqueous complex ionization [7].

The water-rock-gas system is considered as open with respect to atmosphere gases with CO_2 partial pressure equal to 0.001 bar and with oxygen partial pressure corresponding NO_3^- predominance in surface water samples of Ak-Tuz site.

Simulation results of water-solid-gases interaction show the following: micro component concentrations correspond to active water exchange (mass ratio rock to water R/W≈0.0001-0.004) and to positive redox potential Eh~+750-800 mV and pH~8.0 (see table 3). Comparison of simulation and analytical data of macro and micro components of aqueous phase shows that the water exchange when sampling corresponds to R/W≈0.004 (bold shift) and with such water exchange the element concentrations do not exceed pmc. Significant decrease (more than 10 times) in water exchange will lead to pmc excess (at R/W>0.06 only). Model concentrations of micro and macro components might be more accurate if Ak-Tuz rock and tailings compositions are known.

Model concentration of SO_4 is significantly lower than analytical one. It supports the opinion about oxidation sulfides. Surplus of sulfate decreases barium concentration in equilibrium to barite $BaSO_4$. Simulation permits us to say that concentrations of As, Se. Hg with low pmc might be overcome at passive water exchange only.

Table 3. Comparison of results of simulation of water-rock-gas system at R/W=0.0001-0.2 (various water exchange) and results of analytical determination of the same chemical elements (mg/L) of Ak-Tuz surface water samples.

Elements	**Concentrations at various water exchange (R/W model)**						**Analysis**	
	0.0001	0.0007	0.004	0.03	0.06	0.2	min	max
				Macro components				
Ca	7.6	35	36	57	97	205	24	88
Mg	0.5	2.9	5.4	8.5	14.5	30.1	4	29
Na	0.3	1.8	11.1	68.8	171	426	<0.1	26
K	0.04	0.2	1.4	8.6	21.3	49.2	<0.1	0.7
HCO_3	28.4	121	122	108	90.5	73.6	31	256

SO_4	0.02	0.1	0.9	4.4	10.1	23.2	13.6	189
Cl	0.3	1.9	12	73	180	450	4	13
NO_3	0.5	3.2	20	125	310	772	<0.7	9.8
pH	7.3	8.0	8.0	7.9	7.9	7.9	7.3	8.7
				Micro components				
Ag	1E-6	6E-6	4E-5	2E-4	6E-4	1E-3	5E-5	3E-4
Al	3E-6	1E-5	1E-5	1E-5	1E-5	8E-6	-	-
As	2E-5	1E-4	7E-4	4E-3	1E-2	3E-2	-	-
Ba	6E-3	4E-2	2E-1	4E-1	2E-1	1E-1	0.03	0.021
Cr	6E-4	4E-3	2E-2	1E-1	3E-1	8E-1	0.002	0.008
Cu	4E-4	2E-3	1E-2	3E-2	3E-2	3E-2	0.002	0.007
Fe	2E-8	1E-8	1E-8	1E-8	1E-8	1E-8	<0.1	-
Mn	2E-5	1E005	1E-5	2E-5	4E-5	9E-5	-	-
Mo	1E-5	6E-5	3E-4	2E-3	5E-3	1E-2	-	-
Ni	6E-4	4E-3	2E-2	1E-1	4E-1	9E-1	0.0004	0/001
Pb	2E-4	1E-3	8E-3	5E-2	1E-1	3E-1	0.0004	0.0008
Sb	2E-4	1E-3	2E-3	2E-3	2E-3	1E—3	-	0.16
Si	1.3	1.3	1.3	1.3	1.3	1.3	-	-
Sr	8E-3	5E-2	3E-1	6E-1	1	2	0.01	0.3
Ti	7E-5	4E-4	2E-3	2E-3	2E-3	2E-3	0.002	0.03
Th	2E—9	7E-9	7E-9	6E-9	5E-9	4E-9	-	-
U	3E-5	2E-4	1E-3	6E-3	1E-2	4E-2	-	-
Y	1E-4	8E-4	5E-3	3E-2	8E-2	0.2	-	-
Zn	8E-4	5E-3	3E-2	2E-1	5E-1	1.2	0.007	0.02

Simulation allows us to evaluate pmc of Th in 40 samples of surface waters. Simulation was done assuming the presence of chloride, sulfate, hydroxo and carbonate Th-complexes formation. Thermodynamic properties of the complexes were taken from [11]. Simulation shows the predominance of hydroxo and hydroxo-carbonate complex species with total concentration less than **$n\times10^{-8}$ mg/kg H_2O** (table 4) in most aqueous phase samples.

Table 4. Simulated concentrations of Th aqueous species in surface waters of Ak-Tuz 2014-sampling.

Sample	mg/L				Wight. %		
	$Th(OH)_4$	$Th(OH)_5^-$	$Th(OH)_3CO_3^-$	Sum	$Th(OH)_4$	$Th(OH)_5^-$	$Th(OH)_3CO_3^-$
Ak-1	2.54E-10	6.98E-08	1.51E-08	8.51E-08	0.3	81.93	17.77
Ak-2	2.54E-10	7.77E-08	1.60E-08	9.39E-08	0.27	82.71	17.02
Ak-3	2.54E-10	3.05E-08	2.68E-08	5.75E-08	0.44	53.06	46.5
Ak-4	2.54E-10	7.26E-08	1.84E-08	9.13E-08	0.28	79.54	20.18
Ak-5	2.54E-10	1.25E-07	1.39E-08	1.39E-07	0.18	89.78	10.04
Ak-6	2.54E-10	4.98E-08	2.27E-08	7.28E-08	0.35	68.4	31.25
Ak-7	2.56E-10	6.50E-08	1.01E-10	6.54E-08	0.39	99.45	0.15
Ak-8	2.54E-10	6.28E-08	2.00E-08	8.30E-08	0.31	75.61	24.09
Ak-9	2.54E-10	5.68E-08	2.29E-08	7.99E-08	0.32	71.01	28.67
Ak-10	2.54E-10	2.68E-08	3.18E-08	5.89E-08	0.43	45.53	54.04
Ak-11	2.54E-10	4.66E-08	2.22E-08	6.91E-08	0.37	67.49	32.14
Ak-12	2.54E-10	2.03E-07	1.00E-08	2.14E-07	0.12	95.19	4.69
Ak-13	2.54E-10	2.43E-07	9.16E-09	2.52E-07	0.1	96.26	3.64
Ak-14	2.54E-10	1.39E-08	4.49E-08	5.91E-08	0.43	23.55	76.02
Ak-15	2.54E-10	5.87E-08	2.01E-08	7.90E-08	0.32	74.27	25.41
Ak-16	2.54E-10	2.37E-08	3.32E-08	5.72E-08	0.45	41.48	58.08

Ak-17	2.54E-10	2.05E-08	3.64E-08	5.72E-08	0.45	35.94	63.61
Ak-18	2.56E-10	1.28E-08	1.88E-10	1.33E-08	1.93	96.65	1.42
Ak-19	2.54E-10	2.68E-08	3.50E-08	6.21E-08	0.41	43.22	56.37
Ak-20	2.54E-10	3.55E-07	6.96E-09	3.62E-07	0.07	98.01	1.92
Ak-21	2.54E-10	6.82E-08	1.88E-08	8.72E-08	0.29	78.15	21.55
Ak-22	2.54E-10	4.12E-08	1.77E-08	5.92E-08	0.43	69.62	29.96
Ak-23	2.54E-10	5.10E-08	2.37E-08	7.50E-08	0.34	68.09	31.57
Ak-24	2.54E-10	2.03E-08	3.60E-08	5.66E-08	0.45	35.91	63.64
Ak-25	2.54E-10	5.14E-08	2.20E-08	7.36E-08	0.35	69.81	29.84
Ak-26	2.54E-10	2.15E-08	3.74E-08	5.91E-08	0.43	36.35	63.22
Ak-27	2.56E-10	2.08E-08	1.95E-10	2.12E-08	1.21	97.87	0.92
Ak-28	2.54E-10	2.03E-08	3.67E-08	5.73E-08	0.44	35.48	64.08
Ak-29	2.54E-10	2.52E-08	3.60E-08	6.15E-08	0.41	41	58.59
Ak-30	2.54E-10	1.64E-08	5.08E-08	6.75E-08	0.38	24.35	75.27
Ak-31	2.54E-10	1.88E-08	3.88E-08	5.78E-08	0.44	32.52	67.04
Ak-32	2.54E-10	2.13E-08	3.91E-08	6.07E-08	0.42	35.15	64.43
Ak-33	2.54E-10	6.60E-08	2.09E-08	8.71E-08	0.29	75.69	24.02
Ak-34	2.54E-10	2.42E-08	3.53E-08	5.98E-08	0.43	40.5	59.08
Ak-35	2.54E-10	6.94E-08	1.89E-08	8.86E-08	0.29	78.38	21.34
Ak-36	2.54E-10	1.75E-08	4.39E-08	6.16E-08	0.41	28.34	71.24
Ak-37	2.54E-10	1.23E-08	6.00E-08	7.26E-08	0.35	16.98	82.67
Ak-38	2.56E-10	5.93E-08	1.11E-10	5.97E-08	0.43	99.38	0.19
Ak-39	2.54E-10	2.69E-08	3.71E-08	6.42E-08	0.4	41.86	57.75
Ak-40	2.54E-10	2.45E-08	3.23E-08	5.71E-08	0.45	43	56.56

ThO_2 (thorianite) is the main solid limited Th enrichment in aqueous solution. It is fixed in equilibrium simulated mineral assemblages. It means that surface water aqueous thorium concentration, exceeding the values in table 5, is colloids or dispersed species as it is shown for Al for alumosilicates solubility study [1].

3. Summary

1. Statistics of macro and micro component chemical analysis of Ak-Tuz surface waters samples collected in 2014 is done. The results of statistics lead to following conclusions:
- water exchange time of several months proves atmosphere precipitates as main source of Ak-Tuz surface waters;
- sulfides oxidation in ore tailings is estimated,
- small variations of isotopic composition of water samples proves identical precipitate income.
2. The various competing Th-complexes are calculated. The thorium aqueous concentration in surface waters is calculated ($n\times10^{-8}$ mg/kg H_2O). The surpass of this thorium concentration means the presence of colloids and dispersed species of thorium in Ak-Tuz surface waters.
3. Thermodynamic model is created and simulation of water-rock-gas system is done. Comparison of simulation results and analytical determination of macro and micro component aqueous concentrations proves the model to be efficient to predict surface waters composition at other water exchanges.

Acknowledgements
The reported study was funded by according to the research project ISTC #KR-2091.

References

[1] Alekseev V A, Ryzhenko B N and Shvartsev S L 2005 *Geological evolution and selforganization of water-rock system* vol 1 Water-rock system in the earth crust: interaction, kinetics, equilibrium, simulation (Novosobirsk: SB RAN) p 242 [in Russian]
[2] Borisov M V and Shvarov Yu V 1992 *Thermodynamics of geochemical processes* (Moscow:

MSU) p 256 [in Russian]

[3] Ryzhenko B N and Cherkasova E V 2012 Chemical Composition of Natural Waters and Brines as a Result of Hydrogeochemical Processes in Water-Rock-Gas Systems *Geochem. Int.* **50** 1101–1150

[4] Kyrgyz complex hydrogeological expedition *Pollution of ground waters in the east part of Chu valley* Report on ecologo-hydrogeological study on K-43 and other mapsites within KR boundaries for 2000–2005 years [in Russian]

[5] *Permeable mean concentrations of chemical substances in the waters of industrial and potable uses* 2003 Document 2.1.5.1315-03 Chief sanitary officer of RF. 30.04.2003. (Moscow) 78 [in Russian]

[6] Naumov G B, Ryzhenko B N and Khodakovsky I L 1974 *Handbook of thermodynamic data* US Dept. of Commerce PB-226722. (Washington DC)

[7] Ryzhenko B N 1981 *Thermodynamics of equilibrium in hydrothermal solutions* (Moscow: Nauka) p 191 [in Russian]

[8] Smirnov N V, Zotov I A, Stavinsky V A, Isaev I S and Kim V F 1992 *Proceedings USSR Academy of Sciences* **332** 382–385 [in Russian]

[9] Torgoev I A and Aleshin Yu G 2009 *Geoecology and waste of KR mining* (Bishkek: ILIM) [in Kyrgyzstan]

[10] Cherny M, Mashek V, Erochin B and Nurabaev A 2011 *Ecological situation on the mining sites of river Kichikimin (Ak-Tuz part) and rivers Kashka and Berkut (site Kashka)* Final Report GEOMIN (Chech Republik)

[11] Borkowski M and Richmann M 2009 *Comparison of recent Thorium Thermodynamic Data with those used in WIPP FMT_050405* CHEDAT DATABASE. LA-UR 09-07298 (Los Alamos)

[12] Jung H G and Himmelsbach T 2008 *Hydrogeochemical Groundwater Monitoring in Mailuu-Suu, Kyrgyz Republic* Final Report BGR Federal Institute for Geosciences and Natural Resources (Hannover)

[13] Johnson J W, Oelkers E H and Helgeson H C SUPCRT 1992. A software package for calculating the standard molal thermodynamic properties of minerals, gases, aqueous species, and reactions from 1 to 5000 bar and 0 to 1000°C *Comput.Geosci.* **18** 899–947

Modeling the interaction of mine brines with chloride minerals of potassium-magnesium deposits

V V Fetisov

Perm State National Research University, 15 Bukirev St., Perm, 614990, Russia

E-mail: fetisov.v.v@gmail.com

Abstract. The article is devoted to study of dynamics of saturation degree of suprasalt brines with respect to major chloride minerals of salt strata in the initial phase of an accident related to discontinuity of waterproof stratum in the mine of the Verkhnekamskoe salt deposit (Berezniki-3 mine, 1986). Physicochemical modeling has showed that the brines discharged into mine are in equilibrium with halite during all period of observation. At the same time, their degree of saturation with respect to sylvite and carnallite regularly decreases with the increase in inflow of the suprasalt Cl-Na brines. The initial stage of suprasalt brine penetration into mine is characterized by an increase in the saturation degree with respect to the considered chloride minerals, which is showed on the chart presented in the article. However, there are brines oversaturated with respect to halite, which occurs over a brief period. In contrast to the mine brines of different genesis being in equilibrium or close to equilibrium with sylvite, saturation index (SI) for this mineral decreases in the suprasalt brine. This allows one to recommend the use of this parameter in the study of the mine brines to timely detect suprasalt brines entering the mine.

1. Introduction

The Verkhnekamskoe salt deposit (VKSD) is located on the left bank of the Kama river within Perm Krai. By its tectonic nature, the deposit is related to Solikamsk depression of the Fore-Urals Trough. The deposit can be characterized as a multimineral one: sylvinite (raw material for potash fertilizer production), carnallite (artificial carnallite production in magnesium industry) and rock salt are mined there. VKSD is one of the biggest deposits in the world, with 30 % of world potash ore. Deposit salt stratum is of 8 100 km^2, the area of potash deposit is 3 750 km^2 [1]. The deposit is developed by underground mining method, the mines being 100-500 m in depth.

VKSD has solid (more than 500 m) salt strata with the layers (from bottom to top) of underlying rock salt (URS) of 320-400 m, potash deposit (70-100 m) and covering rock salt (CRS) (20 m). Salt stratum is underlied with clay-anhydrite deposits of 200-220 m and covered with salt-marly (SMS), terrigenous-carbonate (TCS) and speckled (SS) strata of the Ufimian strata of the lower Permian (figure 1). A set of salts is related to the Iren horizon of the Kungurian strata of the lower Permian System. Sylvinite (with average thickness of 17.5 m) and carnallite (with average thickness of 54 m) zones comprise potash deposit [1].

The lens-shaped salt strata of the deposit act as a regional aquaclude dividing suprasalt and subsalt waters, the first being fresh, mineralized waters and brines, while the second being highly-mineralized brines. Waters of brine horizon come out from the holes in some areas at the top of salt strata or near it. It has been noted that these waters have Cl-Na composition, and their mineralization achieves 320 g/L [1].

The deposit mines are very often characterized by the technogenic brines - condensate and stowage brines. Salt deposit contains some amount of post-sedimentation (intrasalt) brines. Condensate brines are caused by the interaction of moisture from air and salt rocks. In the mines these brines are everywhere represented by the puddles, small lakes in the degradations and downpour from the cover. Stowage brines are represented by the liquid phase entering the mines together with the salt-processing waste applied in stowing worked-out area and being pressed while solidifying stowage material in mines.

Waterproof stratum (WPS) is defined as a stratum of rocks the upper border of which is the first rock salt layer of intermediate zone (IZ) of salt-marl strata and the lower border of which is the cover of the salt deposit's upper layer being processed (figure 1). With discontinuity of WPS due to natural factors or mining works, the suprasalt waters interact with the rocks of salt deposit and wash away the cracks, which results in the increase in water inflow in the mines and mine flooding. WPS discontinuity and the breakthrough of the suprasalt stratum groundwater have led to the flooding of two VKSD mines (Berezniki-3 mine, 1986; Berezniki-1 mine, 2006) and an accident in Solikamsk-2 mine (2014). WPS outbreaks together with further solution of salt stratum in the areas of suprasalt waters discharging in a mine are accompanied with the sinkholes on the surface.

In case of an accident, the highly-mineralized suprasalt brines of CS leaching are the first to be discharged in the mines. Mining work safety is closely connected with fast detection of these brines among the other mine brines described above.

2. Objects and methods

The research in this paper is supported by the evaluation of saturation degree of accident causing suprasalt brines with respect to chloride minerals of VKSD salt strata on the basis of saturation index (SI) identification. Due to highly-mineralized brines, the activity coefficients in the paper were identified with the Pitzer model [3-6]. Being in agreement with the experimental data, this model is widely used in the papers of the scientists [7, 8].

To calculate brine equilibrium with chloride minerals (halite NaCl, sylvite KCl and carnallite $KMgCl_3*6H_2O$), PHREEQC code was applied [9]. Physicochemical modeling was done with standard atmospheric pressure at 10°C brine temperature.

In our studies of accident causing suprasalt brines connected with the WPS discontinuity, we turned to the accident happened in Berezniki-3 mine in 1986, which was described in detail in the works published [2, 10, 11]. In the present research, we analyzed some measurement data provided in these works, namely, the data on hydrochemical sampling of brines in one of the deposit mines regarding the major ions, Br^-, pH value and density.

Two layers of sylvinite zone ("AB" and "Red-II") till 1979 and later layer "Red-II" were worked out in a mine located at the center of VKSD southern part. On the 11th of January, 1986, brines in the form of separate small streams were discovered from the cover of chamber № 50 of the western half block of the mine. At the same time the brines appeared in the cover of chamber № 52. For the next several days the brines started to flow in the covers of the other chambers as well. Approximate evaluation showed that at the initial stage the overall inflow in a mine was about 10 m^3/hour. On the 6th of February summarized inflow of brines was about 60 m^3/hour, on the 22nd of February – 100 m^3/hour. On the 1st of March the inflow was 150 m^3/hour, in the morning of the 9th of March it was 500 m^3/hour. In the evening of the 9th of March the inflow increased up to 5,000 m^3/hour and increased further. By the end of June all worked-out area of Berezniki-3 mine field was flooded with suprasalt waters, with their calculated overall volume being about 15 mln. m^3 [1].On the night of July 23, 1986, six and a half months after the brine's streams had been found in the cover of the mine, a sinkhole in the ground formed over the place of suprasalt groundwaters breakthrough. The thickness of the suprasalt stratum in the breakthrough area consisting of SS, TCS and SMS is about 350 m (Figure 1). Below, there is covering salt (CS) of 20 m in thickness and productive strata of 70 m in thickness with interlayering rock salt, sylvinite and carnallite. Underlying salt is up to 400 m in thickness. The sinkhole was formed in the area with WPS of more than 100 m in thickness.

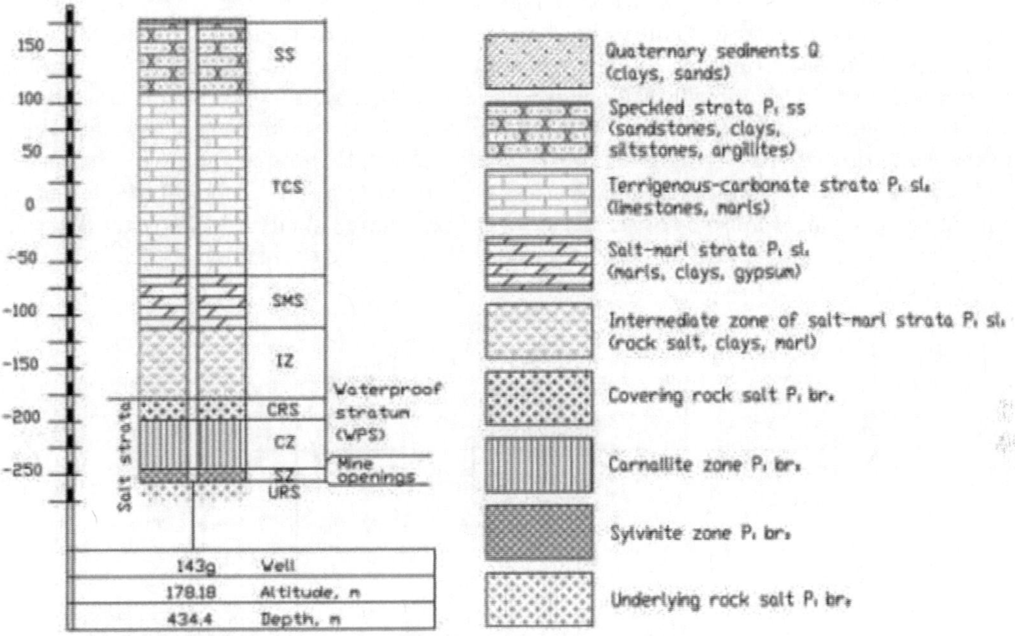

Figure 1. Geological strip log 143g (Berezniki-3 mine).

3. Results of the research

Table and Figure 2 show the dynamics in the composition components of suprasalt brines inflowing the chamber 52 of Berezniki-2 mine from January 11, 1986 till March 08, 1986. The brines were sampled for the chemical analysis from 1 to 6 times within 24 hours. "Jump" in the chart from January 21, 1986 till February 01, 1986 corresponds to the lack of observation data.

Table. Changes in composition components of suprasalt brines entering the mines from January 11, 1986 till March 08, 1986, g/L, density ρ (g/cm^3).

Values	N	pH	ρ	Cl^-	SO_4^{2-}	HCO_3^-	Br^-	Na^+	K^+	Mg^{2+}	Ca^{2+}	TDS
Observation period		**11 Jan – 12 Jan, Cl-Mg type, Q = 10 m^3/h**										
Beginning		5.70	1.282	258.86	0.49	0.33	4.26	12.00	8.26	69.92	11.50	365.62
Ending		5.10	1.260	278.36	0.63	0.18	1.96	27.40	18.10	70.98	8.00	405.61
Average	3	5.35	1.267	263.88	0.57	0.23	2.76	17.86	15.15	70.17	9.08	379.71
		13 Jan – 21 Jan, Cl-Na-Mg type, Q = 100 m^3/h by 22 Jan										
Beginning		5.75	1.230	214.53	1.72	0.16	0.97	58.30	32.40	29.49	6.25	343.82
Ending		6.24	1.216	207.44	1.73	0.15	0.73	73.19	28.00	21.58	4.50	337.32
Average	23	6.14	1.222	210.61	1.60	0.14	0.77	69.59	30.24	23.66	5.05	341.67
		01 Feb – 08 Mart, Cl-Na type, Q = 5000 m^3/h by 09 Mart										
Beginning		6.20	1.210	202.12	2.30	0.15	0.51	90.73	19.05	13.98	3.50	332.34
Ending		6.50	1.204	195.03	3.86	0.18	0.20	114.72	6.25	4.26	1.75	326.25
Average	93	6.54	1.206	195.39	3.42	0.18	0.24	108.79	9.40	6.07	2.33	325.82

IOP Conf. Series: Earth and Environmental Science **33** (2016) 012016 doi:10.1088/1755-1315/33/1/012016

In the initial period of the accident (January 11-12) the discharging brines were characterized with Cl-Mg composition. Their average density was 1.267 g/cm^3, and mineralization was 380 g/L. The 12th of January saw the brines with the highest concentration of chlorides and mineralization (405 g/L) for the period in question. It is possible that the first streams of the brines in the cover of the mine were followed by the initial cavity of solution in carnallite zone (CZ in figure 1) due to the inflow of suprasalt brines. Here the brines were enriched with Mg^{2+} in their interaction with carnallite, Cl^- concentration together with their mineralization and density increased. With further development of the crack (damage) in waterproof stratum (WPS), the brines from the sinkhole flew into the mines.

With the expansion of cross-section area of discharge channels to mines and with the increase in inflow of suprasalt waters with Cl-Na composition, the brines turn into Cl Na-Mg type, on average their mineralization changes up to 342 g/L, density - 1.222 g/cm^3 (January 13-21). Later (February 1 - March 8) the mines receive Cl-Na brines typical for the lower part of SMS, with average mineralization of 326 g/L, density of 1.206 g/cm^3.

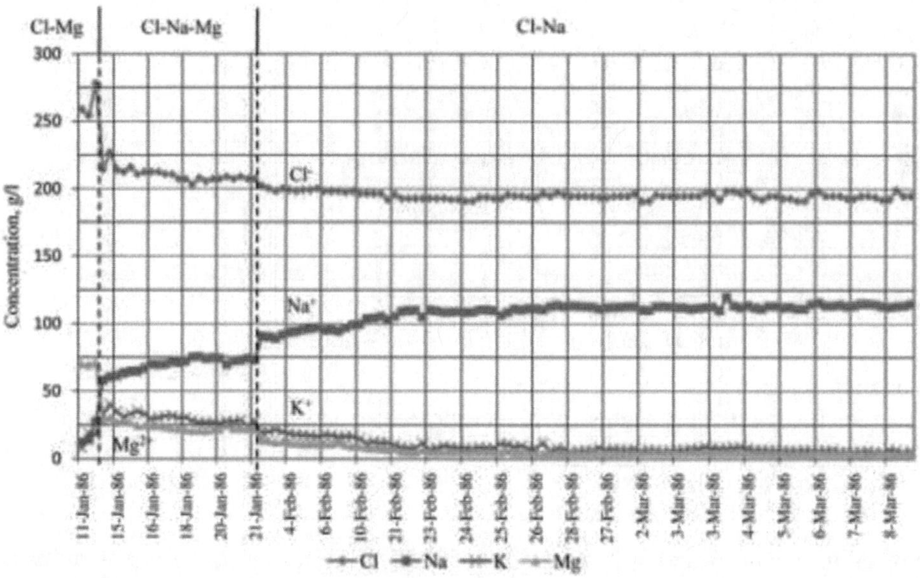

Figure 2. Dynamics of composition components of suprasalt brines discharging in mines (accident in Berezniki-3 mine in 1986).

The chart in figure 3 illustrates changes in saturation degree of suprasalt brines entering the mines with respect to the considered chloride minerals. The chart shows that the inflowing brines over the whole observation period are characterized with the equilibrium state with respect to halite. At the same time, in the initial period of the accident (January 12) the brines oversaturated with respect to halite (SI_{halite} = 0.46), sylvite ($SI_{sylvite}$ = 0.14) and equilibrium with carnallite ($SI_{carnallite}$ = 0) are discharged for a short period of time. Some samples of post-sedimentation (intrasalt) brines met in the mine's cover in the border of carnallite and sylvinite zones ("B"-"V" layers) are noted to be maximum close to chloride minerals by the chemical composition and saturation degree. For example, the brine of sample № 1407 (Solikamsk-2 mine, 2012) is characterized by 407 g/L mineralization, 1.273 g/cm^3 density, content (g/L): Cl^- - 269.52; Na^+ - 24.86; K^+ - 18.23; Mg^{2+} - 60.60. Degree of brine saturation (SI) to chloride minerals is: with respect to halite 0.31; to sylvite 0.06; to carnallite"-0.30". Thus, we can identify the initial cavity of solution in carnallite zone, where inflowing Cl-Na brines are enriched with Mg^{2+} and Cl^- and reach equilibrium with respect to carnallite. With the development of WPS damage before the mine's cover these brines are first to discharge. G.V. Beltyukov [2] holds the same opinion. He suggests that "paleokarst cavities filled with the brines and hydraulically connected with the suprasalt waters" are the initial sources of brines inflow.

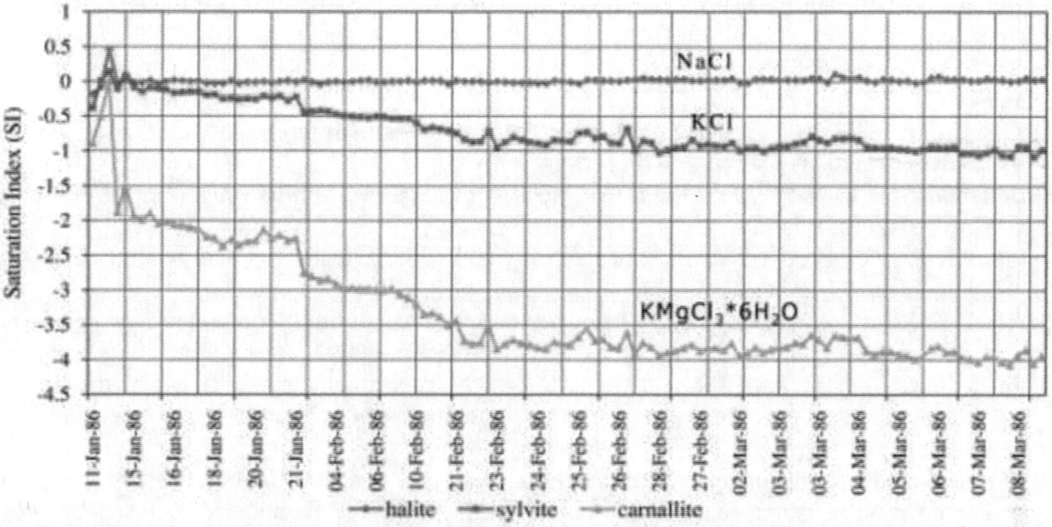

Figure 3. Dynamics of saturation degree of suprasalt brines entering the Berezniki-3 mine with respect to chloride minerals from January 11, 1986 till March 08, 1986. Chamber № 52.

With the increase in inflow of suprasalt Cl-Na brines, their saturation degree with respect to sylvite and carnallite regularly decreases (see figure 3). In observation period corresponding to the discharge of Cl-Na-Mg brines in the mines (January 13-21), saturation degree (SI) with respect to sylvite changed from "-0.11" to "-0.22" (average value "-0.17"), with respect to carnallite from "-1.88" to "-2.25" (average value "-2.12"). With the inflow of Cl-Na brines (1 February - 8 March)b saturation degree with respect to sylvite changed from "-0.44" to "-1.00" (average value "-0.83"), with respect to carnallite from "-2.75" to "-3.95" (average value "-3.66"). In his paper A.I. Kudryashov [1] points out that by 10.03.1986 more than 250 thousand m^3 of "maximum salt saturated solutions" flew into Berezniki-3 mine. Here, considering the above-mentioned facts, we should note: solutions saturated with respect to halite (but undersaturated with respect to sylvite and carnallite).

Analysis of 324 brine samples of different genesis taken in VKSD mines in 2011-2014 shows that mine brines are in equilibrium or in close equilibrium with respect to halite and sylvite [12]. For example, saturation index SI by sylvite ranges between "-0.2" and "0.2" with the prevalence (94 % of samples) of the values from "-0.11" to "0.18". Due to genesis and chemical composition the brines of the mine differ in their saturation degree with respect to carnallite. The present paper shows that in comparison with the mine brines suprasalt Cl-Na brines (the ones which penetration into the mine is caused by WPS discontinuity) are less saturated with respect to sylvite. Therefor, it is advisable to take into account the saturation degree with respect to sylvite expressed in SI, when studying the mine brines, to improve safety in mining.

4. Conclusion

Suprasalt Cl-Na brines connected with leaching of CS and salt layers in the lower part of SMS are in equilibrium with respect to halite. When the suprasalt brines flow into Bereznki-3 mines due to WPS discontinuity, in the initial stage their interaction with carnallite in productive stratum increases the saturation degree with respect to sylvite and carnallite. In this case the brines have Cl Na-Mg composition. Brief inflow of Cl-Mg brines being in equilibrium with carnallite and oversaturated with respect to halite is likely to be connected with the formation of the initial solution cavity in carnallite zone, which development follows the first accident brines in the mine's cover.

With the increase in inflow rate in the mines, the discharging brines change their chemical type from Cl Na-Mg to Cl-Na. Their composition corresponds to suprasalt brines of the SMS lower part; solution of carnallite zone sediments has the minimum impact on the composition.

In comparison with the mine brines of different genesis, suprasalt brines are undersaturated with respect to sylvite.

References

[1] Kudryashov A I 2001 *Verkhnekamskoe salt deposit* (Perm: Mining Institute of Ural Branch of RAS) (in Russian) p 429

[2] Beltyukov G V 2000 *Karst and hypergenic processes in evaporates* (Perm: Perm State University) (in Russian)

[3] Pitzer K S 1973 Thermodynamics of electrolytes. I. Theoretical basis and general equations *J. Phys. Chem* **77** 268–277

[4] Pitzer K S and Simonson J M 1986 Thermodynamics of multicomponent, miscible, ionic systems: theory and equations *J. Phys. Chem* **90** 3005–3009

[5] Pitzer K S, Wang P, Rard J A and Clegg S L 1999 Thermodynamics of electrolytes. 13. Ionic strength dependence of higher-order terms; equations for $CaCl_2$ and $MgCl_2$ *J. Solution Chem.* **28** 265–282

[6] Mironenko M V and Polyakov V B 2009 On algorithm for the calculation of the equilibrium composition of water–salt systems on the basis of the Pitzer model *Geochem. Int.* **47** (10) 1036–1040

[7] Bukaty M B 1999 Equilibrium between underground brines of the Tunguska basin with minerals of evaporite and terrigenous facies *Geology and Geophysics* **40** (5) 750-763

[8] Shvartsev S L 2000 The chemical composition and strontium isotopes of brines from the Tunguska basin: implications for their formation *Geochem. Int.* **38** (11) 1070-1083

[9] Parkhurst D L, Appelo C A J 1999 *User's guide to PHREEQC (Version 2): A computer program for speciation, batch-reaction, one-dimensional transport, and inverse geochemical calculations* (Denver, Colorado: US Geological Survey) p 312

[10] Shiman M I 1992 *Preventing the Flooding of Potash Mines* (Moscow: Nedra) (in Russian) p 176

[11] Andreichuk V, Eraso A and Domínguez M C 2000 A large sinkhole in the Verchnekamsky Potash Basin in the Urals *Mine Water and the Environ.* **19** 2-18

[12] Fetisov V V, Kataeva E P, Fetisova N F 2015 Interaction of natural and technogeneous brines of Verkhnekamskoe salt deposit with basic minerals of salt strata *Bulletin of the Tomsk Polytechnic University* **326** (9) 87-94

Hydrogenous mineral neoformations in Tomsk water intake facility from underground sources

E Dutova[1], I Vologdina[2], D Pokrovsky[3], N Nalivaiko[4], K Kuzevanov[5] and V Pokrovsky[6]

[1,4,5,6] Tomsk Polytechnic University, 30 Lenin Avenue, Tomsk, Russia
[2] Tomsk State University, 36 Lenin Avenue, Tomsk, Russia
[3]Tomsk University of Architecture and Civil Engineering, 2 Solyanaya Square, Tomsk, Russia

E-mail: [1]dutova@sibmail.com, [2]irvalvo@inbox.ru, [3]dsp@sibmail.com, [4]nalivaiko@mail.ru, [5]kuzevanovkk@mail.ru, [6]pokrovskiy.v@gmail.com

Abstract. The article considers study outcomes of hydrogenous mineral neoformations precipitated on deferrization filters of Tomsk water intake facility from underground sources. Compositionally, these precipitations are colloform and polymineral including ferrous, carbonate and aluminosilicate mineral phases. Ferrous phase predominates and embraces ferric hydroxides (ferrihydrate, goethite, hematite and lepidocrocite) and ferrous hydrophosphates (vivianite, strengite, strunzite and rockbridgeit). Carbonate and aluminosilicate minerals are calcite and kaolinite-group, respectively.

1. Introduction

Water supply is one of the most common human activities that is accompanied by the formation of significant precipitation. In many cases problems crop up from operating water intake facilities, such as incrustation, water supply pipe network incrustation, which, in its turn, reduces tap drinking water quality. Researchers consider such a factor as "water intake facility ageing" and identify the following environmental degradation causes at water supply facilities: equipment corrosion, sanding, ochering, deposition of various chemical elements and slimming caused by microorganism growth.

Studies of chemical and mineral composition of the precipitation, its morphology, formation behavior patterns are of both theoretical and practical importance. There are two sides of this issue: on the one hand, such data is necessary in selecting the relevant water treatment technology and further optimization of the water intake equipment operation; on the other hand, up-dated model of hydrogenous mineral formation allows observing the formation of minerals from the first moment of their origin.

The authors have been studying precipitations forming in water supply systems of water intake facilities from underground sources in Tomsk and Tomsk Oblast [1-3]. Studies revealed data on precipitation composition and morphology and the precipitation properties – water aeration regime relation, as well as type of water load.

Research findings proved the fact that such precipitations could be used as raw sources, i.e. sorbents to purify water from petrochemicals and phenols [4], as well as catalysts in the hydrocarbon oxidizing process [5].

2. Study subject

Tomsk water intake facility from underground sources is one of the largest intake facilities in Russia (Figure 1) and it develops Paleogene aquifers being main water supply source for the majority of West-Siberian populated areas. The chemical composition of ground waters has been studied during the entire development period of this ground water basin. According to research data [6], production wells are compositionally hydrocarbonate with various calcium-magnesium ratios; fresh waters with mineralization varying from 196 to 600 mg/dm^3; weak-acid to weak-alkaline (pH: 6.2 – 8.0); very soft to hard (from 0.6 to 7.0 at a mean value of 3.8 mg-equivalent/dm^3), and predominately hard. Concentrations of most analyzed components do not exceed the maximum allowable concentration (MAC). Ferrum and manganese content is higher than MAC, whereas ferrum concentration exceeds the MAC in 100% cases. Proposed water treatment scheme includes simplified aeration followed by filtering through granular loading and decontamination. In general, water treatment shows rather positive results – water through technological treatment responds to the existing regulatory documents.

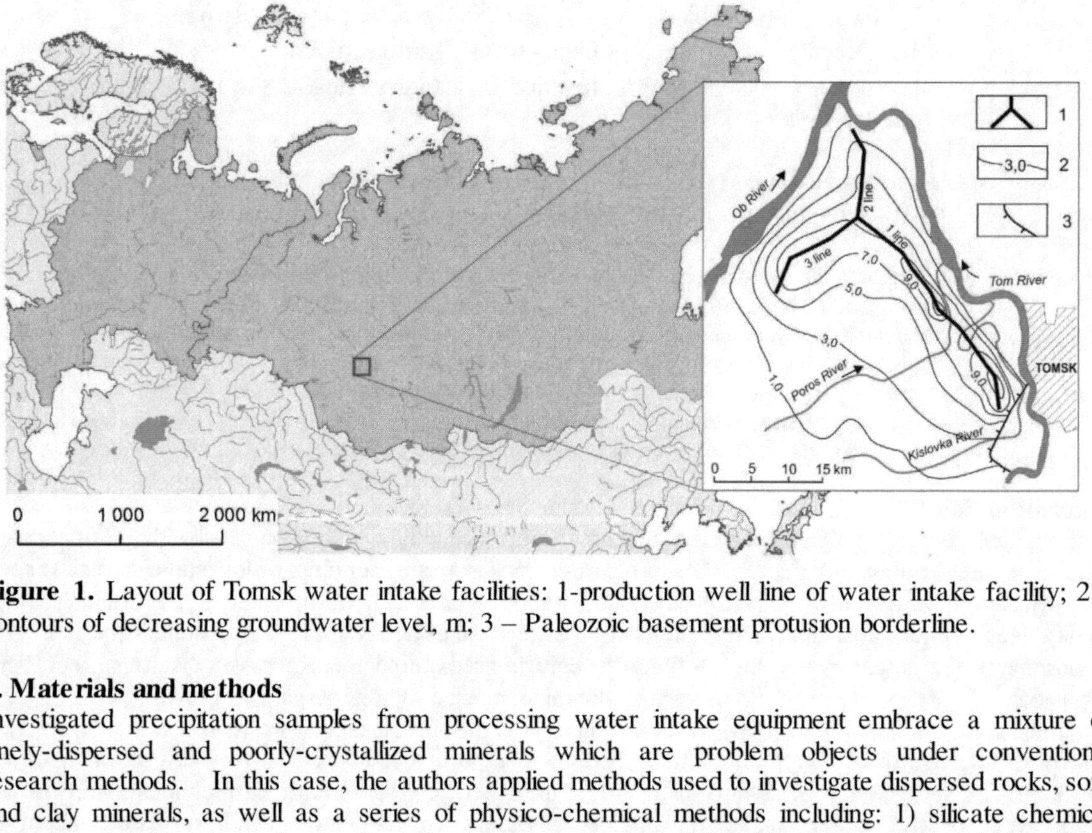

Figure 1. Layout of Tomsk water intake facilities: 1-production well line of water intake facility; 2 – contours of decreasing groundwater level, m; 3 – Paleozoic basement protusion borderline.

3. Materials and methods

Investigated precipitation samples from processing water intake equipment embrace a mixture of finely-dispersed and poorly-crystallized minerals which are problem objects under conventional research methods. In this case, the authors applied methods used to investigate dispersed rocks, soils and clay minerals, as well as a series of physico-chemical methods including: 1) silicate chemical analysis; 2). X-ray analysis; 3) infrared spectroscopy (IS); 4) thermogravimetric analysis; 5) spectrometry – ED (on SEM); 6) microdiffraction and spectrometry via electron transmission microscopy (ETM) and 7) composition analysis via scanning electron microscopy VEGA II LMU, equipped with energy-dispersed spectrometer INCA Energy.

4. Results and discussion

Precipitation samples were selected from the filters of Tomsk intake facility from underground sources including operation time of the filters. Quartz sand and crushed albitophyre were used as a filter bed. Sand grain sizes vary from 0.2 mm to 1.5 mm, while albitophyre grains: 5 – 7 mm.

Suspended particles in the filtered water settle down on the grain surface forming a layered film, which, in its turn, forms rounded brown aggregates related to oolites. The size and specific morphology of oolites depend on filter bed type and operating filter time. In operation time the precipitation film thickness varies from a thinnest layer on filter bed grains to large -sized oolites (3-4 times more than grain size) (Table 1). After air drying, the precipitation structure preserved and only mechanically disintegrates.

Table 1. Morphological characteristics of mineral neoformations.

Filter №	Filter operation time, years	Filter bed composition	Oolites size (diameter) and shape
11	21	quartz sand	5.8 – 6.5 mm (spheroidal)
19	19	quartz sand	3.5 – 4 mm (spheroidal)
7	6	albitophyre	5 – 10 mm (spheroidal -elliptical)
14	1	albitophyre	≈ equal to the filter bed grain size
17	less than 1	albitophyre	≈ equal to the filter bed grain size

Well-developed oolites in the cross-section revealed a bedded structure: if mechanically disintegrated, upper oolite layer (crust) easily laminates from the inner dense layer (core). The crust differs from the core due to its light color and specific bedding embracing alternating dark and light layers (Figures 2). Oolites have such concentrically-layered structure formed in long-term operating filters.

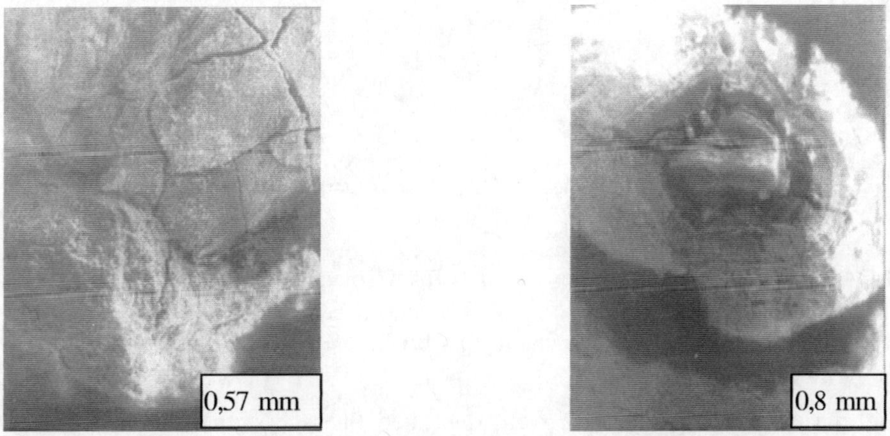

a b

Figure 2. Structure of mineral neoformations from Tomsk water intake facility filters. Oolite is divided into two parts- (1) inner- core, and (2) outer (crust): a) conchoidal-concentric oolite from filter № 7 (cross-section); b) – layered oolite core from filter № 11 (cross-section). (Electron-microscopic analysis was conducted in Electron Microscopy Lab, TSUAB).

Microstructure studies of mineral neoformations indicate the fact that precipitations, as mineral aggregates, have specific microstructures characteristics of which are in direct relationship with the composition and behavior of the hydrogeochemical environment where they are formed. Precipitation structuring formed during the water treatment process is a multi- process as in the natural environment. This, in its turn, is most pronounced in different internal structures of oolites formed on filters of different water intake facilities.

When investigating the precipitations of different ages, it was possible to observe the microstructure evolution itself, practically from its nucleation. Initial stages of the microstructure formation were experimentally investigated. The experiment included: water-cured albitophyre and quartz grains for 48 hours. This water underwent a simplified aeration immediately at Tomsk water intake facility. Scanning and transmission electron microscopes were used to examine the precipitation microstructure and composition. Comparatively different microstructure films formed on albitophyre and quartz were not identified. SEM imaging showed a fibrous structured precipitation film composed randomly of interwoven 0.4 micron fibers and a few micrometers length (Figure 3a). The microstructure is heterogeneous: fibers are interwoven in rope-like bundles which, in its turn, form flaky ultra-microagregates of 3 – 5 μm in diameter by sticking together and inter-twisting. Fibrous microstructure observed in 48-day precipitations is simply a macro- particle aggregate consisting of joint chains. The same fibrous structure was observed in prepared suspended substances. In this case, this fibrous structure included elongated bands and ribbons, laterally about 0.15 μm and 2 μm or more in longitudinal (Figure 3b). The X-ray diffraction pattern of aggregates shows two weak diffusion halos which indicate that the aggregates are small in thickness and are amorphous (Figure 3c). In some sections one can observe twisted and interwoven bands forming micro- dispersed structures.

Six-month aggregates from Filter №17 have the following structure: rounded flakes (particles) of up to 4 μm in diameter (Figure 3c). These flakes (particles) are oriented perpendicular to grain surface of the filter bed (based on SEM results). Particles intertwine laterally forming thin layers. This structure can be characterized as a colloform structure.

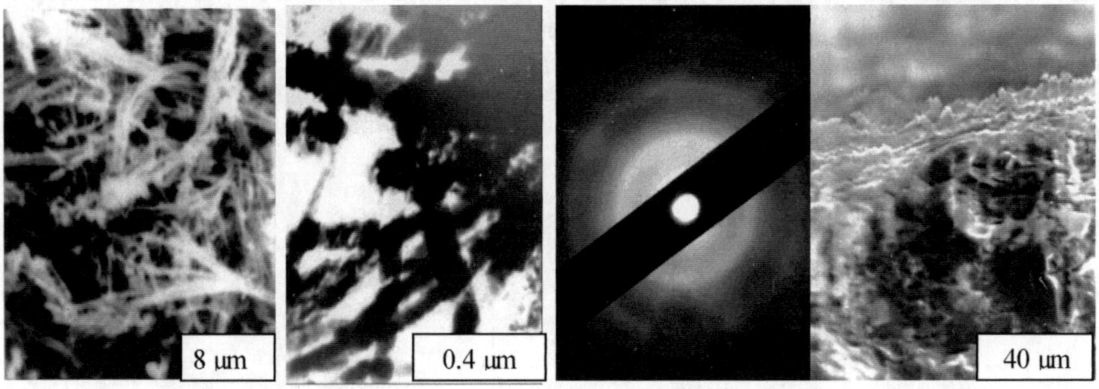

a) b) c) d)

Figure 3. Mineral neoformation microstructures from Tomsk water intake facility filters: a) – fibrous precipitation structure formed during the experiment (SEM imaging); b) – ribbon-like aggregates (TEM imaging); c) – microdiffraction image of ribbons; d) flaky aggregates in the precipitation film from Filter № 17 (SEM imaging).

In all the filters the aggregates from deposited filter bed grains have a colloform structure, differing only in thickness and density of the precipitation itself. Precipitation thickness on albitophyre grains (in Filter №17) is about 16.5 μm. Comparingly, in Filter №14, after a one-year operation, precipitation thickness on albitophyre grains is 0.8-1.2 mm, while in Filter №7, operating over six years, conchoidal-concentric oolites formed on the albitophyre grains (Figure 2a). The thickness of the upper loose precipitation layer (crust) is approximately 0.2 – 0.7 mm, while the thickness of the inner dense layer (core) ranges from 0.8 to 1.2 mm. Oolites formed on the filter bed grains, precipitation thickness of about 2mm during nine-year operation of Filter №19. The research results showed that precipitations have a homogeneous coagulation structure.

Bulk chemical composition of newly-formed oolite particles is given in Table 3 (in analyzing prepared samples filter bed grains were removed). Basically, precipitation composition includes ferrum, aluminum, phosphorus and silicon compounds and precipitations contain a significant amount of water.

Table 3. Chemical composition of precipitations from Tomsk water intake facility filters

Component	Content, %	Component	Content, %
SiO_2	6.34	K_2O	0.42
Al_2O_3	17.44	Na_2O	0.27
Fe_2O_3	28.44	CO_2 carbonate.	1.41
TiO_2	0.06	Ignition loss	29.94
MnO	3.35	Total	95.78
CaO	5.63	H_2O hygroscopic	7.31
MgO	2.70	H_2O inherent	18.65
SO_3	0.12	Humus	4.88
P_2O_5	10.29	$C_{,\ organicr}$	2.83

NOTE: Analysis was carried out in the laboratory of the Biology Institute, Tomsk State University.

The phase and mineral composition of precipitations have been studied via thermogravimetric analysis and infrared spectroscopy and X-ray diffraction methods. It was established that precipitations are a polymineral mixture composed of ferrous, carbonate and aluminosilicate mineral phases. Ferrous phase is predominate (up to 65 %) and includes ferric hydroxides (ferrihydrate, goethite, hematite and lepidocrocite) and ferrous hydrophosphates (vivianite, strengite, strunzite and rockbridgeit). Carbonate and aluminosilicate minerals include calcite and kaolinite-group minerals, respectively. There are some traces of organic matter as iron organic complexes.

Aggregates consisting of disc-like particles (Figure 4a) formed as a result of Gallionella iron bacterial activity were found in the precipitations via transmission electron microscope. The microdiffraction image (Figure 4b) of aggregates is a combination of circular (ferrihydrate: d □ 2.24, 1.97 and 1.5 Å) and point (hematite: d □ 3.62, 257 and 1.18 Å) X-ray diffraction patterns.

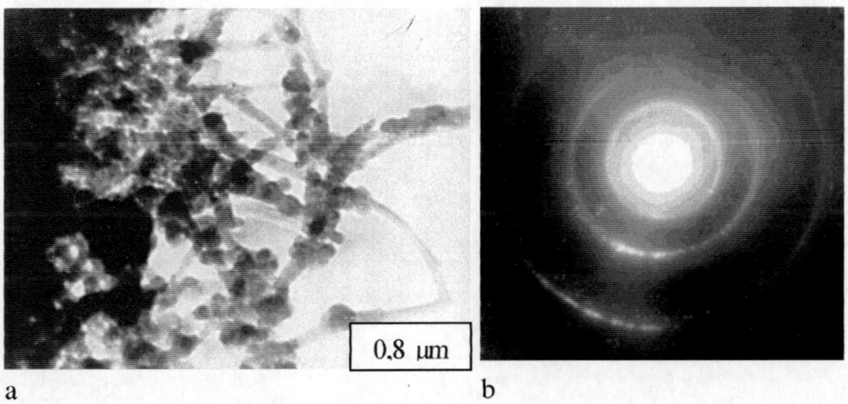

a b

Figure 4. (a) bacterial disco-like formations (b) aggregate of microdiffraction pattern. TEM imaging.

5. Conclusions

During water filtration treatment concentric-layered aggregates- oolites – are formed on the upper layer of the grain filter bed. Initially, fine -filmed fibrous microstructures are formed on bed grains which gradually evolve into a colloform structure. Colloform structure unit is rounded flakes of about 5 μm in diameter and their intergrowth and recompaction resulting in a layered structure. The precipitation thickness and bedding morphology depend on service time of the filter. Newly-formed oolites predominately include ferrum, aluminum, phosphorus and silicon compounds and are a polymineral mixture whose basic the component is ferrous phase – ferrihydrate, goethite and hematite. Iron bacteria are important in the precipitation of ferric hydroxides. Phosphate and carbonate phases are less. There are clay minerals of kaolinite group in the precipitation.

References

[1] Pokrovsky D S, Dutova E M, Vologdina I V and Taylashev A S 2002 Prirodno-tehnogennoe mineraloobrazovanie na fil'trah obezzhelezivanija vodozabora Tomskogo Akademgorodka *Izvestiya Tomsk Polytechnic University* **305** (6) 319–329

[2] Dutova E M, Vologdina I V, Pokrovsky D S and Zamorovskaya L V 2008 Izmenenie gidrogeohimicheskih uslovij pri jekspluatacii Akademicheskogo mestorozhdenija *Izvestiya Tomsk Polytechnic University* **312** (1) 59–63

[3] Pokrovsky D S, Dutova E M and Vologdina I V 2010 Hydrogeochemical Environments and Mineral Neocrystallizations from Subsurface Sources *Izvestiya Higher Educational Establishments Construction* **11–12** 54–61.

[4] Novoselova L Yu and Sirotkina E E 2010 The structure of sorbents based on thermally activated iron-containing water preparation precipitate *Russian Journal of Physical Chemistry A* **84** (6) 1033–1038.

[5] Novoselova L Yu 2013 Composition, structure, and sorbability of the thermally treated water deironing precipitate with respect to carbon monoxide Powder Ttechnology **243** 149–154

[6] Pokrovsky D S, Dutova E M, Rogov G M and et al. 2002 *Mineral'nye novoobrazovanija na vodozaborah Tomskoj oblasti* (NTL Publishing House) p 176

Microflora of drinking water distributed through decentralized supply systems (Tomsk)

A A Khvaschevskaya[1], N G Nalivaiko[2] and A V Shestakova[3]

[1, 2, 3] Tomsk Polytechnic University, 30 Lenin Avenue, Tomsk, 634050, Russia

E-mail: [1]unpc_voda@mail.ru, [2]nng@tpu.ru, [3] stysha@sibmail.com

Abstract. The paper considers microbiological quality of waters from decentralized water supply systems in Tomsk. It has been proved that there are numerous microbial contaminants of different types. The authors claim that the water distributed through decentralized supply systems is not safe to drink without preliminary treatment.

1. Introduction

Tomsk oblast is the area with a plenty of fresh groundwater sources. The raw water distributed through decentralized supply systems is intensively used by individuals for drinking: streams, dug and drilled wells. People assess the quality of this water by sight and taste in the absence of experts in organoleptic properties of water. In compliance with the current national sanitary regulations [1], one of the major criteria to assess drinking water quality is its epidemiological safety. As for centralized drinking water supply, this requirement is met through preliminary treatment and monitoring microbiological quality of source waters. However, these measures are ignored in respect with streams, dug and drilled wells.

In this paper, we study the microbiological quality of waters from decentralized water supply systems in Tomsk as it is the criterion of water safety.

2. Materials and methods

The study object is waters from streams, dug and drilled wells located within Tomsk area and used for drinking and household needs. These decentralized water supply systems are based on string tapping and located in different parts of the town with different technogenic load. Most dug and shallow drilled wells are within the residential area with no centralized water supply and truck patches. The water samples were collected from most commonly used systems.

The samples were collected into sterile glass containers in compliance with procedure guideline "Sanitary Survey and Microbiological Testing of Drinking Water Quality". In the course of the research, 77 samples were collected and investigated. In microbiological quality testing conventional microbiology methods were applied [2]. The microbiological water quality testing included data on physiological groups indicating microbial and chemical contamination, as well as those on self-purification capacity. To identify and determine the number of microorganisms, elective media were created with due regard to all physiological needs of the studied groups.

The waters chemical composition was analyzed to determine its impact on microorganisms habitat. For this purpose, the following methods were applied: titrimetry, potentiometry, colorimetry, conductivity, stripping voltammetry, atomic absorption spectrometry, ion chromatography, and fluorometry.

3. Results and discussion

The chemical composition of the studied groundwater is mainly HCO_3-Ca-Mg or HCO_3-Mg-Ca, neutral or slightly alkaline, fresh, moderately hard. There are also waters of Cl-HCO_3 and Cl-SO_4-HCO_3 anion composition. Water salinity ranges between 0.26 g/L and 1.45 g/L. pH and hardness values in studied waters vary significantly, from 5 to 8 and from 3.6 to 11.5 mEq/L, respectively. In terms of chemical composition, some components, such as organic matter, nitrogen compounds and chloride ion, reach or even exceed MPC for drinking water [1]. Water supply sources contamination depends on ground water level and the territory anthropogenic load [3, 4].

The microflora of samples is characterized by a wide diversity of physiological groups which cause aerobic and anaerobic destruction of minerals and organic compounds. These processes have a significant impact on water quality. The number of bacteria within different physiological groups is shown in the table below while the biogeochemical characteristics and the particularities of interaction with the components of the environment were discussed in detail in our previous studies [3, 5, 6].

Table. Number of microorganisms in waters from decentralized water supply systems in Tomsk.

Physiological groups of bacteria	The number of microorganisms (cell / ml *)		
	Streams	Dug wells	Drilled wells
Mesophilic saprophytes	30-690/70	30-240/100	20-280/50
Psychrophilic saprophytes	30-43000/2830	60-6300/1670	30-1470/590
Oligotrophs	30-74100/8100	390-48050/5820	3200-920000/13500
Oil oxidizing microorganisms	30-57200/1900	30-1620/560	20-4200/730
Denitrifying bacteria	10-10000/1000	10-10000/1000	10-10000/1000
Ammonifying bacteria	10-3200/790	1020-26470/4540	10-350/120
Nitrifying bacteria	10-1000/100	10-1000/100	10-100/10
Uro-bacteria	30-5800/450	30-2340/730	0
Sulfur reducing bacteria	10-10000/1000	30-1000/100	0
Cellulose-decomposing aerobic bacteria	0-100/10	10-1000/100	0
Cellulose-decomposing anaerobic bacteria	0-100/10	10-1000/100	0
Iron bacteria	0-2000	0	0-5000

Note: The values in the table are presented as follows: minimum-maximum / average;
*cells/ml (cells per 1 ml of water).

Organic matter is one of the major factors indicating the presence of microorganisms in water. Therefore, the researchers focused on identification of psychrophilic saprophytes and oligotrophs, as the former signifies the presence of labile organic material (dissolved organic matter) while the latter determines its mineralization rate. The number of saprophytes indicates the rate of ecosystem contamination with both microbes and organic matter. In accordance with [6], the number of psychrophilic saprophytes in clean water should not exceed 100 cells per ml and should be less than the number of oligotrophs [7].

According to the research data, oligotrophs and psychrophilic saprophytes are present in all water sources. The number of psychrophilic saprophytes ranges from 30 to 43000 cells/ml, with the highest values characterizing streams and lowest drilled wells. The psychrophilic saprophytes number is particularly high in water sources located within the areas with old buildings and private houses without centralized water supply and sewerage systems. Oligotrophs are normally predominant over psychrophilic saprophytes in number, which indicates self-purification capacity of waters from the vast majority of sources.

In all studies waters, there are a great number of bacteria associated with the metabolism of nitrogen-containing organic compounds. They are ammonifying, nitrifying, denitrifying bacteria and uro-bacteria.

Ammonifying bacteria obtain carbon and energy from wide range of organic compounds, particularly, proteins, and the metabolic product is mostly ammonia [7, 9]. According to [7], the number of these bacteria in clean water should not exceed a few hundred cells per ml, or 3–5 % of total bacteria number, while in contaminated water this number can be several thousand cells per ml. In 70 % of water sources investigated in the course of the research, the number of these bacteria ranges from a few dozen to tens of thousands of cells per ml, which indicates water contamination and allochthonous (anthropogenic) organic matter presence. The greatest number of ammonifying bacteria is characteristic for dug well water (up to 26470 cells/ml), which is 7 times as many as that in stream water and 30 times as many as that in the waters from drilled wells.

Denitrifying bacteria are present in 75 % of water sources, their number ranging between 1000 and 10000 cells/ml. The greatest number of denitrifying bacteria is characteristic for the water of dug wells. These bacteria cause anaerobic destruction of organic matter obtaining oxygen from nitrates. The metabolic end product is ammonia [10].

Nitrifying bacteria, which remove ammonia through its oxidation to nitrites and nitrates, are present in almost all dug wells, 20 % of streams, and only a few drilled wells. The number of these bacteria is significantly lower than that of bacteria producing ammonia. The number of nitrifying bacteria is relatively high in old districts of the town, where the town development density is high and the layer of anthropogenic soil is thick enough, which increases the amount of carbon dioxide and ammonia in the air.

Uro-bacteria are identified in 30 % of streams and 46 % of dug wells, which are contaminated by sewage runoffs, while drilled well waters are characterized with the absence of these bacteria. The bacteria cause the destruction of carbamides, uric acid, which serve as a source of carbon and nitrogen. The metabolic end product is ammonia. More often than not, the bacteria are present in the areas contaminated by sewage and household-generated polluted runoffs.

Hydrocarbon degrading bacteria are present in 40 % of streams and 10 % of dug and drilled wells. Under favorable conditions, these bacteria destroy oil to gas and water. The great number of hydrocarbon degrading bacteria was identified in the streams which could be contaminated with petroleum: within food services area, key transit and unloading points the amount of these bacteria was more than 70 % of total number. According to previous research data [7], in raw waters clean from petroleum, the amount of hydrocarbon degrading bacteria ranges between 1 and 4 % of total bacteria number. Therefore, these data might indicate stream water contamination with oil, which can be confirmed by the high value for petroleum products in drinking-water ranging from 0.68 to 1.3 mg/l, with PMC for drinking water being 0.3 mg/l.

Sulphate reducing bacteria (SRB) are present in 50 % of streams and almost 100 % of wells. The bacteria produce hydrogen sulfide, which has a negative impact on organoleptic properties of water. However, waters from drilled wells are characterized by the absence of these bacteria.

Cellulose-decomposing bacteria, both aerobic and anaerobic ones, are present in the waters of streams equipped with wooden tapping construction and all dug wells. In summer, when the wooden construction is heated, the number of bacteria is up to thousands of cells. These bacteria presence indicates that soil has got into the water while their metabolic products can contribute to the growth of denitrifying, sulphate reducing, and methanogenic bacteria [7].

Mesophilic saprophytes, which indicate the presence of pathogenic microflora, are identified in all water sources, and their number often exceeds MPC (50 cells/ml) [1]. This means that the waters are not safe to drink.

According to the relevant classification [6], the number of psychrophilic saprophytes indicates that in 46 % of streams the water is contaminated, while in 30% – severely contaminated. The water in 50 % of dug wells is moderately contaminated and in 30 % – contaminated. More often than not, drilled well water is contaminated.

Iron bacteria identified in the studied waters are isolated instances, which are generally caused by metal tapping constructions.

In general, the present research data indicate contamination of all water samples with bacteria, ranging in number and types. In terms of total bacteria number, stream water heads the list, dug well water is ranked second, and drilled well water round out the list. The water from all sources is contaminated with organic matter, sulfur and nitrogen compounds. The contaminants get into the water due to human activities and under microbial conditions and cause the secondary microbial and chemical contamination of water, which has a negative impact on organoleptic properties of water.

4. Conclusions

Summing up the results, it can be concluded that the water from decentralized water supply systems in Tomsk is contaminated with bacteria, significant in number and ranging in types. The number of bacteria depends on the water supply system type and technogenic load of the area. Microbiological characteristics of the water indicate the high rate of contamination with bacteria, which means that the water fails to be safe to drink without preliminary treatment.

Acknowledgements

This work was supported under the state assignment of the Ministry of Education and Science of Russia "Nauka" № 5.1931.2014/K.

References

[1] Sanitary Regulations and Standards 2.1.4.1074-01 *Drinking water. Hugienic requirements to the quality of water from centralized drinking water supply systems. Quality monitoring*

[2] Netrusov A I, Egorova M A, Zakharchuk L M et al. 2005 *Praktikum po Mikrobiologii* (Moscow, ACADEMIA) p 608

[3] Dutova E M and Nalivaiko N G 2011 Chemical and microbiological characteristics of groundwater within Tomsk City *Izvestiya Vysshikh Uchebnykh Zavedeniy. Geologiya i Razvedka* **5** 56–61

[4] Amanatidou E, Adamidou K, Trikoilidou E, Katsiouli F, Patrikaki O and Tsikritzis L 2007 Physicochemical and microbiological characteristics of the potable water supply sources in the area of Kozani, Western Macedonia *Desalination* **213** 1–8

[5] Nalivaiko N G, Kuzevanov K I and Kopylova Yu G 2002 *Atlas bakterial'nykh peyzazhey rodnikov g. Tomska* (Tomsk, SST) p 52

[6] Khvaschevskaya A A and Nalivaiko N G Zhelezobakterii pit'yevykh vod goroda Tomska.- Vodnye i ekologicheskie problemy Sibiri i Tsentral'noy Azii *Trudy Vserossiyskoy Nauchnoy Konferentsii s Mezhdunarodnym Uchastiem* **2** 192–199

[7] Korsh L E and Artemyeva T Z 1978 Uskorennye metody sanitarno-bakteriologicheskogo issledovaniya vody (Moscow, Medicine) p 272

[8] Romanenko V I 1979 Mikrobiologicheskie pokazateli kachestva vody *Water Resources* **6** 138–153

[9] Boer W De and Kowalchuk G A 2001 Nitrification in acid soils: micro-organisms and mechanisms *Soil Biology and Biochemistry* **33** 853–866

[10] Lipponen M T, Suutari M H and Martikainen P J 2002 Occurrence of nitrifying bacteria and nitrification in Finnish drinking water distribution systems *Water Research* **36** p 4319–29

Environmental state and buffering properties of underground hydrosphere in waste landfill site of the largest petrochemical companies in Europe

R Kh Musin[1], N A Kurlyanov[2], Z G Kalkamanova[3] and T V Korotchenko[4]

[1, 2, 3] Kazan Federal University, Kazan, Russia
[4] Tomsk Polytechnic University, Tomsk, Russia

E-mail: [1]Rustam.Musin@kpfu.ru

Abstract. The article examines the waste landfill site of PJSC "Nizhnekamskneftekhim" built 1982. Particular attention is paid to the volume of disposed wastes and peculiarities of landfill operation. It has been revealed that the landfill negatively impacts groundwater. The increase in groundwater level and contamination degree is dependent on recharge from infiltration of precipitation that interacts with the waste in the landfill cells. Groundwater contamination follows the longitudinal distribution pattern, with maximum intensity reaching in the nearest area of the landfill. With increasing distance, concentration of all pollutants sharply reduces. Within three kilometers away from the landfill, groundwater turns to its background values indicating its quality. The landfill discharges oil, phenols, formaldehyde, benzol, toluene, xylene, ethylbenzene, and iron and, to a lesser extent, sulfates, chlorides and barium into the underground hydrosphere. The overlimiting concentrations of other components are caused by intensive leaching from the rocks by aggressive carbonic acid water. The concentrations of hydrocarbonates can reach 8 g/l in the groundwater within the landfill and its nearest area, however, under natural conditions, they do not exceed 0.4 g/l. This is only possible in a case of partial activity of carbon dioxide associated with destruction of organic matter disposed in the landfill. One of the processes that play an important role in groundwater quality recovery is mixing of contaminated groundwater with infiltrating precipitation.

1. Introduction

PJSC "Nizhnekamskneftekhim» is one of the largest petrochemical companies in Europe specialized in the production of synthetic rubber, plastics, ethylene and other products (more than 100 names). Its annual volume is more than 1.5 million tons and the cost is more than 100 billion roubles. The main production facility of the company includes 10 plants and 7 different departments (Department of Science and Technology, Design, and etc.). The main production facilities are located in the city of Nizhnekamsk, Republic of Tatarstan. The company was founded in 1967. Most industrial waste is dumped in the landfill which has been in service since 1982. By 2014, the landfill capacity has been depleted up to 80%. The landfill holds about 504.080.7 m^3 of waste consisting of different compounds. Identification of the environmental impact of the discussed landfill is a rather critical issue.

2. Objects and methods of research

The area of the waste landfill is 0.25 km^2 (0.5 * 0.5 km). It is located within the Volgo-Ural anteclise of the Russian plate, i.e. Kama-Vyatka artesian basin, on the left bank of the river Kama, near the axis of the forested watershed, 8 km to the south-east of Nizhnekamsk. The altitude above sea level of unbroken surface within the landfill site is 186–199 m, with altitude above riverbed being less than 120 m. The closest settlements are situated at a distance of 3 kilometers, with the population not exceeding 400 inhabitants. The gravity drain is of south-eastern direction. The surface slope is not more than 2–3^0 (figure 1). The upper part of geological cross-section (200 m) is composed of subhorizontal facies-irregular complexes of carbonate-terrigenous rock (Urzhum and Kazan stage). Within the landfill site, it is represented by Quaternary loam soil up to 2-4 m in thickness and top soil.

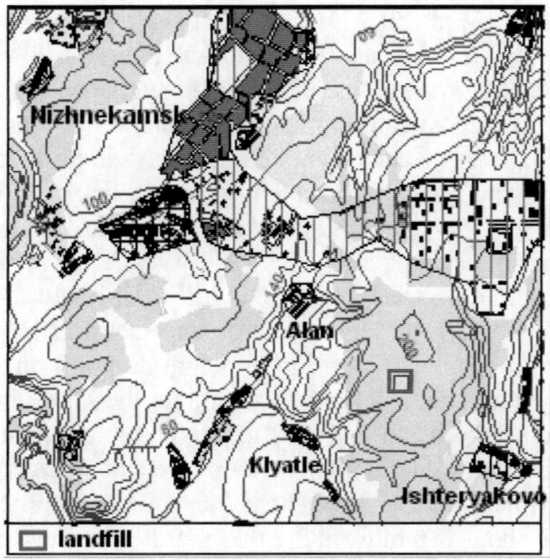

Figure1. The landfill location.

The highest recorded stage (160–205 m) is composed of mottled formations characterized by interbedded sandstones and clay, with thickness of definite layers reaching 3-4 m (greater thickness is much more rare). Sandstones are usually polymistic, fine-grained, and weakly cemented. Clay carries thin variously fractured interbeds of siltstone and fine-grained sandstone. The ratio of pelitic and psammitic rocks is about 2:1.

The underlying Kazan sequence (140-160 m) is of the similar structure. Its basement, being 30-40 m in thickness, is composed of grey-colored clay carrying interbeds of carbonate rocks (1-3 m in thickness). In the landfill site, the Kazan sedimentary complex is underlain by the sequence of so-called "lingual clay", being 14 m in thickness within the hypsometric interval, i.e. 2-16 m. This sequence is extended along the strike and plays hydrogeological role as a regional impermeable horizon.

The groundwater of the described cross-section is interstitial, crevice, and interstitial-crevice. These types form separate aquifers which are basically confined to permeable sandstone and carbonate rocks comprising Urzhum, Upperkazan and Lowerkazan aquifers. The difference in adjacent aquifer levels can reach several meters reflecting the relief of the territory. The groundwater flow is of south-western direction. In addition, downward filtration is identified across the whole cross-section. This explains the fact that some aquifers are hydraulically connected. There no continuous impermeable horizons up to the layer of "lingual clay". Groundwater is recharged from infiltration of precipitation, i.e. rain and snow. The regular annual precipitation is 554.6 mm. Under natural conditions, groundwater above the impermeable clay layer mostly belongs to hydrocarbonate-magnesium-calcium type (here and further the elements are listed in order of increasing mole-percent concentration), mineralization being 0.2–0.8 g/l and total hardness – up to 7-8 mM/l. Mineralization and hardness increase with depth, however, the groundwater quality meets all drinking standards. Brackish water can be found below the impermeable clay layer within the Ufa stage Sheshminsk deposits.

These specific features of groundwater composition are determined by impact of natural factors, more precisely, interaction of atmospheric precipitation and peculiar types of soils and rocks identified within the cross-section. During the rainless period, the depth of the first aquifer is naturally 3-8 m, with free gravitational waters being located within sandstones and Urzhum stage clays, characterized by the average values of filtration capacity, i.e. 1.23 and 0.05 m/day, respectively. It is a free aquifer with the areas of excessive pressure 0.5–1 m [1] which to be referred to henceforth as ground aquifer.

The landfill site comprises 48 separate waste cells, 50*70 m in size and 3–5 m in depth. The waste cell bottom and walls are made of rammed clay; the walls of outermost eastern waste cells are concreted. Approximately 250 types of wastes of III–V hazard classes (solid, liquid and pasty) are disposed of in the landfill. Mainly, these wastes are catalysts, polymer materials, sediments, oil sludge, pipeline and tank sludge, and etc. The landfill holds about 504.080.7 m^3 of wastes which had been dumped by 2014: solid forms – 444717.9; pasty– 51467.8; liquid – 7895.

The landfill is planned to be reconstructed, precisely, the wastes are proposed to be sealed within a waterproof tank. To develop such waterproof tanks, first of all, it is required to remove the waste from fully filled cells. Then, impermeable screen should be installed on the inside of the tank. The cells are planned to be backfilled with the temporarily removed wastes and sealed with the impermeable screen. Seeding the soil to reintroduce vegetation is the final activity. The impermeable screen is proposed to be of geosynthetic materials characterized by extremely low filtration properties – (2–5)*10^{-11} m/c. Eight waste cells equipped only with impermeable screen at the bottom are planned to be on service.

The interception ditch with pipe drainage has been excavated along the external perimeter of the landfill in order to stop the flow of surface waters and part of ground waters. There are also natural impermeable screen, i.e. a clay fence up to 6 m in depth and 5 m in width along the northern and southern borders of the landfill, and pump station which pumps drainage water and water from the waste cells to the water treatment facilities located outside the landfill site. The whole area is enclosed by a concrete fence 2 m in height. In addition, there is a continuous monitoring of air and groundwater quality. The quality of soil around the landfill site, as well as the quality of surface and ground waters at some distance from the site (near the closest settlements), is also controlled on a schedule basis. Groundwater quality is monitored by means of two observation well networks. The first network which includes 12 wells is located close to the landfill site, while the second one comprising 20 wells is about 150 m from the first monitoring network.

The landfill site has been thoroughly studied. The detailed geo-engineering research was carried out in 1975, 1991 and 2011. The official monitoring of surface and ground surface waters, as well as air and soil quality assessment, is regularly conducted. The geo-engineering research involved monitoring of groundwater level and hydrochemical sampling. In addition, hydro-geoenvironmental research were carried out within the landfill site and outside in 2004, 2011, and 2012. Thus, there are a number of hydroisobath maps made up at different time and various hydrochemical data including the following parameters of groundwater: organoleptic properties, permanganate index, COD, BOD; mineralization, pH, total hardness, CO_3^{2-}, HCO_3^-, SO_4^{2-}, Cl^-, NO_3^-, NO_2^-, PO_4^{3-}, F^-, Ca^{2+}, Mg^{2+}, Na^+, K^+, NH_4^+, SiO_2; microcomponents: Al, B, Ba, Sr, Be, As, F, Br, Fe, Cd, Mn, Cu, Mo, Ni, Hg, Pb, Se, Zn, Cr, Co; organic substances: petroleum products, phenols, formaldehyde, benzene, toluene, xylene, ethylbenzene, detergents, pesticides (DDT, 2,4-D, γ-HCH; HS); α- and β- activity.

To determine the impact of the landfill on the ground hydrosphere, the authors conducted hydrodynamic and balance calculations using analytical method and modelling approach (Processing Modflow) aimed at defining: intensity of infiltration recharge of groundwater and underlying intermediate water; flow rate of contaminated groundwater in the landfill "outlet"; flow rate of pure groundwater within the second and the third aquifer systems in the landfill "inlet"; efficiency of drainage system. In addition, hydrogeochemical data have been processed.

3. Results and discussion

The data of hydro-geoenvironmental research show that groundwater is increasingly negatively affected by the landfill. During the first decade of landfill operation, there was a rise in groundwater level of about 1 to 5 meters just below the waste cells. The groundwater anionic composition changed from hydrocarbonate to hydrocarbonate-chloride. The increase in groundwater mineralization from 0.2–0.4 to 4–5 g/l was identified. The total hardness changed from 4–8 to 40–70 mM/l. Today, mineralization of groundwater within the landfill and its nearest areas (observation wells of the first monitoring network) can reach 7–12 g/l; hardness can reach 70–135 mM/l; the concentration of the most characteristic pollutants (mg/l) are as follows: petroleum products – up to 500–982; phenols – up to 13.9; total iron – up to 153 (water is mainly of chloride-hydrocarbonate sodium-calcium type). However, the concentrations of most pollutants decrease by one or two orders of magnitude at a relatively close distance from the landfill, i.e. the second monitoring network of observation wells

(Figure 2). According to previous geo-engineering research and environmental investigations, the changes in pollutants concentration were stipulated by liquid waste seepage.

The results of the current research provide the grounds for the following conclusions:

1. The fact that the landfill has become a major source of recharge to the groundwater stipulates the rise in groundwater level within the landfill site. At the same time, it should be noted that groundwater is mainly recharged not by liquid wastes, but by precipitation of different types. Within the landfill site, the over-the-surface flow is disturbed, therefore, most precipitation accumulates (excluding evaporation) within this area. Currently, the intensity of total infiltration within the waste-cell area is 2.8 higher than its standard values, precisely, it reaches 208.2 mm/year (37.5% of normal precipitation). The present water balance within the landfill (m^3/day): inflow (332.75): precipitation – 324.12; liquid industrial waste – 8.63; discharge (332.75): evaporation – 184.69; infiltration – 129.98; runoff – 5.46; pumping from waste cells to the drainage station – 5.05; waste cell flooding – 7.57.

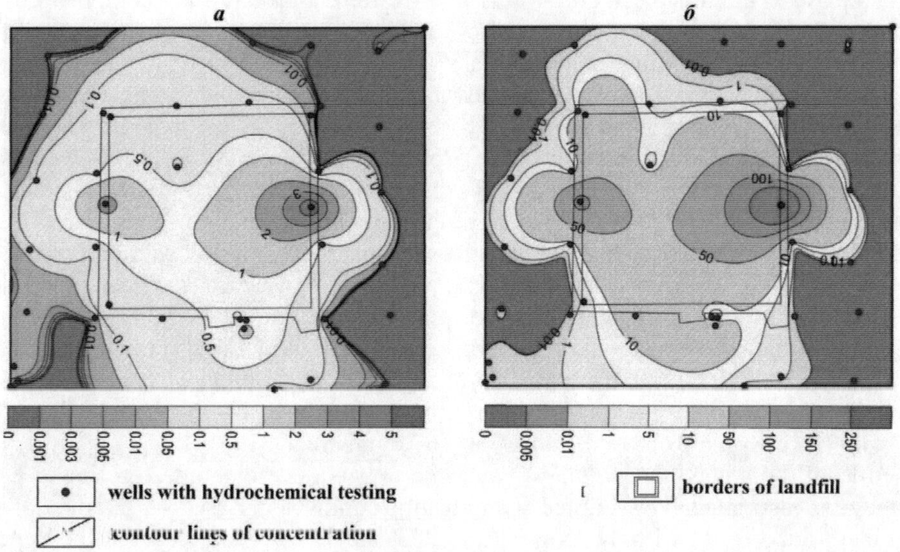

Figure 2. Isoconcentration contour maps (mg/l): phenols (a) and oil (b) in groundwater, sampling data 2011.

2. Groundwater contamination is caused by infiltration and down filtration of precipitation that is altered by waste (to a lesser extent liquid wastes) interaction with water.

3. The flow rate of extremely polluted filtrate through the waste cell bottom is 129.98 m^3/day, including 8.63 m^3/day of liquid waste; the remaining volume is of altered precipitation. In the groundwater flow, the flowrate of filtrate is 121.67 m^3/day (the remaining filtrate volume, i.e. 8.31 m^3/day, recharges the underlying Upper Kazan aquifer system). The total flowrate of groundwater flow is 172.7 m^3/day, including 64.31 m^3/day of water that runs into the underground drainage system. The flow rate of the contaminated ground flow that is not drained is 116.7 m^3/day, including 108.39 m^3/day of groundwater flow and 8.63 m^3/day of deep underground drainage.

4. The landfill wastes can be the reason for the presence of oil products, phenol formaldehyde, benzene, toluene, xylene, ethylbenzene, iron and, to a lesser extent, sulphates, chlorides, barium (high concentrations that 10 times exceed maximum allowable concentrations for drinking water are a specific feature of iron and other highly correlated organic matters). The high concentrations of the above-enumerated elements are basically identified within the waste cell area of the landfill. However, at the distance of 150 m from this area (the second network of observation wells), these concentrations can reduce by factors of 10 and more. The concentration of most inorganic microelements (Al, B, As, F, Br, Cd, Mn, Mo, Hg, Pb, Cr) demonstrates a negative correlation with the distance up to the waste cell area. It should be noted that maximum allowable concentrations can be exceeded in the wells of the first monitoring network, with concentrations of 3-5 being a rare case. This fact can be adequately

explained by intensive leaching both of microelements, and basic cations of mineral matrix by aggressive carbon-dioxide waters with concentration of hydrocarbonate exceeding 800–1000 mg/l (HCO_3^- content can reach 4–8 g/l, while in natural environment it does not exceed 400 mg/l). The hydrocarbonate groundwater is mainly found within the waste cell area of the landfill. This is only possible in a case of partial activity of carbon dioxide [2, 3] associated with destruction (oxidation) of organic matter disposed in the landfill. From the above-mentioned microelements, the presence of F, Br, and probably, As and Hg is stipulated by the disposed wastes, while the rest microelements are held in the cross-section rocks. The landfill operating procedures do not affect the content of nitrates, nitrites, phosphates, fluorides, bromides in groundwater. In addition, the landfill does not alter the content of Co, Cu, Ni, Se, Sr, Zn, Be, as well as synthetic surface active substances and pesticides (DDT, 2.4 D'; γ-HCH; HS).

5. In general, groundwater contamination follows the longitudinal distribution pattern (observation wells located close to the waste cell area would show different groundwater composition). This is due to the fact that each waste cell is designated for dumping a definite type of waste, therefore, each of them is characterized by a peculiar composition of the filtrate. The soil which divides the waste cells demonstrates its own composition of the groundwater. Thus, the groundwater table within the landfill site looks like a "patchwork quilt" in terms of hydrogeochemistry. When separate water currents move in south-western direction as a part of the main ground flow, they can hardly mix, with the boundary water layers being diffused due to lateral hydrodispesion. The low rate of the described processes defines the longitudinal distribution pattern of groundwater contamination.

6. The reduction of groundwater mineralization and hardness, as well as the decrease in pollutant content with increasing distance from the landfill, is caused by various processes and phenomena, precisely, sorption and diffusion, lateral hydrodispesion, pollutant destruction (particularly petroleum products and phenols), precipitation of Fe, Mn, etc. as pH and Eh conditions change. The most significant factor that affects pollutant concentration is natural dilution of contaminated waters with pure waters (in case of south-western filtration, groundwater is recharged from each square meter of the Earth's surface in volume of 1.9*10-4 m3/day (67.6 mm/year).

7. When landfill filtrates (8.63 m^3/day) reach the upper- and sometimes the lower Kazan aquifer systems, they are also intensively diluted with clean groundwater. Based on the data of geofiltration modeling, the ground flows with the flow rate of 253 and 131 m^3/day recharge the upper and lower Kazan complexes from the eastern part of the landfill, i.e. underground watershed. These data are proved by local and spotted distribution pattern of high petroleum products concentrations in waters of the upper Kazan complex in the southern part of the landfill.

7. The identified processes of "self-purification" are obviously correlated with the fact that composition of ground and surface waters near the settlement Klyatle (at a distance of 3 kilometres from the landfill) demonstrates no significant changes.

8. The local occurrence of groundwater contamination is a characteristic feature of many other industrial waste landfills, with organic matter being dumped. The most vivid examples of waste landfills which have been thoroughly studied are as follows: the Kohtla-Jдrve landfill (Estonia) that was started in 1938 and holds $8*10^7$ m^3 of liquid oil shale mining residues [4]; and closed domestic and industrial waste disposal site in Hamilton (Canada) of 50 ha [5].

References

[1] Musin R Kh and Musina R Z 2014 About the influence of industrial waste landfills to hydrolithosphere *J. Nedropolzovanie XXI* **87–89** 99

[2] Shvartsev S L 1982 "*Osnovyi gidrogeologii. Gidrogeohimiya*" (Basics of hydrogeology. Hydrogeochemistry) (Novosibirsk: Nauka) p 288

[3] Roscoe Moss Company 1990 *Handbook of Ground Water Development* (John Wiley & Sons) p 493

[4] Vallner L, Gavrilova O and Vilu R 2015 Environmental risks and problems of the optimal management of an oil shale semi-coke and ash landfill in Kohtla-Jдrve, Estonia *Science of*

The Total Environment **524–525** 400 – 415
[5] Barker J F, Barbash J E and Labonte M 1988 Groundwater contamination at a landfill sited on fractured carbonate and shale *J. of Contaminant Hydrology* **3** 1–25

Groundwater microflora of the Aptian-Cenomanian deposits at the Igolsko-Talovoe field in Tomsk Region

N G Nalivaiko[1], E M Dutova[2] and T S Spiridonov[3]

[1, 2, 3] Tomsk Polytechnic University, 30 Lenin Avenue, 634050, Tomsk, Russia

E-mail: [1] nng@tpu.ru, [2] dutova@sibmail.com

Abstract The authors have studied the microbiological composition of the groundwater of the Aptian-Cenomanian deposits in the territory of the Igolsko-Talovoe field in Tomsk Region. The detected diversity of the physiological groups of bacteria can be a corrosive component for waters used in the reservoir pressure maintenance system. The research findings have allowed making conclusions about the need to study the contribution of all microorganisms inhabiting the waters of the Aptian-Cenomanian deposits to corrosion.

1. Introduction

The groundwater of the Aptian-Cenomanian deposits in the territory of the Igolsko-Talovoe field in Tomsk Region has an original chemical composition and is used in the reservoir pressure maintenance system, but is characterized by enhanced corrosiveness that is traditionally connected with the activity of sulfate-reducing bacteria. However, the basis of microbial communities in the groundwater under study is formed by saprophytic, oligotrophic and nitrate-reducing bacteria. Sulfate-reducing microorganisms are present in these waters in small amounts and are characterized by low metabolic activity. Along with that, nitrate-reducing bacteria that cover the same environmental niche with sulfate-reducing bacteria are available here in considerable amounts and are able to take part in corrosive processes. Therefore, there is a need to study the corrosive activity of all groups of microorganisms living in the groundwater of the Aptian-Cenomanian deposits in order to develop effective means of suppressing corrosive processes.

2. Materials and methods

2.1. Subject of research

The administrative location of the Igolsko-Talovoe field is Kargasoksky District of Tomsk Region, its southwestern part (Figure 1). It was discovered in 1977 and has been developed by VasyuganNeft. The settlements being closest to the field are such towns as Nefteyugansk, Nizhnevartovsk and Pyt-Yakh, as well as Maysk stationary village located at the distance of 25 km to the northeast of the field. Igol village has been constructed in the area of the field.

As for the regional structural and tectonic characteristics of the oil field, it is located in the southern part of the Nyurol depression. It is structurally confined to the Igolsko-Talovoe local elevation being an element of the big Igolskaya dome-like structure and having the central position in it.

The cross section of the Igolsko-Talovoe field is formed by the deposits of the pre-Jurassic foundation that is uncomfortably covered by the deposits of a thick sedimentary mantle. The productivity of the Igolsko-Talovaya area is connected with the regionally oil-bearing J_1 horizon of the

IOP Conf. Series: Earth and Environmental Science **33** (2016) 012020 doi:10.1088/1755-1315/33/1/012020

Vasyugan Formation. The Igolsko-Talovoe field has several big oil pools. The groundwater of the Aptian-Cenomanian deposits is very deep and covered by thick regional impermeable beds that reliably protect it against pollution from the surface [1–4].

The clusters of water wells used in the reservoir pressure maintenance system are located on both banks of the Chertala River at the distance of about 3.5 km (cluster No.9) and 4.5 km (cluster No.36) from the river bed. The complex of facilities using the groundwater of the Aptian-Cenomanian deposits is part of the pressure maintenance system of this oil pool. The areas of the water extraction installations are combined with the oil production facilities of the Igolsky oil field area.

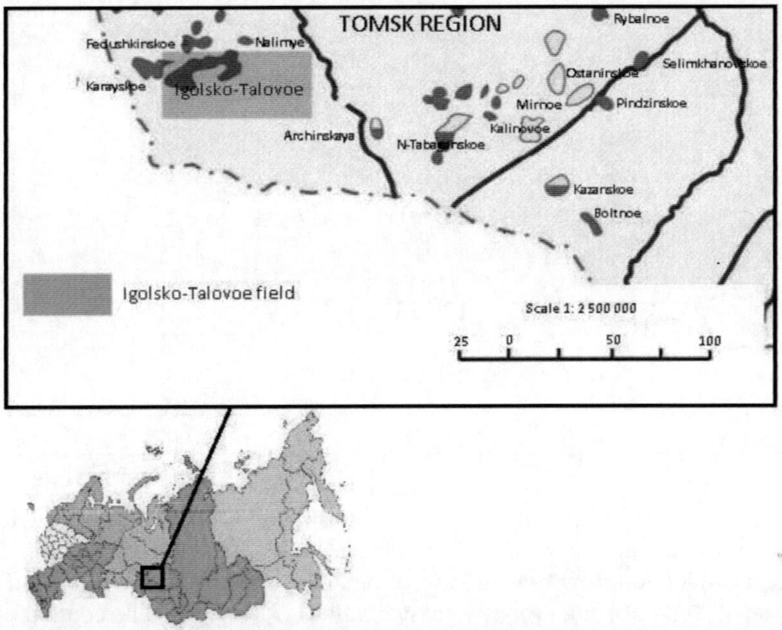

Figure 1. Map of the studied area.

2.2. Sampling and research methods

For research purposes, water samples were taken from production and injection wells once a quarter. Sampling was performed at the water wells from the sampler (valve) after water had been drained for several minutes. Water samples for the microbiological analysis were taken into sterile glass bottles while the rules of asepsis were observed. In the waters under consideration, the authors were determining a wide range of organotrophic microorganisms that can have this or that impact on the chemical composition of the waters in this aquifer system and oil quality, as well as on the technical condition of oil production equipment.

Identification of and quantitative accounting of microorganisms was carried out in accordance with classical techniques common in Microbiology, which are outlined in some of our publications [4–7]. Simultaneously, the authors were examining the chemical composition of the waters as a habitat of the microorganisms.

3. Hydrochemical characteristic

3.1 Microcomponents and pH

The Aptian-Cenomanian aquifer system of the Igolsko-Talovoe field is characterized by horizontal hydrogeochemical zoning being typical of the region in general. First of all, it means enhanced mineralization, presence of macro- and microcomponents in the northwestern direction, progressively with layer deepening.

IOP Conf. Series: Earth and Environmental Science **33** (2016) 012020 doi:10.1088/1755-1315/33/1/012020

Table 1. Components of groundwater chemical composition in the Aptian-Cenomanian deposits of the Igolsko-Talovoe field.

Indicators and components	Cluster and well No.		
	Cluster 9, well 1c	Cluster 9, well 2c	Cluster 9, well 3c
Sampling interval, m	985-1,290	1,130-1,581	1,430-1,757
Exit temperature, °C	48	39	39
pH	6.8	7.0	7.1
Ca^{2+}, mg/dm^3	674	662	687
Mg^2+, mg/dm^3	121	117	121
K+, mg/dm^3	37	38	37
NH^+_4, mg/dm^3	55.0	23.0	67.0
NO_2, mg/dm^3	<0.05	<0.05	<0.05
NO_3^-, mg/dm^3	0.1	0.1	0.1
CO_3^{2-}, mg/dm^3	<0.10	<0.10	<0.10
HCO_3, mg/dm^3	183	159	207
SO_4^2,mg/dm^3	<2	<2	<2
Cl^-,mg/dm^3	12,723	12,723	12,723
Na+, mg/dm^3/	-	5,000	-
Fe, mg/dm^3	4.1	6.3	4.8
Si, mg/dm^3	0.10	0.31	0.31
Mineralization, g/dm^3	20.93	20.92	20.95

The groundwater in the aquifer system under study at the Igolsko-Talovoe field has somewhat enhanced mineralization vs. other oil fields of Tomsk Region (except the Olenye field). Its mineralization has been staying within a narrow interval from 20.31 to 23.64 g/dm^3 over the many-year period of water production from the Aptian-Cenomanian deposits of the Igolsko-Talovoe field. Its pH value is neutral (pH 6.8-7.2), sometimes weakly alkaline (7.30-7.77). The content of chloride ion is 97.7-99.3, sodium ion – 77.4-98.3 %-eq. The content of sulfate ion is low (0-50 mg/dm^3), which is typical of the whole section of the lower hydrogeological stage of the West Siberian Basin. The prevailing type of waters is chloride sodium [1]. The proportion of these components and the type of water coincide with the waters of productive Jurassic deposits. The waters of this aquifer system contain a wide range of microcomponents: iodine, bromine, boron, rubidium, manganese, gold, etc.

The composition of the water-dissolved gases is predominantly methane-based (Table 2). Apart from methane, they are also characterized by the significant amounts of nitrogen. They also have a small content of carbon dioxide, oxygen and ethane.

Table 2. Component composition of water-dissolved gas.

No.	Components	Unit of measure	Content
1.	Gas density	kg/m^3	0.685
2.	Oxygen (O_2)	vol.%	0.341
3.	Nitrogen (N_2)	vol. %	3.222
4.	Carbon dioxide (CO_2)	vol. %	0.762
5.	Methane (CH_4)	vol. %	95.662
6.	Ethane (C_2H_6)	vol. %	0.010

The content of dissolved oxygen in the waters under study varies from 0.10 to 0.31 mg/dm^3, which does not exceed the permissible level of 0.5mg/dm^3. Dissolved hydrogen sulfide has not been found in the waters.

IOP Conf. Series: Earth and Environmental Science **33** (2016) 012020 doi:10.1088/1755-1315/33/1/012020

4 Results and discussion

4. 1 Microbiological composition of waters

The total number of microbes in the water of the wells perforating the aquifer of the Cenomanian deposits at different depths amounts for one to 2.5 million cells/ml and grows with depth (Table 3). At the same time, the number of living bacteria drops with depth. Viable cells comprise somewhat more than a half of the total amount in the deepest well. It means that the aquifer has factors being unfavorable for the activity of microbes.

Table 3. Groundwater microflora in the Aptian-Cenomanian deposits of the Igolsko-Talovoe oil field.

Physiological groups of microorganisms	Cluster and sample No.		
	Cluster 9, well 1c	Cluster 9, well 2c	Cluster 9, well 3c
Total number of bacteria, cells/ml	1,120,000	1,690,000	2,440,000
Number of living cells, %	84	76	56
Saprophytic, cells/ml/%	200,000/21.2	420,000/32.8	80/0.0001
Oligotrophic, cells/ml/%	57,500/6.4	317,200/24.8	600,000/44
Heterotrophic iron-oxidizing bacteria, cells/ml	660/0.05	870/0.05	12,000/0.8
Oil-oxidizing, cells/ml/%	4,200/0.45	3,000/0.23	0
Pentane-oxidizing, standard units	300	300	290
Hexane-oxidizing, standard units	350	390	390
Heptane-oxidizing, standard units	290	300	320
Octane-oxidizing, standard units	270	350	300
Nonane-oxidizing, standard units	440	440	350
Decane-oxidizing, standard units	400	250	270
Benzene-oxidizing, standard units	450	450	400
Toluene-oxidizing, standard units	420	450	400
Phenol-oxidizing, standard units	300	320	290
Naphthalene-oxidizing, standard units	360	490	430
Nitrate-reducing, cells/ml/%	$11.5*10^4/10$	$9.5*10^4/7.3$	$1.5*10^5/11$
Nitrifying	0	0	0
Sulfate-reducing, cells/ml/points	0	10/6	10/6
Thionic, cells/ml	0	10	100

Viable cells form microbial communities of a specific composition being typical of each sampling interval. Despite the fact that the first two wells are characterized by the prevalence of saprophytes, their amount makes only 21 or 33 % of the living cells, whereas they are practically absent in well 3c.

The titer of oligotrophic microorganisms grows with depth. They prevail over saprophytes and make 44 % of the total number of living cells in the water of the deepest well. The titer of oil-oxidizing bacteria drastically drops with depth to make from several thousand to single cells per 1 ml. However, at all depths there is the intensive oxidation of individual hydrocarbons being part of the oil composition, even where oil-oxidizing microbes are practically absent. This fact is testified to by the presence and activity of bacteria oxidizing different hydrocarbons. Apparently, hydrocarbons and oil are oxidized in this aquifer with the help of bacteria from all physiological groups and having an adaptive inducible fermentative system [2]. The water of all wells contains heterotrophic iron-oxidizing bacteria that deposit iron hydroxide on the surface of cells. Their titer pronouncedly grows with depth. Nitrate-reducing bacteria ($N*10^5$ cells/ml) are found ubiquitously in significant amounts (7-11 % of the total number of living cells). Sulfate-reducing bacteria are present in the aquifer as single cells with low metabolic activity. Thionic bacteria have a very low titer. Nitrifying bacteria have not been found.

Thus, the physiological groups of bacteria in the Aptian-Cenomanian water make from 38 to 65 % of the total number of living cells. As for obvious factors, the quantitative content of specific groups of microbes in the aquifer is influenced by the depth of a sampling interval. The total number of microbes in the aquifer grows with depth, whereas the amount of viable cells drops. These quantitative parameters characterize the microbial community of the aquifer under study as oligotrophic. Oligotrophic microorganisms are present in water from all wells, and they are practically the only inhabitants of the deepest well.

There are two problems connected with microbiologic activity for the aquifer under study: enhanced content of mechanical particles in water and their enhanced corrosive activity

The content of mechanical impurities (particles content) in the waters under study amounts to 8.4 to 10.1 mg/dm^3 at the permissible levels of up to 3mg/dm^3. It is found that the composition of these mechanical impurities comprises iron sulfide and oxygenated iron, clay particles, quartz, carbonates produced from the formation. Over 50% of these particles are up to 20 μm in size. A significant part of the mechanical impurities is formed by the fragments of biofilms and mineral-microbial conglomerates.

Despite the fact that sulfate-reducing bacteria in the waters under study are found occasionally, have a low titer (1-10 cells/ml), whereas the main product of metabolism, hydrogen sulfide, is present in these waters in small concentrations (0.3-0.5 mg/dm^3), the corrosive activity of these waters is enhanced (6 out of 10 possible points) and is characterized by the depth of corrosive penetration of 0.1 to 0.5 mm/year, the speed of corrosion amounting to 0.09-0.45 g/m^2.

The discrepancy between the small number and activity of sulfate-reducing bacteria and corrosive activity of the waters in the Aptian-Cenomanian deposits can be reasoned by the participation of metabolites of microorganisms from other physiological groups inhabiting these waters in the corrosive process.

In particular, all samples from the waters under study are characterized by the presence of nitrate-reducing bacteria amounting to 10^4 -10^5 cells/ml. Nitrate- and sulfate-reducing bacteria cover the same environmental niche in rock formations [2, 3]. Both of them can successfully live and grow both in wells and in the aquifer, if they have enough sources of carbon in the form of short-chain fatty acids. There is little information on the participation of nitrate-reducing bacteria in metal corrosion. However, it is proved that the corrosive process accelerates in the presence of denitrifying microorganisms, which is connected with the generation of aggressive corrosive products of metabolism: carbon dioxide and ammonia that can lead to electrochemical metal corrosion [3].

To determine the role of nitrate- and sulfate-reducing bacteria of the Aptian-Cenomanian waters at the Igolsko-Talovoe field in metal corrosion, the authors have made an experiment with steel samples No.3 [3]. The corrosive activity of water and enrichment cultures of microorganisms was tested in compliance with GOST 9.905-82 "Corrosion test methods" and RD 39-3-973–83 "Methods of controlling microbiological contamination of oil field water and evaluating the protective and bactericidal action of reagents." The study implied the use of flat samples made of steel No.3: three samples per test. The value of sample surface roughness was up to 6.3 μm. The weight of the samples under test varied from 7,420 to 800 mg. The samples were suspended in laboratory flasks on a fishing line. To test the corrosive activity of nitrate-reducing bacteria, the flasks were filled in with an enrichment culture with the Giltay medium and the bacteria titer of 10^3 cells/ml. To test the corrosive activity of sulfate-reducing bacteria, the samples were doped with an enrichment culture with the Postgate medium and the titer of sulfate-reducing bacteria amounting to 10^3 cells/ml. The experimental flasks were placed to a heated magnetic mixer. The test temperature was 20-22 °C. The test time made 6 hours. To determine the weight loss of the samples, their surfaces were cleaned. The film made of corrosion products was removed with a solution not interacting with the main metal, washed with water and defatted. Then the samples were dried and weighted. The speed of corrosion was determined with a gravimetric method on the basis of the difference between the initial weight of a sample and its weight after the test. Metal samples in the abacterial Giltay and Postgate media were used for control purposes. Weight losses were converted into the losses of units of area per annum. The values of weight loss for the samples differed insignificantly in the fourth sign and, on average, amounted to 0.002 mg for sulfate-reducing bacteria and 0.0025 mg for nitrate-reducing bacteria over the six-hour period. On conversion to the depth of penetration, the possible corrosive activity of the microflora under given conditions will amount to 0.15-0.2 mm annually. These results characterize the corrosive activity of both types of bacteria as enhanced. Thus, the corrosive activity of nitrate-reducing bacteria is comparable with that of sulfate-reducing bacteria. Corrosion was practically absent in the control experiments without bacteria. It is necessary to take the laboratory conditions of the test into account when interpreting the results of this experiment.

Apparently, the participants of corrosive processes in the Aptian-Cenomanian waters can be microbes from all groups that are present here. Further research is needed to determine the degree of their participation in corrosive processes. This problem is rather relevant now, as corrosion-related losses are rather significant. To suppress bacterial corrosion, specialists use a wide range of bactericides, but they are not always effective enough and there are no multipurpose agents.

5. Conclusions

The present study allows concluding that the total number of microorganisms in the aquifer system of the Aptian-Cenomanian deposits in the territory of the Igolsko-Talovoe field is from one to 2.5 million cells/ml and grows with depth. Living cells comprise 56-84 % of the total number and their amount drops with depth. The composition of the microbial communities differs in the number of microbes and the diversity of physiological groups. The prevalent type in the composition of the microbial community of this aquifer system is oligotrophic microorganisms. Sulfate-reducing microorganisms in the waters under study are found in insignificant amounts and have low functional activity. The enhanced corrosive activity of the waters in this aquifer system might be reasoned by the presence of nitrate-reducing bacteria that cover the same environmental niche with sulfate-reducing bacteria. Following the quantitative accounting results, it is possible to conclude that saprophytic, oligotrophic, iron-oxidizing and oil-oxidizing microorganisms in the composition of the microflora under study can have an impact on corrosive processes.

References

[1] Ivanov V G and Laminsky A I 2003 Chemical Composition of Groundwater in the Aptian-Albian-Cenomanian Deposits of Tomsk Region and their Complex Use in Oil Production Areas *Problems of Exploration and Environmental Geochemistry of Siberia* (Tomsk: Publishing office of TPU) 159–161 [in Russian]

[2] Netrusov A I, Egorova M A, Zakharchuk L M et al. 2005 *Practical Guide on Microbiology* (M.: ACADEMA) p 608 [in Russian]

[3] Nalivaiko N G and Pasechnik E Yu 2011 Microflora of the Aptian-Cenomanian Waters at Tomsk Region Oil Fields *J. Prospect and Protection of Mineral Resources* 2 [in Russian]

[4] Dutova E M, Nalivaiko N G, Kuzevanov K I and Kopylova J G 1998 The chemical and microbiological composition of urban groundwater, Tomsk, Russia *Groundwater in the Urban Environ.: Proc. of the XXVIIth Congress IAH* (Nottingham, UK) **2** 125-30

[5] Pokrovsky V D, Dutova E M, Kuzevanov K I, Pokrovsky D S and Nalivayko N G 2015 Hydrogeological Conditions Changes of Tomsk, Russia *IOP Conf. Seri.: Earth Environ. Sci.* **27** 12031

[6] Savichev O G et al. 2015 Microbiological composition of river waters in the Ob' basin (West Siberia) and its associations with hydrochemical indices [Electronic resource] *IOP Conf. Seri.: Earth Environ. Sci. Problems of Geology and Subsurface Development* **27** 6

[7] Akhmed-Ogly K V et al. 2104 Parameter Calculation Technique for the Waste Treatment Facilities Using Naturally-Aerated Blocks in the Bog Ecosystems [Electronic resource] *IOP Conf. Seri.: Earth Environ. Sci.: XVIII Int. Sci. Symp.in Honour of Academician M. A. Usov: Problems of Geology and Subsurface Development* Tomsk, Russia **21** 6

Environmental monitoring of natural waters in Krasnodar and Stavropol Territories

T S Glyzina [1], E G Matugina [2], B M Bagamaev [3], Yu M Tokhov [4], Yu V Kolbysheva[5], E V Gorchakov [6], T V Sotnikova [7] and A S Shilova[8]

[1, 2, 5, 7, 8] Tomsk Polytechnic University, 30 Lenin Avenue, Tomsk, 634050, Russia
[3, 4, 6] Stavropol State Agrarian University, 12 Zootechnicheskiy Ln, Stavropol, 355017, Russia

E-mail: [1]inosine@mail.ru

Abstract. The environmental monitoring of natural waters in Krasnodar (Uspensky and Novokubansky districts) and Stavropol (Kochubeyevsky District) Territories was conducted. In the course of study, various elements and compounds harmful to animals and humans, which exceed maximum permissible concentrations, were identified.

1. Introduction

In the territory of large enterprises such as the Stavropol SRPS included in the United Energy System (UES) of the South, sites of mining and processing industries, and cattle-breeding complexes as well, the geochemical redistribution of elements and compounds takes place. It leads to their accumulation in natural waters, soils, and plants. It is known that the regions of the South are strategically important and of great priority for developing the agriculture and livestock industry in Russia. Plants, accumulating harmful elements from natural waters and soils, are used as feed for animals; therefore, these harmful elements get in crop and livestock products of the country.

2. Materials and Methods

The aim of this study is to assess the negative environmental impact of a number of factors on animals and humans. Accurate data of the study would be obtained only over a long period. For this purpose, the degree of precipitation in different seasons, different types of waters (drinking water, river water, water from underground sources), air pollution, and soil condition have to be identified and assessed. These data are essential for the monitoring, assessment and prospective measures to reduce emissions from anthropogenic sources as well as to support further ecological balance within the recommended standards of sanitary control in the border areas of Krasnodar and Stavropol Territories of Russia.

In Stavropol Territory there are a lot of small and large rivers, lakes, water reservoirs (i.e. artificial surface watercourses and reservoirs), whose water is used for drinking and household needs, industry and agriculture. The river system is unevenly distributed across Stavropol Territory. The main source of Stavropol rivers' inflow is the seasonal rainfall and snow melting. The rivers originating in the mountains are fed by high-melting snow and glaciers. Rivers of Stavropol and Krasnodar Territories have a mixed type of nutrition. There are many different rivers, reservoirs, and lakes in Stavropol Territory. They are used in agriculture and industry as well as for drinking and domestic purposes. The Kuban River is the largest river of Stavropol Territory and the Northern Caucasus. It rises from glaciers of the western slope on Mount Elbrus. The Kuban River has its source at the confluence of the Rivers Ullukam and Uchkulan and is 970 km in length. Within Stavropol Territory it flows across Kochubeyevsky District and along the border of Novoalexandrovsky District near Krasnodar Territory. The main inflow sources of the river are the following: rain – 60%, melting snow and glaciers – 24%, groundwater – 16%. As snow and glaciers play a key role in the recharge of these rivers, it explains the high and long summer flood. The water of the Kuban River refers to a drinking one and can be diverted for irrigation as well. However, it is very polluted by wastes from enterprises. The highest flood levels are observed usually in late July – early August. The flood fall occurs slowly and ends in late September; but sometimes it lasts till the first ice [1].

Taking into account the atmospheric conditions, sampling was carried out in different seasons. The list of study substances had been proposed by Regional Committee of Nature Resources. The analysis was carried out in accord with state-approved methods and certified techniques [2-4]. JSC "Nevinnomyssk Azot", located in Kochubeevsky District of Stavropol Territory, called Nevinnomysskaya Industrial Zone, was chosen for the emission control. Samples were taken near the enterprise. The samples analysis gave the following results presented in Table 1.

Table 1. Emissions of pollutants in the area of JSC "Nevinnomyssk Azot".

Determined substances	Emitted, tons/ year	Maximum permissible emission, tons/ year
Sulphur dioxide (SO_2)	6624	27853
Carbon oxide (CO)	2946	3830
Nitrogen oxides (NO, NO_2)	18098	15078
Hydrogen sulphide (H_2S)	91	1.2
Ammonia (NH_3)	2085	1974
Phosphoric anhydride (P_2O_5)	9.3	42.7

The presented data show that the excess of permissible emissions limit is observed in such indicators as nitrogen oxides, hydrogen sulphide and ammonia. The determined compounds were in the form of aerosol. It can be due to the intense movement of air masses in different seasons, especially in spring and summer.

The greatest excess was found in the samples taken in "Armavir corridor". It is connected with the average movement of air masses along the corridor: east wind - 58%, west wind - 14%, north wind - 16% and south wind - 12%. "Armavir corridor" is clearly visible on the physical map (Figure 1). It is ridges location that leads to a fixed annual movement of air masses.

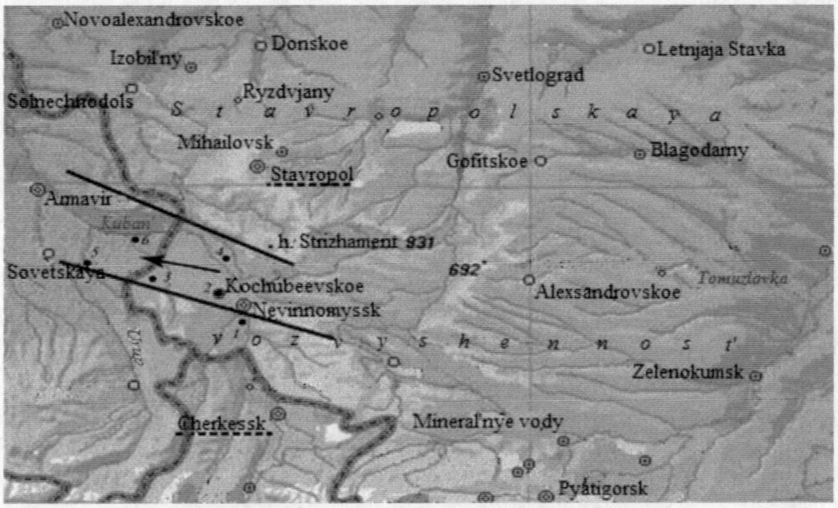

Figure 1. Location of «Armavir Corridor».

IOP Conf. Series: Earth and Environmental Science **33** (2016) 012021 doi:10.1088/1755-1315/33/1/012021

3. Results and Discussions

The determined compounds are toxic. They cause changes in the blood circulation, effect badly the respiratory system and skin, and the nervous system as well. They can also lead to metabolic disorders and allergic skin lesions.

Sulfuric anhydride is formed by the gradual oxidation of sulfur dioxide in the air, involving the light. The final reaction product is sulfuric acid aerosol in the air, rain water and clouds. When precipitating, it acidifies the soil, forming the corresponding metal salts, and exacerbates the respiratory diseases. Chemical enterprises emit the sulphuric acid aerosol into the air. This emission is concentrated in the low clouds and high humidity. Plants around such enterprises are usually dotted densely with small necrotic spots. It proves the presence of sulphuric acid aerosol in the atmosphere in a great amount. Pyrometallurgical enterprises of nonferrous and ferrous industry, as well as TPP, annually emit dozens of millions of tons of sulfur dioxide [5]. Nitrogen oxides such as NO, NO_2, N_2O_3, N_2O_5 are easily dissolved in water contained in the air. They form the aerosol of nitric and weak nitric acid. Aerosol and water sampling was carried out along the corridor border. Sampling areas are shown in Table 2.

Table 2. Sampling areas in «Avamir corridor».

Point number	Settlement	Distance from starting point, km
1	Ivanovskoe	7
2	Kochubeevskoe	10
3	Vrevskoe	32
4	Barsukovskaya	24
5	Urupskiy	57
6	Uspenskoe	46

Results of the determinations of volatile substances are shown in Table 3.

Table 3. Amount of volatile substances in aerosol emissions, mg/l.

Determined substances / Point number	Sulphur dioxide (SO2)	Carbon monoxide (CO)	Nitrogen oxides (NO, NO2)	Hydrogen sulfide (H2S)	Ammonia (NH3)	Phosphoric anhydride (P2O5)
1	0.07	0.27	0.43	2.06	4.12	0.25
2	0.09	0.47	0.92	2.49	4.35	0.35
3	0.08	0.45	0.78	2.48	4.25	0.27
4	0.08	0.44	0.78	2.46	4.22	0.27
5	0.07	0.39	0.73	2.24	4.05	0.23
6	0.08	0.27	0.45	2.10	4.02	0.25
MPC, mg/dm^3	0.02	0.01	0.05	0.08	0.02	0.05

Results of the determinations of substances in water bodies are shown in Table 4.

Table 4. Amount of substances in water bodies, mg/l.

Determined substances / Point number	Sulphur dioxide (SO_2) in form SO_4^{-2}	Nitrogen oxides (NO, NO_2) in form NO_2^- and NO_3^- respectively	Hydrogen sulfide (H_2S)	Ammonia (NH_3) in form NH_4^-	Phosphoric anhydride (P_2O_5) in form PO_4^{-3}
1	700	0.33-43	2.06	4.12	25
2	900	0.46-92	2.49	4.35	3.5
3	800	0.42-78	2.48	4.25	2.7
4	800	0.44-78	2.46	4.22	2.7
5	700	0.45-73	2.24	4.05	2.3
6	800	0.14-45	2.10	4.02	2.5
MPC, mg/dm^3	500	0.1-45	0.05	2.5	3.5

According to the received data, at the Nevinnomyssk SRPS heating boilers operating on heavy fuel oil have been replaced with the boilers using gaseous fuels. The repeated sampling of aerosol and water was carried out in 2013 at the same border of the corridor. Sampling areas are shown in Table 2. Results of the determinations of substances in water bodies are shown in Table 5.

Table 5. Amount of substances in water bodies (research data, 2013), mg/l.

Determined substances / Point number	Sulphur dioxide (SO_2) in form SO_4^{-2}	Nitrogen oxides (NO, NO_2) in form NO_2^- and NO_3^- respectively	Hydrogen sulfide (H_2S)	Ammonia (NH_3) in form NH_4^-	Phosphoric anhydride (P_2O_5) in form PO_4^{-3}
1	350	0.03-5	0.06	1.12	2.5
2	500	0.06-11	0.49	1.35	3.5
3	400	0.02-9	0.48	1.25	2.7
4	400	0.04-9	0.46	1.22	2.7
5	400	0.05-8	0.24	1.05	2.3
6	400	0.04-5	0.10	1.02	2.5
MPC, mg/dm^3	500	0.1-45	0.05	2.5	3.5

The presented data show that the modernization of the Nevinnomyssk SRPS has led to the reduction of toxic emissions. Monitoring data demonstrate a gradual improvement in the ecological situation.

4. Summary and Conclusions

The urgency of the production development in the agricultural areas of Krasnodar and Stavropol Territories of Russia to maintain the dynamics of the national economy is out of question. However, a considerable damage to the environment of these areas, which are recognized as a national treasure, requires changes in the existing approaches to organizing these enterprises. It is necessary to study and

IOP Conf. Series: Earth and Environmental Science **33** (2016) 012021 doi:10.1088/1755-1315/33/1/012021

propose solutions to this problem through the adoption and consolidation of norms and rules for the operation of enterprises. Thus, the principles of enterprises' activities should be defined. The implementation of these principles should be based on the nature protection activities. Because of the negative impact on the nature of Krasnodar and Stavropol Territories, the rational approach to solution of environmental problems with the involvement of all levels of authorities is required. It affects not only the environment, but also, agriculture, which is of major priority for Stavropol and Krasnodar Territories.

References

[1] *Stavropolregion* URL: http://ecology.gpntb.ru/ecolibworld/project/regions_russia/north_caucasus/Stavropol
[2] HS 2.1.5.1315-03 - *Maximum allowable concentrations (MACs) of chemicals in the water of water objects used for drinking and domestic-recreation purposes*. Approved by G.G. Onishchenko URL: http://www.etch.ru/norma.php?art=4
[3] Baturina F M, Chizhova L N and Mishvelov E G 2001 Ekologicheskaya otsenka territorii Kochubeevskogo rajona *Vestnik veterinarii* **18** 74 – 78 (in Russia)
[4] Matugina E G, Glyzina T S, Kolbysheva Yu V, Klyuchnikov A S and Vusovich O V 2015 Compensation for oil pollution damage *IOP Conf. Ser.: Earth Environ. Sci.* **27** 012070
[5] *Sulphuric anhydride* URL: http://refsru.com/referat-25349-1.html

Ecological risk assessment of Tomsk region groundwater used for drinking purposes

N V Konchakova[1], N S Ushakova[2] and T Yu Aikina[3]

[1, 2, 3]National Research Tomsk Polytechnic University, Russia, 634050, Tomsk, Lenin Avenue, 30

E-mail: [1]konchakova.nata@mail.ru

Abstract: The present paper is devoted to the chemical composition analysis of Neogene-quaternary and Paleogene groundwater widely used for drinking in the territory of Tomsk region. It has been shown that groundwater under study contains iron and manganese in excessive concentration. Consequently, this water can negatively affect human health. The ecological and human health risk assessment of Tomsk region groundwater used for drinking has been conducted. According to the calculations, it has been defined that in the overwhelming majority of cases there is a great risk to use groundwater of Tomsk region for drinking purposes.

1. Introduction

A peculiarity of water consumption in Tomsk region is priority groundwater use for drinking water supply. Tomsk region groundwater is rather effectively protected from surface contamination by low-permeability argillaceous deposits. However the water quality in its natural condition does not meet the sanitary regulations and standards (SanR&S) 2.1.4.1074-01 "Drinking Water. Hygienic Requirements to Water Quality in Centralized Drinking Water Supply Systems. Quality Control" as evidenced in publications by many outstanding scientists, for example, V.S. Kustovskiy, I.M. Ermashova, Yu.K. Smolentsev, N.M. Rasskazov, S.L. Shvartsev, D.S. Pokrovskiy and others. However, these works do not focus on the risks of using such water for drinking purposes.

In this regard, the objective of the present paper is ecological risk assessment of using groundwater for drinking purposes by population in the territory of Tomsk region.

Human health risk assessment is a quantity and/or quality characteristic of harmful effects which develop or can develop as a result of current or possible environmental factors impact on a definite group of people under specific conditions determined by regional features, whereas "ecological risk" is understood as probability of negative changes in condition of environment and/or natural objects due to certain factors [1]. With regard to impact on human health, the term "ecological risk" can be defined as a ratio of possible health hazard value due to the influence of a harmful ecological factor within a definite time interval to a standardized variable of this factor intensity [2]. The risk assessment data serve as guidelines and are used to justify and make decisions on risk management.

Ecological risk assessment is used in cases when it is necessary to give response about possible impact of chemicals in air, water and soil that is natural environment on human condition and health. The issues related to ecological risks assessment are actively being studied in a number of developed countries, for example, USA, Germany, Japan, the Netherlands, and etc. [3, 4, 5].

2. Ecological risk assessment methodology

The ecological risk assessment of fresh groundwater used for drinking purposes in the territory of Tomsk region comprised several steps:

1. Groundwater sampling.
2. Chemical analysis of water samples from 90 settlements in the territory of Tomsk region.
3. Analysis of groundwater chemical composition in the region under study and selecting priority substances for ecological risk calculation of groundwater used for drinking purposes.
4. Application of the approach to the risk assessment for Tomsk region population including non-cancerogenic ecological risk calculations of drinking water contamination for population health.
5. Interpretation of the obtained results.

3. Characteristics of groundwater chemical composition

Neogene-quaternary and Paleogene groundwater is used for public supply in Tomsk region in the vast majority of cases. According to the scientists [6, 7, 8, 9, 10, 11], the quality of such water in natural conditions does not conform to the regulation requirements upon a number of indications.

The chemical composition analysis of Neogene-quaternary groundwater has indicated that the water mineralization in the eastern and northern regions is basically within the range of 0.15-0,4 g/l. Mineralization value reaches 0.6-0.7 g/l in the right south-west of Tomsk region. Water is hydrocarbonate-sulfate, about neutral, pH ranges between 6.5 and 8.05.

Groundwater is smooth and very smooth over a large area of the region under study. Moderately hard water is almost commonly developed in the southern regions. Calcium and magnesium ratio rises sharply only in Tomsk and Bakcharsky regions where water becomes hard and very hard, its hardness ranges from 7 to 13 mg-equ/l.

Organic substances content (in terms of permanganate value) varies over a wide range between 2.6 and 10 mgO/l. According to the data collected by N.A. Ermashova [6], they are generally represented by humic substances (humic acids and fulvoacids).

Substantially all the water in the region under study contains iron in excessive concentration. Iron concentration is 3-10 mg/l in 70 % of the analyzed water samples. Oxidation-reduction potential (Eh) and dissolved organic solid are the major factors which control iron content in water (Fig. 1) [10, 12, 13, 14]. Our research is in accordance with the data obtained by other authors; in particular, iron is presented in water under study in two forms: dissolved bivalent form and bonded with organic substances [6-10, 12-16].

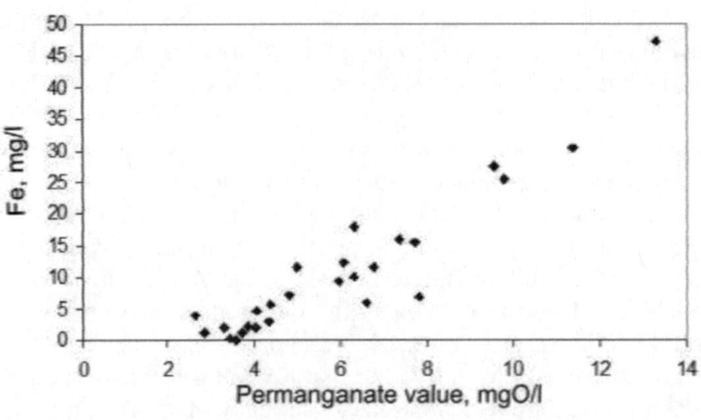

Figure 1. Iron content in the analyzed Neogene-quaternary water in dependence to permanganate value.

The presence of iron combined with organic substances in water has been studied with extractive method in chloroform [10].

The results of the study have shown that the share of iron bonded with organic substances in water samples ranges between 15% and 37% in the area under study. This value increases in the northern direction; it depends on the content and type of the organic substances presented in water. The main problem is that it is very hard to destroy iron combined with organic substances and remove it from water [10].

Manganese is rather widespread in the explored water. The presence of this element in concentration exceeding 0.1 mg/l is 95% with the most typical concentration of 0.1-0.5 mg/l (recorded in 58% water samples); its maximum concentration is 1.1 mg/l.

Paleogene water is most widely used for drinking purposes in the region. It is characterized by about neutral medium, pH ranges between 6.2 and 7.5. They demonstrate basic regularities in distribution of mineralization and its components specific to overlying formation. Water mineralization ranges between 0.09 and 0.5 g/l.

Water is mainly hydrocarbonate-calcium. Water is smooth in the northern and eastern areas; it is mainly moderately hard and occasionally smooth in the central and southern areas.

Organic substances content is almost twice lower than in Neogene-quaternary water. Average oxidability value is 3.6 mgO/l. This value increases in the northern and eastern directions with a decrease in water mineralization in boggy areas and reaches 7.2 mgO/l (which 1.5 times exceeds the maximum permissible concentration).

Iron concentration in the water under study reaches 15.4 mg/l, and is on the average 4.5-5.0 mg/l. Iron concentration in the analyzed Paleogene water samples commonly exceeds the maximum permissible concentration. In the overwhelming majority of cases, Paleogene water contains iron in bivalent dissolved form [6, 9]. Manganese content in water reaches 1.1 mg/l with an average value of 0.3-0.4 mg/l. Its content exceeds the maximum permissible concentration in a large part of the territory.

The chemical composition analysis of the groundwater used for drinking supply in the territory of Tomsk region has indicated that iron and manganese are the major components deteriorating its quality. Rural communities are facing this acute problem as drinking water treatment is exceedingly imperfect there.

4. Assessment of ecological risk level

Iron and manganese that occur in drinking water refer to non-carcinogens. Non-carcinogens are substances which cause negative changes in health condition, in particular, the rise in sickness and mortality rate that can be determined by both short-term and long-term exposure. Non-carcinogenic effects comprise: respiratory system irritant action; toxicity to liver, kidneys and other vital organs; central nervous system changes; fertility disorder and death [1, 16-21].

Non-cancerogenic health risk determination from the presence of chemicals in drinking water has been carried out according to the risk coefficient at per oral route with drinking water (CO_{per}) using the formula [2]:

$$CO_{per} = \frac{C_{per}}{RfD}$$

where C_{per} — substance concentration in drinking water, mg/l;

RfD — reference doses in case of chronic intake (0.3 mg/l for iron (the source is NCEA), 0.14 mg/l for manganese (the source is IRIS). Table 1 presents the calculation data and their interpretation.

The data presented in table 1 show that 60 % water samples conform to extremely high hazard level and 25 % - to high hazard level in terms of iron concentration. 25 % water samples conform to high hazard level in terms of manganese concentration. Extremely high and high hazard levels of drinking water are inappropriate for population. When such water is used for drinking purposes, it is necessary to take emergency curative and other measures to ensure health risk reduction.

Table 1. The data of hazard coefficient and risk level evaluation in case of using groundwater for drinking.

Risk level	Hazard coefficient	Iron concentration limits, mg/l	Sample ratio by iron, %	Manganese concentration limits, mg/l	Sample ratio by manganese, %
Extremely high	Higher than 10.0	3.23-15.4	60	-	-
High	5.0 — 10.0	1.65-3.0	25	1.1-0.86	25
Average	1.0-5.0	0.3-0.82	15	0.59-0.17	65
Low	0 — 1.0	-	-	0.13-0.11	10

15 % water samples by iron and 65 % water samples by manganese conform to average hazard level, which means that such a hazard level is inappropriate for population. This risk requires developing and taking scheduled corrective measures as well as continuous monitoring in population settlements.

Only 10 % water samples by manganese have low hazard level, which means that such water conforms to the conditionally appropriate (acceptable) risk zone; the majority of foreign hygienic regulations and those recommended by international organizations for population in general are established precisely at this level.

5. Conclusion

The present research has shown that in most cases Neogene-quaternary and Paleogene water used for drinking purposes in the territory of Tomsk region contains iron and manganese in excessive concentration. This made it possible to define substances which can have negative impact on human organism when they are ingested with drinking water.

The ecological risk assessment made in the current research has indicated that Tomsk region groundwater in its natural condition used by population for drinking is characterized by extremely high or high hazard level in the vast majority of cases, thus, it required to implement urgent and effective methods to purify groundwater from iron and manganese.

References

[1] Revich B A 2004 *Environmental Epidemiology* (Moscow: Academy) p 384

[2] Onischenko G G 2004 *Human Health Risk Assessment from Environmental Chemicals* (Moscow)

[3] *Introduction to Environmental* Management (2010) (NY: CRC Press) p 266

[4] *Environmental health Criteria 210: Principles for the Assessment of Risk to Human Health from Exposure to Chemicals* (1999) (Geneva: World Health Organization International Program on Chemical Safety) p 322

[5] *Environmental health Criteria 214: Human Exposure Assessment* (2000) (Geneva: World Health Organization) p 422

[6] Ermashova N A 1998 *Geochemistry of groundwater of active water exchange zone in Tomsk Oblast in connection with water supply and conservation* (Tomsk: Tomsk Polytechnic University) p 44

[7] Vidyaikina N V 2010 *Ensuring environmental safety in the use of groundwater by the rural population for drinking purposes (in Tomsk Oblast and Khanty-Mansi Autonomous District)* (Tomsk: Tomsk Polytechnic University) p 163

[8] Kraynov S R, Ryzhenko B N and Shvets V M 2004 *Geochemistry of Ground Waters. Theoretical, applied and environmental aspects* (Moscow: Science) p 677

[9] Ivanova I S, Lepokurova O E, Shvartsev S L and Pokrovskii O S 2014 Iron-containing

groundwater in the upper hydrodynamic zone in the central part of West-Siberian artesian basin *J Water Resources* **41** 163–177

[10] Serikov L V, Tropina E A, Shiyan L N, Frimmel F H, Meterveli G and Delay M 2009 Iron oxidation in different types of groundwater of Western Siberia *J. for Soils and Sediments* **9** 103–110

[11] Levina O, Pasechnik E and Baranova A 2014 Change in Chemical Composition of Natural Waters in Response to Technogenic Load (the Case of Tomsk City, Russia) *J. Procedia Chemistry* **10** 137–144

[12] Ginzburg II 1960 Migration of metals in waters *J. Principles of Geochemical Prospecting* chapter 9 186–215

[13] Perminova I V, Frimmel F H, Kudryavtsev A V, Kulikova N A and et al. 2003 Molecular weight characteristics of humic substances from different environments as determined by size exclusion chromatography and their statistical evaluation *J. Environ. Sci. Technol.* **37** 2477–85

[14] Shchukin E D, Pertsov A V, Amelina E A and Zelenev A S 2001 *Colloid and Surface Chemistry* (Elsevier Science B V) p 734

[15] Linnik P N and Nabivanets B I 1986 *Forms of Migration of Metals in Fresh Surface Waters* (Leningrad: Gidrometeoizdat) p 270

[16] Moiseenko T I, Dinu M I and Gashkina N A 2012 Metal Speciation in Natural Waters and Metal Complexing with Humic Matter *J DAN* **442** 691–695

[17] Fritsch E, Allard Th and Benedetti M F 2009 Organic complexation and translocation of ferric iron in podzols of the Negro River watershed. Separation of secondary Fe species from Al species *J. Geochim. et. Cosmochim. Acta.* **7317** 1813–25

[18] Fujii M, Ito H and Rose A L 2008 Superoxide mediated Fe(II) formation from organically complexed Fe(III) in coastal waters *J. Geochim. et. Cosmochim. Acta.* **72** 6079–89

[19] Pham N and Waite D 2008 Oxygenation of Fe(II) in natural waters revisited: Kinetic modeling approaches, rate constant estimation and the importance of various reaction pathways *J. Geochim. Et. Cosmochim. Acta.* **72** 3616–30

[20] Dekun Hou, Jiang He, Changwei Lü, Limin Ren, Qingyun Fan, Jinghua Wang, Zhilei Xie 2013 Distribution characteristics and potential ecological risk assessment of heavy metals (Cu, Pb, Zn, Cd) in water and sediments from Lake Dalinouer, China *J. Ecotoxicology and Environmental Safety* vol 93 135-144

[21] Wu Shan, Hu Yue-heng and Zuo Dan 2011 Discussion on Parameter Choice for Managing Water Quality of the Drinking Water Source *J. Procedia Environmental Sciences* **11** 1465–68

IOP Conf. Series: Earth and Environmental Science **33** (2016) 012023 doi:10.1088/1755-1315/33/1/012023

Conditions of contaminant distribution in the wetland water of Western Siberia (the Russian Federation)

O G Savichev[1], I A Matveenko[2] and E V Ivanova[3]

[1, 3]Department of Hydrogeology, Engineering Geology and Hydrogeoecology, Natural Resource Institute, Tomsk Polytechnic University, Tomsk, 634050 Russia
[2] Foreign Languages Department, Natural Resource Institute, Tomsk Polytechnic University, Tomsk, 634050 Russia

E-mail: [1]osavichev@mail.ru, [2]mia2046@yandex.ru

Abstract: The models of pollutants' distribution in the wetland waters of Western Siberia (Tomsk Oblast, Russia) have been considered. The models' application allows describing the observed changes in the chemical composition of active layer of wetland waters with satisfactory accuracy. With these models, the most significant impact of sewage discharge on the local wetland water composition is observed in wetland edge area of up to 150–300 m in width. Even in this area it is possible to minimize the impact of sewage discharge on the wetlands provided the concentration in sewage water is 1.14-1.3 time more than the background concentration.

1. Introduction

The lowlands of the Ob River basin are characterized by high bogginess, which is 30 % from catchment area of the river or even more in some regions. Besides, at present, further wetland formation is in progress, particularly, in the subzones of the south and middle Taiga (the Ob River's middle course), where a number of authors pointed out the mean rate of vertical growth in the peat deposit thickness of 0.39–0.80 mm/year at maximum to 2.62 mm/year, whereas the wetland area increases by about 100 km^2/year [1, 2]. Besides, in this area a major portion of the Russian Federation's oil and gas production is located [4]. It necessitates the adaptation of regional economic activity to the high bogginess, and, consequently, conditions the relevance of the research in the wetland ecosystem of Siberia, the key properties of which are connected with the excessive humidity (the thickness in the research area is 3-5 m, sometimes – up to 11 m). In the present paper, the research results on one of the aspects of the problem, namely, ways and conditions of pollutant distribution in wetland water have been considered. The research performed with participation of workers of Tomsk Polytechnic University (the Russian Federation, Tomsk) within 1997–2014 was taken as the initial data [3, 4].

2. Materials and Methods

The following research sites were chosen: 1) the north-eastern part of the Vasugan bog located in the watershed area of the Iksa and Bakchar Rivers (tributaries of the Chaya River – the Ob River's left tributary); 2) the oligotrophic bog located in the suburb of Strezhevoy (Strezhevoy bog); 3) the eutrophic Obskoye fen near the Mel'nikovo settlement (Figure 1). The choice of wetlands is determined by the following factors.

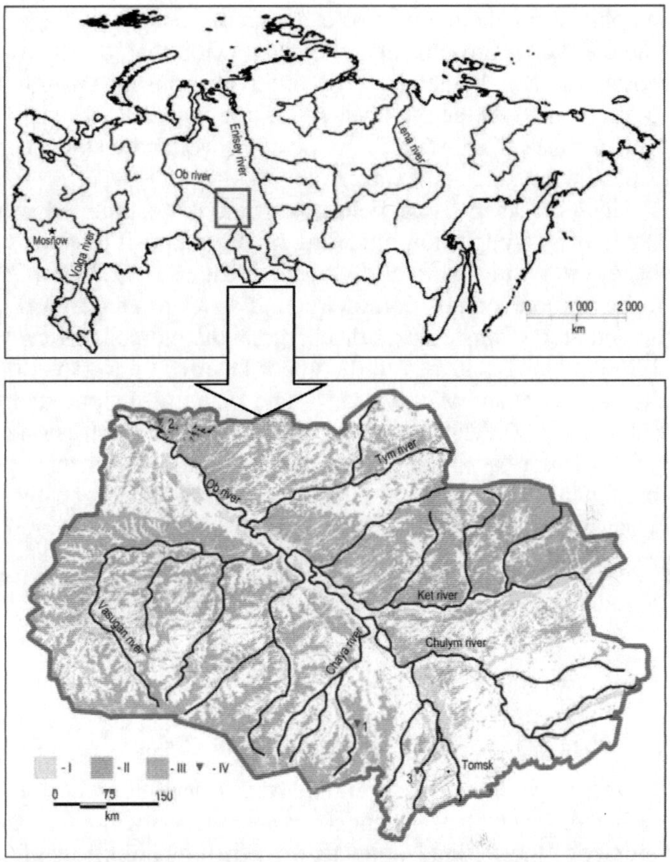

Figure 1. Map of the research areas: I – eutrophic swamps; II – mesotrophic swamps; III – oligotrophic bogs; IV – the research bog areas: 1 – the Vasugan bog; 2 – the Strezhevoy bog; 3 – the Obskoye fen.

The Vasugan bog is one of the largest bogs in the world. According to the different estimates, it has an approximate area ranging from 53.000 to 55.000 km^2 [1, 3]. The research area is located in the north-eastern part of the wetland and corresponds to the drainage basin of the Klutch River (an element of the Klutch, the Bakchar, the Chaya and the Ob Rivers as well as the Kara Sea) and is characterized by the predominance of the oligotrophic ridge-hollow complexes, sphagnous-pine-shrubs, sphagnous-sedge and sphagnous-sedge-cottongrass, pine-sphagnous-shrubs and hummock-ridge-lake bog ecosystems [3]. In general, the research area is distinguished by the following average values: the peat deposit thickness is 1.95 m, peat decay degree – 22 %, peat ash content – 6 %, peat moisture – 90.1 %. The Strezhevoy bog is oligotrophic and located within the subzone of the middle Taiga on the right bank of the Ob River basin. The peat deposit's average thickness in the research pine-sphagnous-shrub area is 1.75 m; average values of peat decay degree – 24 %, ash content – 4 %, moisture – 90.0 %. The Obskoye fen is of the eutrophic type and covers the left bank of the River Ob valley, a belt from 1.5 to 7 km in width and 104 km in length. Peat deposits in the lowland have the average thickness of 3.2 m. The research area is located near Mel'nikovo settlement and characterized by the average values: peat ash content is 28.7 %, peat decay degree – 34 %; moisture – 83.7 % [5].

The research in wetland peat deposit structure includes investigation of an active layer that demonstrates a transition layer from the peat deposit to the vegetation surface or dense root matting with intense water exchange. Peat deposits are a natural layering of definite peat types from the surface to the mineral wetland bottom. The active layer's thickness in the Vasugan and Strezhevoy bogs is 0.32 m and in the Obskoye fen is 0.52 m thick [6]. The Strezhevoy bog serves as a sink for commercial waste water (after the Strezhevoy system's iron removal plant for the clean ground water supply). The Obskoye fen is a sink for the domestic waste water of the Mel'nikovo settlement. At present, the Vasugan bog's research area is not exposed to apparent anthropogenic effect, but in the future it can be used for settled iron ore production and sewage disposal formed in the production process [7]. According to K.Ye. Ivanov [6]: 1) the motion of wetland waters mostly occurs within the peat deposit's upper active layer h_{al} (the usual thickness is 0.2–0.6 m; relatively high values of filtration coefficient and presence of oxygen are observed in the upper layer 0.2-0.5 m. The exchange processes can, certainly, take place, but their scale is insignificant in comparison with those in the upper layer. Besides, within the bog one can distinguish the sites with different transferring properties not only throughout the whole thickness, but also in the upper layer); 2) the wetland water table gradient J_{bw} approximately coincides with that of the wetland surface J_{bs}; 3) the inner wetland ecosystems are characterized by relatively constant magnitudes of surface gradients and permeability coefficients in the active layer k_f. On the whole, in terms of changes in water exchange intensity, the flow intensity of substances dissolved in wetland waters decreases from the surface to the bottom with its peak within the active layer.

Thus, the analysis of field observations on polluted wetlands in Western Siberia as well as the results of geo-migration modeling of peat deposit thickness have shown that: 1) short-term inflow of waste water with the total dissolved substances (TDS) of 7211.8 mg/dm^3 does not exert a significant impact on ion composition of wetland water limited, mostly, by the active layer; 2) a consequence of the continuous waste water discharge is the more significant change in water composition, the highest pollution being confined to the upper layer of up to 0.5–1.0 m in thickness [4]. In this regard the two alternate solutions of the steady-state plane diffusion equation are used to model the distribution of pollutants in wetland waters. In the first case, the diffusion equation's numerical solution is computed in rectangular coordinates under the assumption of primary substance distribution due to the advection component in longitudinal direction of the wetland (waste – wetland) water flow and diffusive one – in transverse direction [4, 8]:

$$V\frac{\partial C}{\partial x} = D\frac{\partial^2 C}{\partial y^2} + f(C) \tag{1}$$

$$V = k_f\, J_{bw} \tag{2}$$

$$D = D_m + k_D\, V \tag{3}$$

where C is the average value of hydrochemical indicator (concentration of the studied substance) in the active layer of peat deposit; x and y are streamwise and crosswise spatial coordinates; V is the velocity of the wetland water flow (if there is an inflow of waste water as well as surface and ground waters from the adjacent dry slopes, the resultant velocity of waste-bog water is calculated for each control point x as an weighted average of the wetland water flow's square and the section's water inflow [x–1; x] under the condition of initial mixing of the water inflow from dry slopes, wetland, and waste waters); D is the hydrodispersion coefficient (it is roughly taken as equal to the coefficient of longitudinal hydrodispersion); D_m is the coefficient of molecular diffusion; k_D is the dispersion coefficient; $f(C)$ is the function of substance inflow and transformation in wetland waters.

Typically, in hydrochemical modeling based on the hydrodynamic approach, the function $f(C)$ has the following forms:

$$f(C) = k_C\left(C_e^{\alpha} - C^{\beta}\right) \tag{4}$$

$$f(C) = k_C\left(C_e - C\right) \tag{5}$$

$$f(C) = k_C\, C^{\beta}, \tag{6}$$

$$f(C) = k_C\, C \tag{7}$$

$$f(C) = 0 \tag{8}$$

$$f(C) = a_0 + a_1\, x + a_2\, x^2 + \ldots, \tag{9}$$

where k_C is the constant of the chemical reaction; C_{eq} is the substance concentration in the solution corresponding to the local equilibrium in "water – organic substance – gas – rock" system; α and β are the coefficients conditionally corresponding to the stoichiometric coefficient in the chemical reaction occurring in the wetland waters and significantly defining the rate of substance concentration in the water; a_0, a_1, a_2,... are the coefficients defined by the least square method by the known values x and the difference between measured and calculated values of the studied hydrochemical indicator at $f(C)=0$. To solve the equation (1), the A. V. Karaushev's technique is used [8] based on the finite-difference approach (10) without transporting through the stream borders and the ratio of computation cell dimensions (11):

$$C_{x+1,y} = 0.5\left(C_{x,y+1} + C_{x,y-1}\right) + f\left(C_{x,y}\right)\Delta x / V \quad (10)$$

$$\Delta x = V\,\Delta y^2 / (2\,D) \quad (11)$$

where $C_{x,y}$ is the substance concentration in a cell with coordinates x and y; Δx and Δy are the increments along x and y. The second alternate allows an analytical solution of the diffusion equation in the cylindrical coordinates in the peat deposit active layer for the function $f(C)$ defined by the equation (7):

$$C_{\max} = C_w \exp\left(-k_C\, r^2 / \left(2\left(D - q_w / \varphi\, h_*\right)\right)\right) \quad (12)$$

where C_{max} is the maximum substance concentration in the wetland waters in a control point; C_w is the substance concentration in waste waters; r is the distance from the inflow source to a control point; φ is the substance distribution sector angle; q_w is the sewage flow rate; h_* is a half of the peat deposit active layer thickness. The active layer's thickness and other hydrologic characteristics of wetlands are estimated according to the observation data [6]. When applying the second solution, the coefficient k_C is defined by a selection with the help of the golden section technique (like other coefficients in the formulas (4–9)) and back calculation from (12) according to the observation data with the subsequent determination of dependence of k_C on the distance to sewage discharge point or wetland boundary. To check the accuracy of calculation, the Nash–Sutcliffe coefficient C_E is used [9] for $C_E > 0.5$.

3. Results and Discussion

According to the O. A. Alyokin's classification [10], due to the absence of any direct anthropogenic impact, the wetland waters of research region are generally stated as fresh waters including: sphagnum bogs with very low mineralization as well as fens with medium and sometimes high mineralization (Table 1). On average, local wetland waters are HCO_3^-/Ca^{2+} according to their chemical composition, in all cases they contain high concentrations of organic substances and some by-products (for example, ammonium ions), as well as a number of metals including iron and manganese. In terms of pH value, waters of sphagnum bogs are acidic on average, but the waters of fens are weakly acidic [1, 3]. The observation data have shown that in natural conditions sufficient differences in the spatial distribution of dissolved substance concentrations can be observed in the wetland waters even in the adjacent areas of the same wetland. This is explained by the discrete character of substance distribution over the bog territory, the conditions of the wetland and ground water interaction, and the peat deposit properties.

The influence of inner hydrodynamic and geochemical conditions is observed in the polluted wetland areas. In addition, the influence noticeably varies during the year. This is defined by the changes in the peat filtration properties and intensity of biochemical processes at air temperature below zero. A general tendency to decrease in TDS is noticeable in polluted wetland waters when moving farther from a sewage discharge point. According to the calculations performed, this tendency is satisfactorily described by the diffusion plane equation. To calculate the transformations of polluted wetland waters, both the numerical and analytical solutions are applicable for the diffusion equation ($C_E > 0.5$), but the analytical solution are often easier to use. Therefore, the equation (12) can be used for the evaluation of dissolved salt migration in the polluted wetland water in terms of TDS values as well as the substances discharged in significant amounts into wetlands with the relatively homogeneous filtration properties of peat deposits. The numerical solution is more appropriate for calculating and predicting the migration of dissolved substances for wetland waters in long wetland areas, for which two or more wetland microlandscapes are typical.

Table 1. Wetland waters' average values (A) of hydrochemical indicators and errors of their determination (δ^1) within 2000–2014.

Indicator	Units	Vasugan bog, N= 41–57		Strezhevoy bog, N=8		Obskoye fen, N=8–21	
		A	δ	A	δ	A	δ
pH	pH units	4.78	0.17	5.75	0.25	7.41	0.08
TDS[2]	mg dm^{-3}	58.5	10.8	65.8	14.0	1105.6	81.0
NH_4^+	mg dm^{-3}	5.18	0.42	2.11	0.55	29.96	8.07
Fe	mg dm^{-3}	3.16	0.58	2.28	1.05	29.16	13.48
COD	mgO dm^{-3}	191.0	21.2	112.3	17.2	340.3	101.4
Oil hydrocarbons	mg dm^{-3}	0.248	0.083	0.028	0.015	0.325	0.162

Note: [1] error of the arithmetic average is defined by the formula: $\delta = \sigma \cdot N^{-0.5}$, where σ is the mean square deviation; N is the number of samples; [2] total dissolved substances (TDS); the value is calculated as a sum of principle ions (Ca^{2+}, Mg^{2+}, Na^+, K^+, HCO_3^-, SO_4^{2-}, Cl^-); the techniques for determination of chemical composition are presented by Ekstein, Savichev, Pasechnik [3]); COD – chemical oxygen demand

The analysis of the field observation data on disturbed wetlands and the results of the sewage discharge modeling at the wetland boundary show that increase in dissolved substances' concentrations as compared to the geochemical background is associated mainly with the zone of 150 to 300 m. In the absences of anthropogenic impact in Western Siberia, such areas are usually referred to as the transition zone between a wetland and non-wetland area. In this case, the groundwater inflows with a significant dissolved salt content and an interaction between acidic wetland waters and mineral particles play a major role in the geochemical condition's development. Hence, the observed effect of anthropogenic activity in the research area has not resulted in sufficient changes in the large wetland hydrochemical balance. On the other hand, the additional (continuous) flow of dissolved salts in waste waters can result in shifting the existing phytocenosis towards increase in peat formation in the inner wetland ecosystems. As a result, fens are being formed in the place of upper bogs [4] with an elevated concentration of dissolved salts.

The latter circumstance defines the necessity for the evaluation of acceptable pollutant absorption into wetland water in which no statistically significant changes in the wetland ecosystem are observed. For this purpose it is recommended ($C_{w,\lim}$) to define the acceptable pollutant concentration in waste water, which is based on the comparison of two samples of volume N in the conditionally background (C_b) and disturbed (C_x) states. C_x is determined by the formula (13), where the error δ of determining C_b – by the formula (15):

$$C_x = C_b + f(C)\left(C_w - C_b\right)/k_n\,, \tag{13}$$

$$k_n = \left(C_w - C_b\right)/\left(C_{\max} - C_b\right), \tag{14}$$

where k_n is the multiplicity of dilution (an indicator of how many times the substance concentration decreased in waste waters in the research area); C_w is the substance concentration in waste waters. Supposing that values C_b and C_x are distributed in terms of Gauss's law and dispersions σ_b^2 and σ_x^2 are known to be equal to ($\sigma^2=\sigma_b^2=\sigma_x^2$), it is possible to use the Student test in the form:

$$Z = \left|C_x - C_b\right|/\left(\sigma \cdot \sqrt{2/N}\right). \tag{15}$$

Then, taking into account (13, 14), at the specified critical value Z_α, the assumption (15) for $C_{w,\lim}$ can be formulated in the form:

$$C_{w,\lim} \le C_b\left(k_n/f(C) - k_n + 1\right) + \left(k_n/f(C)\right)Z_\alpha\,\sigma\,\sqrt{2/N}\,. \tag{16}$$

Taking into account calculations of the equations (1–12) this relation amounts to 1.3–6.4 for Tomsk Oblast at the average value 3.0 [16]. The models of dissolved salts distribution were developed for the undisturbed and polluted wetland waters of Western Siberia based on the numerical and analytical solutions of the diffusion equation on a plane. Their application allowed us to describe the observed changes in TDS and other indicators in active layer of wetland water with different anthropogenic load with satisfactory accuracy.

4. Conclusion

The analysis of observation data and modeling results showed that the most significant impact of sewage discharge on the wetland water's chemical composition is observed at the wetland edge, up to 150–300 m in width. In the given distance from the wetland boundary, there occurs a sufficient (usually several times) decrease in dissolved salt concentrations in the wetland waters. The approach to defining the admissible concentration of pollutants in the waste waters $C_{w,\lim}$ is considered on the basis of comparing two samples in the conditionally background and disturbed states with the use of the Student test. Its study shows that even in the zone of the waste water impact (up to 150–300 m in width) it is possible to minimize the adverse effects on the wetland, if the concentration of pollutants in waste water is 1.14–1.3 time more than their background concentrations in wetland waters.

References

[1] Bazanov V A, Berezin A Ye, Savichev O G and Skugarev A A 2009 The phytoindication method for mapping peatlands in the taiga zone of the West-Siberian Plain *International Journal of Environmental Studies* **66** 4 473 – 484

[2] Pologova N N and Lapshina E D 2002 Carbon accumulation in the peat formations of The Great Vasyugan Mire *The Great Vasyugan Mire; present state and processes.* (Ed. M.V. Kabanov). Tomsk Atmospheric Optics Research Institute, Siberian Branch of Russ. Acad. of Sci., Tomsk, 174–244

[3] Eckstein Y, Savichev O G and Pasechnik E Yu 2015 Two decades of trends in ground water chemical composition in The Great Vasyugan Mire, Western Siberia, Russia *Environmental Earth Sciences* **1** 3 – 15 DOI 10.1007/s12665-014-3908-z

[4] Savichev O G 2015 Distribution of inorganic pollutants over the depth of upper peat deposit *Contemporary Problems of Ecology* **8** 1 118–124

[5] Schipper A M, Zeefat R, Tanneberger F, J P van Zuidam Hahne, Wulf S, Sebastiaan A, Loos S, Bleuten W, Joosten H, Lapshina E D and Wassen M J 2006 Vegetation characteristics and eco-hydrological processes in a pristine mire in the Ob River valley (Western Siberia) *Plant Ecol* December DOI 10.1007/s11258-006-9253-x

[6] Ivanov K E 1975 *Water Exchange in Marsh Landscapes* (Hydrometeoizdat, Leningrad)

[7] Savichev O G, Domarenko V A and Mazurov A K 2014 Estimation of allowable influence on water objects by development of sedimentary iron ores of Bakchare unit (Tomsk area) *Geoinformatika* **3** 56–63

[8] *Methodical Bases of Evaluating Anthropogenic Impact on Quality of Surface Waters* 1981 Ed. A V Karaushev (Hydrometioizdat, Leningrad)

[9] Nash J E and Sutcliffe J V 1970 River flow forecasting through conceptual models part I – A discussion of principles *Journal of Hydrology* **10** (3) 282–290

[10] Alyokin O A 1970 *Bases of Hydrochemistry* (Hydrometeoizdat, Leningrad)

Dust pollution of snow cover in the industrial areas of Tomsk city (Western Siberia, Russia)

A V Talovskaya[1], E A Filimonenko[2], N A Osipova[3], E G Yazikov[4] and L V Nadeina[5]

[1, 2, 3, 4, 5] National Research Tomsk Polytechnic University, 30 Lenin Avenue, Tomsk, 634050, Russia

E-mail: [1]talovskaj@yandex.ru

Abstract This article describes the results of long-term monitoring (2007-2014) of snow cover pollution in the territory of Tomsk city. Snow samples were collected in the territory of Tomsk. Determination of dust load level was carried out by comparing with the background and reference values. It has been determined that the north-east and central parts of Tomsk are the most contaminated areas, where brickworks, coal and gas-fired thermal power plant are located. The analysis of long-term dynamics showed a dust load decrease in the vicinity of coal and gas-fired thermal power plant due to upgrading of the existing dust collecting systems. During the monitoring period the high dust load in the vicinity of brickworks did not change. The lowest value of the dust load on snow cover was detected in the vicinity of the petrochemical plant and concrete product plants. The near and far zones of dust load on snow cover were determined with the reference to the location of the studied plants.

1. Introduction

Today, the study of the snow cover is one of the most affordable and reliable methods to obtain information about the intake of pollutants from the atmosphere. Snow cover is successfully used by many researchers from different regions and countries to indicate air pollution because snow is an excellent natural absorber for many substances from the air [1-9].

The chemical composition of snow is determined by several processes: formation of snowflakes around condensation nuclei (dust, drops, etc.) in the cloud, sorption of impurities during the passage of snowflakes through the atmosphere; inflow of pollutants from the air (aerosols, dust, ash, smoke, exhaust fumes and other).

Air pollution assessment in Tomsk region is of special interest as there are specific industries within the area. Volumes of solid pollutants emitted into the ambient air from stationary sources of Tomsk region increased more than 2.5 times for the period from 2005 to 2011. South-Eastern part of Tomsk region is the most urbanized because Tomsk city and Seversk town are located there, with the population more than half of the region's population. Also the northern areas, where oil and gas complexes are located, make a significant contribution to the pollution of the region [10].

The study of snow cover as an indicator of anthropogenic air pollution was conducted in the territory of Tomsk city and its surroundings [11-13].

The present study of snow cover was carried out in the industrial areas, with brickworks, coal and gas-fired thermal power plant, petrochemical plant and concrete product plants. The industrial enterprises emit annually between 35.5 and 37.1 thousand tons of pollutants into the ambient air in

IOP Conf. Series: Earth and Environmental Science **33** (2016) 012024 doi:10.1088/1755-1315/33/1/012024

Tomsk [10]. The main sources of emission of solid pollutants in Tomsk are coal and gas-fired thermal power plant and brickworks. A petrochemical plant operates in the northern part of Tomsk as well; it is the largest Russian enterprises for the production of polyethylene and polypropylene.

The main objectives of the present study were to (a) determine dust load value on snow cover in the territory of Tomsk-city, (b) identify the most snow polluted areas in the city and (c) reveal the changes of dust load between 2007 and 2014 in the vicinity of brickworks, coal and gas-fired thermal power plant, petrochemical plant and concrete product plants.

2. Methods

Snow sampling was conducted on a regular survey grid with a step of 500-600 m in the entire territory of Tomsk in March 2007. Just in the city 69 samples were collected. From 2009 to 2014 we collected snow samples annually in the zone affected by some large industrial enterprises of the city, such as coal and gas-fired thermal power plant, petrochemical plant, brickworks, and concrete product plant. In addition, private housing development and local boilers are found in the zone. Five samples were collected at the distance of 200 to 1500 m north-eastwards from the boundaries of the plants. The total number of samples during 6 years was 120.

Snow samples were collected at the sites with undisturbed snow cover for all its depth except 5 cm above the ground, the mass of each sample was 15-18 kg. While sampling, the pit area from which the samples were taken and the time from the formation of a stable snow cover till snow sampling time moment were checked and fixed.

Snow samples were melted at 20-22° C, and the snow water was filtered through a pre-weighed filter paper. The precipitated insoluble snow residual on the filter were passed through a sieve with a mesh size of 1×1 mm and then it was weighed with the error of 0.01%. Insoluble snow precipitate contains insoluble aerosol particles.

The calculations of a daily dust load (L_{dust}, $mg/m^2 \times day$) were made using equation:

$$L_{dust} = M / (S \times t),$$

where M is the mass of insoluble residue in the snow sample (mg), S is the pit area (m^2), t is the time from the moment of stable snow cover formation till the snow sampling (day) [14].

3. Results and discussion

According to the results of areal snow geochemical survey in the territory of Tomsk in 2007, sites of increased dust load were determined.The first site was revealed in the north-eastern part of the city and it was related to the brickworks location, the second one is located in the city center and is caused by emissions from coal and gas-fired power and heating supply plant. In addition, we also detected the contamination in the residential areas with private low-rise buildings and local boiler houses. The average value of the dust load in the territory of the city was 63 mg/m^2×day, it conforms to a low level of contamination (less than 250 mg/m^2×day according to [14]), by the background of 7 mg/m^2×day.

The average value of the dust load in the territory of Tomsk is 2 times lower than for the territory of the South of West Siberia (135 mg/m^2×day), Seversk (153 mg/m^2×day), Omsk (132 mg/m^2×day) and Rubtsovsk (106 mg/m^2×day); it is 5 times lower than for the territory of Mezhdurechensk (316 mg/m^2×day).

The Table presents the results of the dust load calculations. The analysis of long-term monitoring data shows that high values of dust load in the vicinity of brickworks remain. During the monitoring period, the dust load exceeds the background from 12 to 103 times. There is a decrease tendency of dust load amount by 35% from 2009 to 2012, which is caused by modernization of new dust collecting systems. Our data confirm the data of these plants about emission decrease. This decrease is determined by measures of the Department of natural resources and environmental protection of Tomsk region in response to the complaints of citizens in April 2009 about «red» snow in residential areas, in the territory of schools and kindergartens, located near the plants. In 2013, the dust load dramatically increased.

The contamination level is medium and high according to the reference and the dust load for the city exceeds 5.3 – 11.0 times in the near-field impacted area of brickworks. In the far-field impacted area of brickworks, the amount of dust contamination corresponds to low level of the air contamination after the normative scale and 3 times higher than the average dust contamination for the city. Such variation dynamics of the dust load value with distance from the plants can be explained by the fact that the material composition of the dust is mainly represented by various types of mineral particles with a heavy specific weight (quartz, feldspars, carbonates, clay minerals), which deposit with atmospheric fall-outs in the territory near the emitters. On the other hand, while loading and transporting dust is transported to the residential areas located near the plants.

Table. Dust load on snow cover in the industrial areas of Tomsk region, mg/m^2×day.

Industry	Distance from plant, m	2009	2010	2011	2012	2013	2014
coal and gas-fired thermal power plant	300	152	99	100	87	46	105
	600	219	162	131	38	63	42
	900	84	53	66	21	31	44
	1200	70	70	65	98	38	28
	1500	52	65	48	41	44	35
	mean value	115±31	90±20	82±15	57±15	44±5	51±14
brickworks	200	571	275	215	693	450	216
	400	584	720	455	100	526	405
	600	292	300	255	191	217	246
	800	256	259	134	96	221	106
	1000	122	123	162	86	249	95
	mean value	365±91	335±101	244±57	233±116	333±65	213±56
petrochemical plant	300	45	51	50	44	28	51
	600	54	45	63	33	54	52
	900	73	112	80	52	29	48
	1200	69	74	76	62	58	48
	1500	85	55	90	37	28	62
	mean value	65±7	67±12	72±7	46±5	39±7	52±3
Concrete plant	200	127	149	120	67	36	45
	400	139	212	145	54	84	89
	600	108	56	95	34	54	80
	800	82	81	85	71	34	78
	1000	94	n.d.	80	57	96	142
	mean value	110±10	125±35	105±12	57±6	61±13	87±16

Note: n.d. – no data; the average dust load for Tomsk-city is 63 mg/(m^2×day), the background is 7 mg/(m^2×day).

The similar tendency to the change of dust load volume with distance from the plants and dynamics in the impacted area of brickworks is presented in the vicinity of the construction industry enterprises, specializing in the production of reinforced concrete constructions. In general, all recorded data of the dust load in this territory exceed the regional background 15 - 18 times, and double the average amount for the city. It should be noted that formation of dust contamination in the vicinity of these plants is caused by the emissions from local boiler-houses, private houses and small private enterprises, located near the plants.

According to the study in the vicinity of coal and gas-fired thermal power plant, the dust load level reduced about by 45 % from 2009 to 2013. This reduction occurs despite the fact that the volume of burning coal at the station increased during these 5 years and the main part of coal (up to 80 – 90 %) is burnt from November to March. The decrease in dust load can be explained by the fact that, according to the official data, in 2010 ash collector of boiler was reconstructed and two additional ash collectors were installed. Moreover, ash collectors are annually repaired and tuned to improve the quality of dust emission cleaning; it provides complete large-sized particle trapping.

Dust load in the near-field impacted area exceeds the background from 17 to 27 times, and the excess of the average value for the city is 2-3 times. The increased values of the dust load in the near-field area can be connected with influx of dust by wind from coal storages, located in the power plant area, or during coal unloading. In the far-field impacted area dust load complies with the average value for the city.

The most significant dust precipitations in the area of 2 km from 100 m chimneys of coal and gas-fired thermal power plant are in the form of large-sized particles, despite permanent cleaning system for ash emissions to the atmosphere. The velocity of these particle precipitation develops several tens centimeters per second. An approximate assessment of such high precipitation rate is caused by simple kinematic features. For example, if average wind speed is 10 m/s, a particle must travel for a distance of 1 km in just 100 seconds and reach the earth's surface from the height of 100 meters. It points to the strong dependence of dust particles washing-out processes in the composition of ice grains, formed by freezing of water vapour in the smoke stream of power plants in winter. This effect was investigated and confirmed at the coal-fired power station in Kyzyl [15].

The dust load volume does not significantly change by years in the vicinity of a petrochemical plant. It is worth noting that this dust load level could be caused by transport of dust emissions from the plans located nearby.

4. Conclusion

As a result, we identified the most contaminated areas of snow cover in the territory of Tomsk and Tomsk region. According to the data of our snow survey in the territory of Tomsk in 2007, areas with high dust load accords to the position of the brickworks and coal and gas-fired thermal power plant. In addition, the results of long-term observations (2009-2014) of dust load from surrounding plants of Tomsk are ranked as follows: brickworks - 304 $mg/m^2 \times day$, plants for the production of concrete products - 88 $mg/m^2 \times day$, coal and gas-fired power and heating supply plant - 78 $mg/m^2 \times day$ and petrochemical plant - 58 $mg/m^2 \times day$. We discovered that most part of dust deposited on the snow cover at the distance of 300-600 m, there is a tendency of distance decrease from 600 to 1500 m from the borders of the studied plants. The obtained results can be used for air monitoring scheme optimization and undertaking detailed assessment of long-term atmospheric pollution in winter.

Acknowledgements

This study was financially supported by Russian Ministry of education and science under President Grant for Russia young scientists support and BP Exploration Operating Company Limited.

References

[1] Baltrenaite E, Baltrenas P, Lietuvninkas A, Sereviciene V and Zuokaiteet E 2014 Integrated evaluation of aerogenic pollution by air-transported heavy metals (Pb, Cd, Ni, Zn, Mn and Cu) in the analysis of the main deposit media *Environ. Sci. Pollut.* **21** 299-313

[2] Elİk A 2002 Monitoring of heavy metals in urban snow as indicator of atmosphere pollution *Int. J. Environ. Anal. Chem.* **82** 37-45

[3] Azimi S, Ludwig A, Thevenot D R and Colin J L 2003 Trace metal determination in total atmospheric deposition in rural and urban areas *Sci. Total Environ.* **308** 247-56

[4] Calvo A I, Alves C, Castro A, Pont V, Vicente A M and Fraile R 2013 Research on aerosol sources and chemical composition: Past, current and emerging issues *Atmos. Res.* **120** 1-28

[5] Cereceda-Balic F, Palomo-Marin M R, Bernalte E, Vidal V, Christie J, Fadic X, Guevara J L, Miro C and Gil E P 2012 Impact of Santiago de Chile urban atmospheric pollution on anthropogenic trace elements enrichment in snow precipitation at Cerro Colorado, Central Andes *Atmos Environ* **47** 51-7

[6] Farahmandkia Z, Mehrasbi M R and Sekhavatjou M S 2011 Relationship between

Concentrations of Heavy Metals in Wet Precipitation and Atmospheric Pm10 Particles in Zanjan-Iran *Iran J. Environ. Healt.* **8** 49-56

[7] Sakai H, Sasaki T, and Saito K 1988 Heavy-Metal Concentrations in Urban Snow as an Indicator of Air-Pollution *Sci. Total Environ.* **77** 163-74

[8] Samara C and Tsitouridou R 2000 Fine and coarse ionic aerosol components in relation to wet and dry deposition *Water Air Soil Poll.* **120** 71-88

[9] Viklander M 1999 Substances in Urban Snow. A comparison of the contamination of snow in different parts of the city of Lulea, Sweden *Water Air Soil Poll.* **114** 377-94

[10] *State report on environmental protection in Tomsk region in 2014* http://www.green.tsu.ru/dep/

[11] Talovskaya A V, Filimonenko E A, Osipova N A, Lyapina E E and Yazikov E G 2014 Toxic elements (As, Se, Cd, Hg, Pb) and their mineral and technogenic formations in the snow cover in the vicinity of the industrial enterprises of Tomsk *IOP Conf. Ser.: Earth Environ. Sci.* **21** 012042

[12] Osipova N A, Yankovich E P, Yazikov E G and Talovskaya A V 2012 Heavy metals in the air and their adverse effects on human being health *Proc. of the 7th Int. Forum on Strat. Tech. IFOST* **1** 153-6

[13] Osipova N A, Filimonenko K A, Talovskaya A V and Yazikov E G 2015 Geochemical Approach to Human Health Risk Assessment of Inhaled Trace Elements in the Vicinity of Industrial Enterprises in Tomsk, Russia *Hum. Ecol. Risk Assess.* **21** 1664-85

[14] Saet Yu E, Revic B A, Yanin E P and Smirnova R S 1990 Environmental geochemistry *Nedra Moscow* **335** (in Russian)

[15] Belyaev S P, Beschastnov S P, Khomushku G M, 1997 Some regularities of environmental pollution as the result of coal combustion as an example of Kyzyl *Russ. Meteorol. Hydro.* **12** 54-63 (in Russian)

Ground and Intermediate Water Equilibrium with Water-Bearing Rock Minerals (Moldova) under Anthropogenic Impact

A N Timoshenkova[1], C Ye Moraru[2], Ye Yu Pasechnik[3], O G Tokarenko[4] and V A Butoshina [5]

[1,2] Institute of Geology & Seismology of ASM, MD-2028, Kishinev, Republic of Moldova
[3, 4,5] Tomsk Polytechnic University, 30 Lenin Avenue, Tomsk, 634050, Russia

E-mail: [1]atimoshencova@gmail.com, [2]cmoraru@yahoo.com, [3]tog@tpu.ru, [3]paseyu@yandex.ru, [5]butoshina@list.ru

Abstract. The calculation results of ground water equilibrium with the major water-bearing rock minerals (Moldova) are presented under the condition of anthropogenic impact. As a calculation model the HydroGeo software is used. It is shown that both "ground water-rock" and "intermediate water-rock" systems are in equilibrium with a number of minerals.

1. Introduction

Ground water in Moldova is exposed to intensive anthropogenic impact. The representative examples of strong anthropogenic input are groundwaters of Kishinev, where an excess of nitrate pollution norms is observed, and the most examined Lower-Sarmatian aquifer, which is used for central water supply of residential areas and industrial sites.

The "water - rock" system is a global one on the Earth and its geological evolution leads to development of various geochemical types of ground and surface waters, different secondary mineral neogenesis including weathering crust, hydrothermally-transformed rocks, mineral deposits, etc. [1]. To understand the evolution mechanisms of underground hydrosphere composition is generally impossible without detailed research in thermodynamic equilibrium condition of groundwater with major water-bearing rock minerals. Therefore, the main goal of the article is to reveal the basic features of equilibrium in "water-rock" system and, hence, to determine the typical factors of water formation under the anthropogenic impact.

2. Materials and methods

2.1. Subject of research

The research subject is groundwater of Kishinev city and water of Lower-Sarmatian aquifer in the north-central part of Moldova (figure1).

IOP Conf. Series: Earth and Environmental Science **33** (2016) 012025 doi:10.1088/1755-1315/33/1/012025

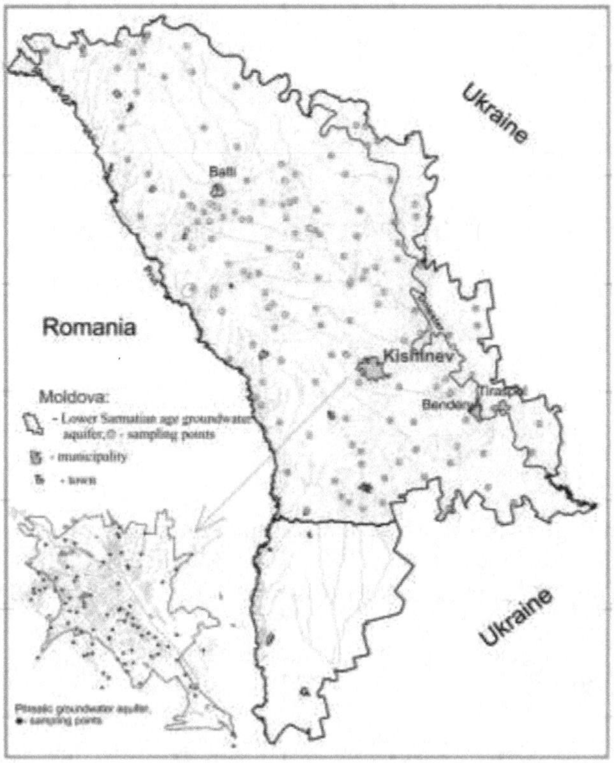

Figure 1. Sampling points.

2.2. Sampling and research methods

The research is based on the results of groundwater sampling (Kishinev city) that is performed in summer of 2012-2013 from 85 sources of temporary water supply [2]. To study Lower-Sarmatian aquifer 148 water samples are collected. The chemical analysis of water samples is performed in the accredited Laboratory of Hydrogeology and Engineering Geology, Geology and Seismology Institute (Academy of Science, Moldova) using the up-to-date analytical techniques such as titrimetry and ion chromatography.

Equilibrium calculation is carried out by common methods developed on the basis of hydrogeochemical process thermodynamics [3] using HydroGeo software [4]. The data on free energy of carbonate, aluminosilicate mineralization and dissolved elements are taken from the work [3]. To establish the rate of water equilibrium with definite minerals, we used the mineral stability field mapping technique developed by R.M. Garrels and Ch.L. Christ [5], as well as saturation index (si), which is equal to: $si = lg\ Q/K$, where Q – reaction quatant; К – reaction constant. While saturating with some mineral, the solution saturation index increases to zero (equilibrium state). In case of oversaturation, its values become positive. The calculations are carried out at temperature 25°C.

3. Geological and Hydrogeological Structure of the Study Area

Geologically, the study area is located in the south-west of East-European Precambrian platform within the Moldovan plate. Crystalline basement is located at the depth of 0 – 100 m (north) and approximately 1600 – 1800 m and more (south) and consists of Archean and Lower-Proterozoic strongly dislocated rocks of magmatic and metamorphic origin. Sedimentary section is of sedimentary formation of Upper Proterozoic, Lower- and Middle Paleozoic, and Cainozoic eras. The upper part of sedimentary section is composed of marine deposits of Middle Paleogene (Eocene), Neogene (Baden and Sarmatian Stages), and Quaternary deposits. For Neogene deposits the formation of three reef ridges is typical, the second of which is Middle-Sarmatian one crossing Kishinev city [6].

Hydrogeologically, the study area is south-western part of the Black Sea artesian basin of the first order. N.F. Frolov [7] distinguishes the area between the Prut and Dniester Rivers as a separate artesian basin of the second order called “Moldovan artesian basin”.

Depending on hydrodynamic and genetic features of geological conditions, there are two types of groundwater. The first type is groundwater of active water exchange zone occurred mainly at the depth of 0-15 m, in rare cases deeper. Water-bearing rocks are inequigranular sand and sandy clay. The aquifer bottom consists of dense blue-gray and green-gray clays of different age, the surface of which is washed out in some areas. Groundwaters are fed by precipitation infiltration, water inflow from the terraces above flood-plain, in high water season – due to flood water. In the territory of Kishinev city four unconfined aquifers can be distinguished: 1) aquifer in the recent alluvial deposits; 2) that in ancient alluvial deposits; 3) that in alluvial-talus deposits; 4) that in eolian-talus deposits [8].

Intermediate water is a complex of Sarmatian aquifers.

IOP Conf. Series: Earth and Environmental Science **33** (2016) 012025 doi:10.1088/1755-1315/33/1/012025

Groundwater in Lower-Sarmatian deposits is distributed everywhere and well-studied over the territory of Moldova, with the exception of extreme south, where rocks of this age are absent. Groundwater reservoir is fractured limestone, which alternates with sand and clay loam in the Central Transdniester.

Lower-Sarmatian aquifer is separated from Upper-Cretaceous aquifer of Podolskian argillaceous sand formation that acts as an aquiclude between them.

Head waters: the pressure increases from north to south reaching 2 -40 m in the north and in the south more than 300 m, then decreasing towards south-west and south-east up to 7 - 10 m. Groundwater is discharged into the Black Sea; first it appears to flow into overlying Middle-Sarmatian deposits, from the latter it flows into the sea (S.T. Vznuzdayev, 1969). Besides, there is a discharge in the river valley, in the tectonic zones [6].

4. Results and discussion

As for their chemical composition, groundwaters of Kishinev city are mostly hydrocarbonate or hydrocarbonate-sulfate (in terms of prevailing anion and cation content) with different ratios of magnesium, calcium, sodium (Table 1). There are also sulfate-hydrocarbonate and sulfate waters. Waters are mostly weakly alkaline, fresh, seldom weakly brackish. Water mineralization varies in the wide range – from 0.39 to 2.63 g/L.

Table 1. Chemical composition of groundwater in Kishinev city (the number of samples – 85).

Component	Concentration, mg/L		
	Min	Max	Average
TDS, g/L	0.4	2.6	1.1
pH	7.0	8.4	7.6
HCO_3^-	244.0	902.8	526.1
Cl^-	13.6	408.7	110.8
SO_4^{2-}	8.0	1488.0	270.4
NO_3^-	0.4	994.4	116.5
Ca^{2+}	14.2	383.2	121.6
Mg^{2+}	7.8	222.5	90.4
Na^+	28.0	305.0	124.9
K^+	0.1	125.2	6.4
Fe^{2+}	0.1	1.0	0.3
Sr^{2+}	0.5	6.8	1.8
SiO_2	1.9	35.0	9.7

The distinguishing feature of groundwater in Kishinev city is elevated concentration of nitrate (the maximum concentration is 994.4 mg/L), the source of which is agricultural activity [2, 8].

In terms of chemical composition water of Lower-Sarmatian aquifer is mostly hydrocarbonate sodium or sodium-magnesium (Table 2).

Table 2. Chemical composition of drinking intermediate groundwater of the north-central part of Moldova (the number of samples – 148).

Component	Concentration, mg/L		
	Min	Max	Average
TDS, g/L	0.48	4.64	1,38
pH	6.80	8.98	7.95
Na^+	27.00	1400.00	307.33
Mg^{2+}	0.50	215.00	33.28
Ca^{2+}	1.50	343.00	36.50
K^+	0.50	20.00	5.12

HCO_3^-	317.20	2656.00	691.06
Cl^-	7.00	574.00	68.16
SO_4^{2-}	2.00	1422.10	228.16
SiO_2	0.50	44.20	12.48

There is also hydrocarbonate-sulfate and sulfate-hydrocarbonate waters with different ratios of sodium, magnesium, calcium. In 5 % of cases there are hydrocarbonate-chloride sodium-magnesium waters. Water mineralization ranges from 0.48 to 4.64 g/L. Water is mostly alkaline (the average value of pH = 7.95), weakly brackish and fresh, seldom brackish, at the average salinity value 1.38 g/L.

Hence, water is mostly alkaline and the major salt-forming components of that water are hydrocarbon-ion and sodium. Taking this fact into account, one may state that it is typical alkaline sodium water sometimes distinguished by high mineralization.

4.1 Equilibrium with carbonate minerals.

Thermodynamic calculations have shown that, despite low mineralization values and calcium content, the major part of studied groundwaters of Kishinev achieves the equilibrium with such carbonate mineral as calcite (figure 2). The long distance between data points shows the state of water oversaturation with calcite. This is true for waters of all chemical types. Insignificant part of water still remains undersaturated with calcite. The average mineralization value of this water is 0.4 g/L, the content of hydrocarbonate ion – 329 mg/L, calcium – 16 mg/L at pH = 7.7.

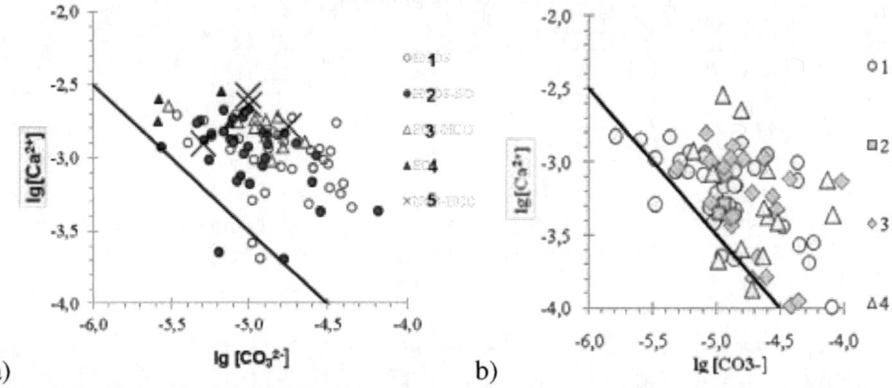

a) b)

Figure 2. Diagram of equilibrium a) groundwater in Kishinev (chemical types of water: 1 – HCO_3; 2 – HCO_3–SO_4; 3 – SO_4–HCO_3; 4 – SO_4; 5 – NO_3–HCO_3) and b) Lower-Sarmatian ground water of the north-central part of Moldova (chemical types of water: 1 – HCO_3; 2 – HCO_3–Cl; 3 – HCO_3–SO_4; 4 – SO_4–HCO_3) with calcite at 25°C.

In case of intermediate water, the group of calcite unsaturated water includes some waters of hydrocarbonate and sulfate-hydrocarbonate types with the average mineralization value – 1 g/L, hydrocarbonate-ion content – 512.2 mg/L, calcium – 39.38 mg/L at pH=7.37, whereas the average values of the above mentioned parameters of calcite saturated waters do not differ significantly.

4. 2 Equilibrium with aluminosilicate minerals.

The diagrams of saturation with aluminosilicate rock minerals are shown in figure 3. All groundwaters of Kishinev and intermediate waters of north-central part of Moldova (with the exception of several samples) are not in equilibrium with endogenous aluminosilicates (albite, anorthite) and are in equilibrium with secondary minerals (aluminum, kaolinite, montmorillonite oxides). Some samples are in equilibrium with gibbsite, potassium, and feldspar.

IOP Conf. Series: Earth and Environmental Science **33** (2016) 012025
doi:10.1088/1755-1315/33/1/012025

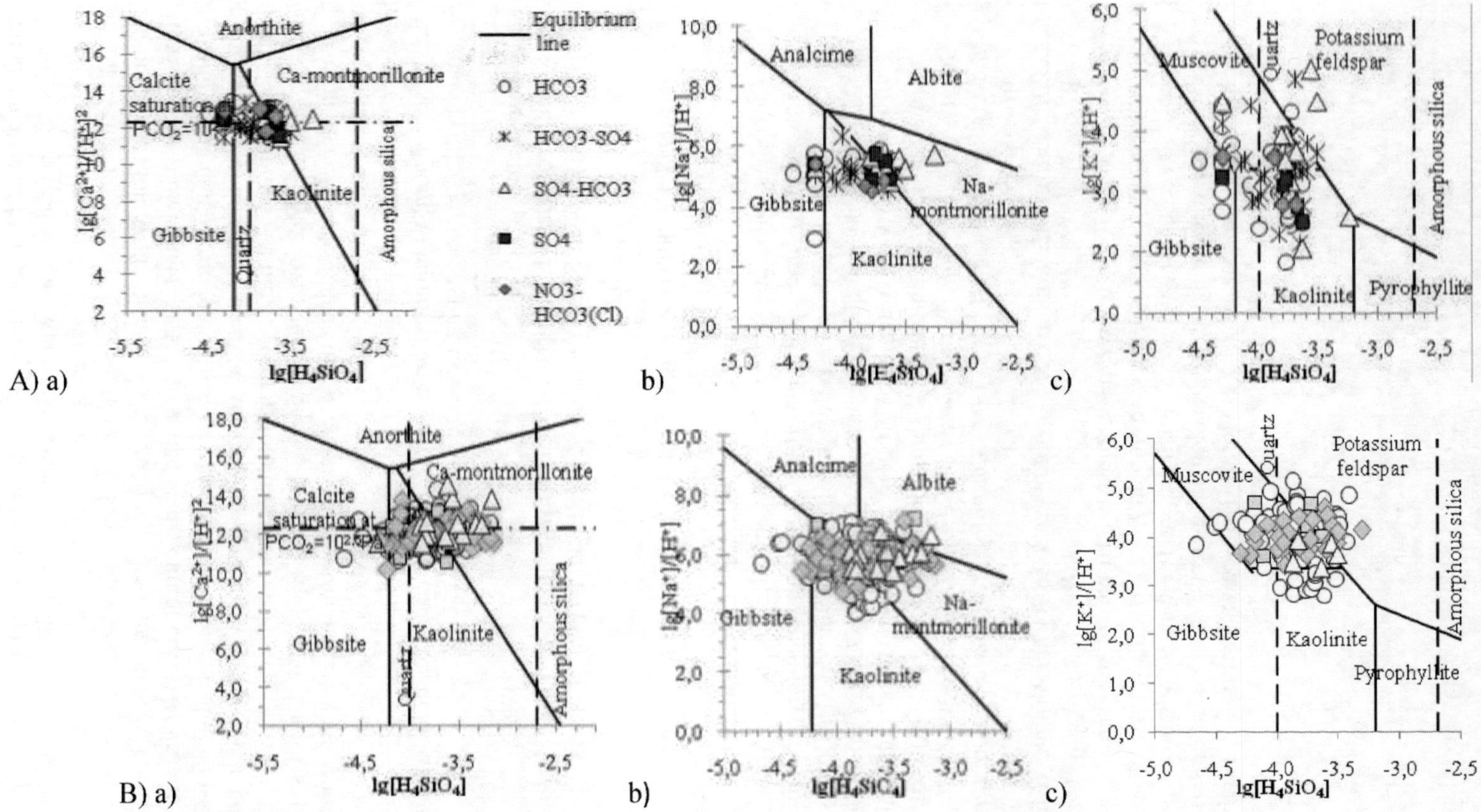

Figure 3. Equilibrium diagrams A) ground water of Kishinev city and B) Lower-Sarmatian ground water of north-central part of Moldova with aluminosilicate minerals at standard condition: a) $HCl–H_2O–Al_2O_3–CO_2–CaO–SiO_2$ system; b) $HCl–H_2O–Al_2O_3–CO_2–Na_2O–SiO_2$ system; c) $HCl–H_2O–Al_2O_3–CO_2–K_2O–SiO_2$ system.

5. Conclusions

All studied waters of Moldova territory are in equilibrium with carbonate minerals, even groundwaters of Kishinev, despite the rapid water exchange and short time of groundwater interaction with rock. The observed equilibrium of Kishinev groundwater with calcite proves the early stage of groundwater evolution which is accompanied with rather low mineralization values (less than 1 g/L) due to hydraulic connection with surface water and precipitations. It is just in those points where water equilibrium with carbonate shifts towards undersaturation. All groundwaters of Kishinev and intermediate waters of north-central part of Moldova (with the exception of several samples) are not in equilibrium with endogenous aluminosilicates (albite, anorthite) and are in equilibrium with secondary minerals (aluminum, kaolinite, montmorillonite oxides).

In the territory of Moldova agriculture is widely developed that has a great impact on water composition. For example, elevated concentration of nitrite is a typical feature for Kishinev that allows distinguishing so-called "exotic" type of water – NO_3–HCO_3–Mg–Ca, sometimes with increased chloride ion.

All these factors point to equilibrium-non-equilibrium nature resulted from interactions in the "water-rock" system and a definite evolution stage.

Acknowledgements

Authors thank Prof. Dr. S L Shvartsev for his helpful advice and suggestions. The investigations were supported by The Russian Foundation for Basic Research (Project No 14-35-50510).

References

[1] Shvartsev S L 2008 Geochemistry of fresh groundwater in the main landscape zones of the Earth *Geochem. Int.* **46** 3 128–1398

[2] Timoshenkova A N, Pasechnik E Y and Tokarenko O G 2014 Equilibrium of Groundwater with Carbonate Minerals of the Water-Bearing Rocks under Anthropogenic Impact (by the example of Kishinev, Moldova) *IOP Conf. Ser.: Earth Environ. Sci.* **21** 1 12-24

[3] Alekseev V A, Ryzhenko B N, Shvartsev S L, Zverev V P, Bukaty M B, Mironenko M V, Charykova M V and Chudaev O V 2005 *Geological evolution and water-rock self-organization* V.1. Water-Rock System on the Earth Crust: interaction, kinetics, equilibrium, simulation. (Novosibirsk: Punlishing House SB RAS) p 244 [in Russian]

[4] Bukaty M B 2002 *Bulletin of Tomsk Polytechnic University* **305** 6 348–65 [in Russian].

[5] Garrels R M and Christ Ch L 1965 *Solutions, Minerals and Equilibria* (New York: Harper & Row) p 450

[6] *USSR Geology. Moldavian SSR. Moldavskaya SSR* 1969 Vol. 45 (Moscow: Nedra) p 456 [in Russian]

[7] Frolov N M 1961 Groundwaters of the western part of the Black Sea artesian basin (Moscow: Academy of Science of the USSR) p 230

[8] Moraru C Ye 2012 Hydrogeochemistry of active water exchange zone in the southwest of the Black Sea artesian basin (geoecological aspects): abstract of dissertation … Doctor of Geol.-Mineral. Sciences: / C Ye Moraru (Moscow: Institute of Environmental Geoscience of Russian Academy of Science) p 52

Mineral-organic formations in Berezitovy deposit (the Amur region, Russia)

E A Vakh[1,2], A S Vakh [1,2], V I Petukhov[1], T V Nikulina[3] and I A Tarasenko[1,2]

[1] Far Eastern Federal University, 8 Sukhanova St., Vladivostok, 634050, Russia
[2] Far East Geological Institute, Far East Branch of Russian Academy of Science, 159, Prospekt 100-letiya, Vladivostok, 690022, Russia
[3] Institute of Biology and Soil, Far East Branch of Russian Academy of Science, 159, Prospekt 100-letiya, Vladivostok, 690022, Russia

E-mail: [1]Adasea@mail.ru, [2]vakh@fegi.ru, [3]petukhovv@mail.ru, [4]nikulinatv@mail.ru, [5]tarasenko_irina@mail.ru

Abstract. The article examines the structure and composition of mineral-organic formations within the hypergenesis zone of Berezitovy deposit (the Amur region). The detailed study has shown that these recent formations are represented by algae identified as *Trentepohlia jolithus* (Linnaeus) Wallroth. The process of macro and micro element accumulation in these formations is likely to have a complex sediment-chemogenic-organogenic nature and results from the flow of the suspended and dissolved substances formed within the hypergenesis zone of sulphide ores. It is also assumed that some elements accumulated in the formations were previously absorbed by algae from the mineralized water environment.

1. Introduction

In hypergenesis zone of sulphide ore deposits, the original sulphide minerals exposed to the atmosphere and hydrosphere are intensively dissolved due to various sulphurous processes, which, in its turn, results in the development of geochemical anomalies in the element-rich water environments. The recent studies aimed to analyze the element concentration in water environments within the hypergenesis zone of sulphide ore deposits have shown that metal content in acid waters can be extremely high, several times exceeding natural background values [1, 2, 3, 4]. In addition, formation of acid and high-mineralized water systems characterized by complex hydrochemisty is always accompanied by the processes of authigenic mineral formation [5], which leads to formation of sulphate, aluminium fluoride and other complex minerals [6].

2. Materials and methods

2.1 Research object

The research was carried out in Berezitovy gold-polymetallic deposit, the Amur River region (Figure 1), located within the Selenga-Stanovoy terrane [7]. The deposit consists of two joined, steeply dipping and hopper-shaped fluid-explosive bodies identified within the Paleozoic granitoids [8]. These bodies are composed of tourmaline-garnet-quartz-muscovite metasomatites which carry stringer-porphyry sulphide mineralization of primarily pyrite, pyrrhotite, false galenite, and galena composition [8]. The concentrations of macro- and microelements have been identified in granites, ore-bearing metasomatites, surface and ground waters of the ore field [7, 8]. The conducted research has revealed that acid (pH 3.0–4.5) sulphate waters with total salt content up to 10 g/l are the basic reason for anomalously high metal content in surface waters [1].

The hydrochemical study of the surface waters in the Konstantinovsky streamlet, within the head of which Berezitovy deposit is located, has revealed abundant mineral-organic formations recently occurred across the whole valley of the streamlet due to the action of acid sulphate waters. This can be explained by the fact that such formations have not been identified in other water streams, beyond the zone of action within Berezitovy ore field. The mineral-organic formations cause development of significant number of dark-brown and reddish-brown crusts on the surface of block deluvial streamlet deposits up to 1-2 mm in thickness (Figure 2). The crusts are of fragile and loose texture. Most of them contain fine-grained filamentous aggregates of white and light-grey colors. The upper layer is a thin film composed of dark-brown aggregates. The crusts do not dissolve in water.

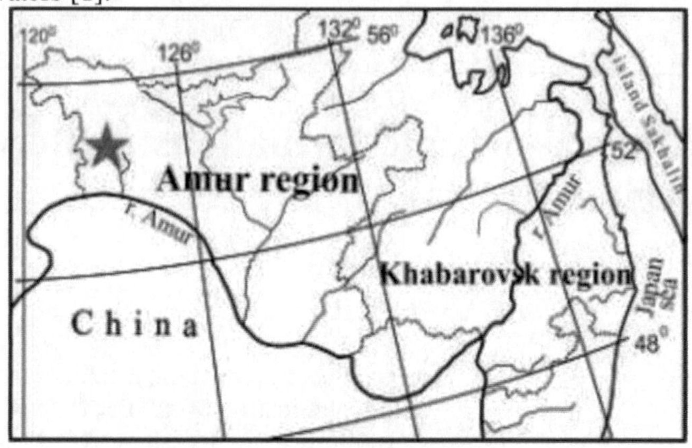

Figure 1. Location map of the study area.

2.2 Research methods

The samples of the mineral-organic crusts were taken from the middle part of the Konstantinovsky streamlet (area of exploratory adit № 2) during three years under various climatic conditions of the summer months. The chemical and microelemental composition was studied on the basis of plasma spectrometry using quadrupole plasma mass spectrometer Agilent 7500 series. In order to identify mineral phases in the crust composition, infrared spectroscopy, X-ray diffraction analysis, and scanning electron microscopy were applied. Infrared absorption spectrum of the samples were recorded by the multifunctional infrared Fourier spectrometer Nicolet 6700 (Termotechno, USA) over the frequency range of 4000-50 cm. The morphological study of mineral aggregates was carried out using analytical scanning electron microscopy, precisely, microscope JSM 6490 equipped with energy dispersive x-ray spectroscopy INCA Energy 350. X-ray diffraction meter (XRD) MiniFlex II manufactured by RIGAKU (Japan) was used for X-ray diffraction analysis. In order to identify the species of algae, light microscopes «Axioskop 40» (Zeiss, lens 40x/0.65 and 100x/1.25 oil) and «Alphaphot–2 YS–2» (Nikon, lens 40x/0.65 and 100x/1.25 oil) were applied. The species of algae were identified in accordance with the procedure proposed by N.A. Moshkova and M.M. Gollerbakh [10]. All studies were carried out on the basis of the Center for Elemental and Isotopic Analysis, Far East Geological Institute (FEGY), Far East Branch of Russian

Figure 2. The recent mineral-organic formations on the surface of the Konstantinovsky streamlet block deluvial deposits.

Academy of Science and Institute of Biology and Soil, Far East Branch of Russian Academy of Science (Vladivostok).

3. Geochemistry of mineral-organic formations

3.1 Formation composition

As initially proposed, the studied crusts were formed as the result of the recent autogenic mineralization. They are basically composed of iron oxide and iron hydroxide formed during chemogenic precipitation from the ore waters. As a rule, such aggregates are well represented in the zone of oxidized sulphide ores. However, the detailed study of the crusts sampled from the Konstantinovsky streamlet has shown that they have more complex mineral-organic origin.

3.2 Biological studies

The crusts analysis by means of scanning electron microscopy has revealed that most of light-grey aggregates are composed of microsponge elongated organic formations of complex structure (Figure 3). More detailed analysis gives grounds to believe that these recently formed aggregates are represented by green algae known as *Trentepohlia jolithus* (Linnaeus) Wallroth (class Ulvophyceae, order Trentepohliales, family Trentepohliaceae) [10]. *T. jolithus* forms thick mats or tussocks 1-2 mm in thickness which are rust colored when wet. As they dry, they dim and turn to dusty-vinous or greyish-green. The tussock fibers abundantly and randomly ramify. The cells are of elliptical and cylindrical shape, slightly tighten near the traverse septums. Cell thickness ranges between 14–27.5 µm, the length - 25–45 µm. The cell membranes are thick and obliquely laminated (figure 3). It is the first time when *T. jolithus* was found in the south of the Russian Far East [9, 11].

Having porous structure, crust organic formations contain a large number of fine-grained mineral granules. Mineral aggregates are composed of hypogene rock-forming minerals, ore, and secondary recently-formed minerals found within the deposit.

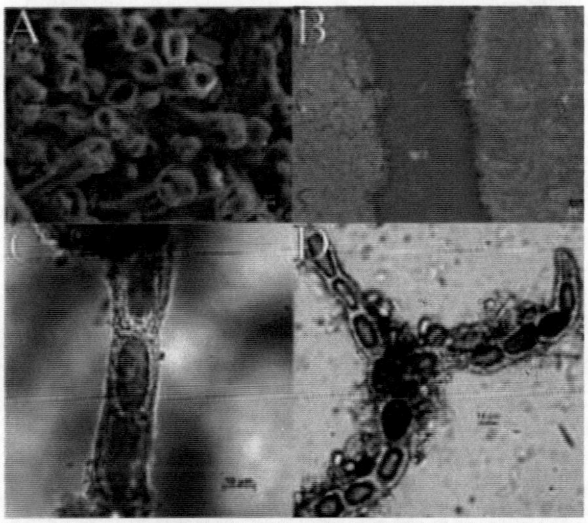

A, B – morphology of organic formations (scanning electron microscope EVO-50XVP light microscope); C – thallus fragment, D – lamination of cell membrane (light microscope).

Figure 3. Trentepohlia jolithus.

3.3 Mineral composition

The infrared spectroscopy results have shown that hydromica, kaolinite, quartz, and goethite are the basic minerals that compose the crusts. Based on the X-ray diffraction analysis of heat-treated crusts (at 1100 C°), the presence of quartz, plagioclase, muscovite (1M and 2M), biotite, and hematite in the crusts has been identified.

Based on the X-ray microanalysis carried out by means of energy dispersive x-ray spectroscopy INCA Energy 350, the following single aggregates (not more than 10-20 µm in size) of hypogene minerals have been identified: melanterite, plumbojarosite, zinc- melanterite, and halloysite.

3.4 Chemical and micro element composition

Chemical analysis of mineral-organic formations (table1) shows that they are composed mainly of silicon oxide, aluminum, and iron. The percentage of carbonates themselves is rather small. The mineral-organic formations are characterized by high loss-on-ignition values. This fact proves the leading role of organic compounds in their composition. High concentrations of ore elements (Zn, Pb, Cu, As and Cd) are identified in the micro element composition of the crusts. This is a characteristic feature of the contaminated surface water, the Konstantinovsky streamlet [1].

It is noteworthy that the concentrations of micro elements in crust samples collected at different time periods and places across the streamlet are rather constant (table 1).

Table 1. Chemical and micro element composition of the recent mineral-organic formations found in the Konstantinovsky streamlet (Berezitovy deposit).

Elements	Sampling years		
	2009	2010	2011
SiO_2 (mas. %)	17.71	23.05	19.32
TiO_2	0.22	0.31	0.25
Al_2O_3	6.09	7.80	6.96
Fe_2O_3 (tot.)	2.52	3.07	2.81
MnO	0.06	0.09	0.07
MgO	0.60	0.71	0.66
CaO	0.68	0.91	0.80
K_2O	1.54	1.91	1.73
Na_2O	0.79	1.03	0.89
P_2O_5	0.15	0.20	0.16
loi[a]	69.50	60.42	65.85
Total	99.87	99.51	99.50
Li (g/t)	6.16	7.54	6.70
Be	0.44	0.64	0.62
Sc	3.09	4.03	3.62
V	21.50	28.30	23.90
Cr	15.10	18,52	17.01
Co	3.60	4.21	3.99
Ni	8.70	11.50	10.50
Rb	67.65	80.98	75.47
Sr	106.22	137.40	118.33
Y	8.70	11.90	10.00
Nb	4.51	6.32	5.27
Cs	2.25	2.48	2.44
Ba	286.30	386.40	327.20
Hf	0.45	0.60	0.49
Ga	7.00	8.94	8.12
Mo	4.02	5.14	5.08
Sn	1.18	1.43	1.26
W	4.13	5.72	5.53
Cu	85.70	87.90	84.00
As	66.60	51.04	50.46
Pb	526.10	796.50	608.70
Zn	726.20	998.70	860.70
Ag	3.56	4.15	3.74
Cd	3.58	4.74	3.72
Th	6.76	8.56	7.61
U	3.09	2.80	3.07

[a.] loi – loss-on-ignition. Determination of element concentration in water samples was carried out by ISP-MS Agilent 7500 (by M. G. Blokhin, E. V. Elovskiy).

This proves the fact that being a basic constituent of mineral-organic formations, algae can significantly contribute to the concentration of elements absorbing them from the Konstantinovsky streamlet.

4. Results and discussion

Thus, it can be stated that the recent mineral-organic crusts found in the Konstantinovsky streamlet are composed mainly of organic formations (supposedly, algae) which were favorable for accumulation of fine-grained minerals suspended, finely dispersed, and dissolved in the surface water of the streamlet. Inorganic material identified in the crusts is a complex combination of chemically and physically weathered host rocks and ores. The area where mineral-organic formations were found is confined to the zone of acid sulphide surface waters of the Konstantinovsky streamlet characterized by high concentrations of heavy metals [4].

The processes of micro element accumulation in biologicals have been previously studied. It has been stated that some types of water plants such as algae and macrophytes are capable of accumulating definite chemical elements including toxic heavy metals [12]. The level of element concentration is basically dependent on hydrochemical composition of water environment and severity of water contamination with heavy metals. Therefore, it is possible to conclude that the organic formations, i.e. algae, found in the Konstantinovsky streamlet significantly contribute to minimizing the impact of contaminated mining runoffs on the environment.

5. Conclusion

The recent mineral-organic aggregates, widely-spread within the valley of the Konstantinovsky streamlet, formed under the impact of acid ore waters of Berezitovy deposit. The process of macro and micro element accumulation in these formations is likely to have a complex sediment-chemogenic-organogenic nature and is the result of the flow of the suspended and dissolved substances formed within the hypergenesis zone of sulphide ores. It is also assumed that some elements accumulated in the formations were previously absorbed by algae from the mineralized water environment.

The specific mineral-organic formations, firstly found in the Russian Far East and confined to the area of the surface waters contaminated with the products of the sulphide ore hypergenesis zone, can be regarded as a new indicator of geo-ecological condition of modern technogenic landscapes.

Acknowledgments

The investigations were partially supported by the Russian Foundation of Fundamental Research (№ 15-05-00809), as well as by the Ministry of Science and Education of the Russian Federation (as part of the research “Improvement of Constructive and Geoecological Safety, Energy Efficiency of Vessels and Facilities at Continental Shelf of the Far Eastern and Arctic Seas”, NIR 543.2014).

References

[1] Vakh E A, Vakh A S and Kharitonova N A 2013 The presence of REE in the waters of the hypergenetic zone of the sulfide ores, the Beresitovy deposit (Upper Priamurie) *Russian Journal of Pacific Geology* Vol. 32 **1** 105–115
[2] Tabaksblat L S 2004 Sostav tekhnogennykh drenazhnykh vod rudnykh mestorozhdeniy *Izvestiya Vysshikh Uchebnykh Zavedeniy*: *Geologiya i Razvedka* **4** 43–48 [in Russian]
[3] Chudaeva V A and Chudaev O V 2011 Accumulation and Fractionation of Rare Earth Elements in Surface Waters of the Russian Far East under the Conditions of Natural and Anthropogenic Anomalies *Geochemistry Int.* **3** 523–549
[4] Protano G and Riccobono F 2002 High contents of rare earth elements (REEs) in stream waters of a Cu-Pb-Zn mining area *Environmental Pollution* **117** 499–514
[5] Smirnova O K, Plyuskin A M and Khazheeva Z I 2013 Sovremennoe mineraloobrazovanie v mestakh skladirovaniya otkhodov gornorudnoy promyshlennosti *Otechestvennaya geologiya* **3** 104–111 [in Russian]

IOP Conf. Series: Earth and Environmental Science **33** (2016) 012026 doi:10.1088/1755-1315/33/1/012026

[6] Zamana L V 2013 Geochemistry of acidic drainage water of gold ore field of the East Trans-Baikal *Water: chemistry and ecology* **8** 92–97

[7] Vakh A S, Moiseenko V G, Stepanov V A and et al. 2009 The Berezitovy Gold–Base Metal Deposit *Doklady Akademii Nauk* Vol. 425 **2** 204–207

[8] Vakh A S, Avchenko O V, Kiselev V I, Sergeev S A and Presnyakov S L 2013 U-Pb isotopic geochronologic study of zircons from granites and ore-bearing metasomatites of the Berezitovoe gold-polymetallic deposit (Upper Priamurye, Russia) *Russian Journal of Pacific Geology* Vol. 32 **6** 20–39

[9] Medvedeva L A and Nikulina T V 2014 *Katalog presnovodnykh vodorosley yuga Dal'nego Vostoka Rossii* (Vladivostok: Dal'nauka) p 271 [in Russian]

[10] Moshkova N A and Gollerbakh M M 1986 *Zelenye vodorosli. Klass Ulotriksovye. Poryadok Ulotriksovye. Opredelitel' presnovodnykh vodorosley SSSR* (Moscow-Leningrad: Nauka) **10** Part 1 p 360 [in Russian]

[11] Liu G, Zhang Q, Zhu H and Hu Z 2012 Massive Trentepohlia-Bloom in a Glacier Valley of Mt. Gongga, China, and a New Variety of *Trentepohlia* (Chlorophyta) // PLoS ONE. 2012. V.7. Is.7. e37725

[12] Khristoforova N K and Malinovakaya T M 1995 Soderzhanie metallov v fukusakh bukhty Kraternoy (Kuril'skie ostrova) v svyazi s usloviyami sushchestvovaniya *Russian Journal of Marine Biology* **21** 77–82

Public Health Risk Conditioned by Chemical Composition of Ground Water

E Yankovich[1], N Osipova[2], K Yankovich [3]and I Matveenko[4]

[1, 2, 3, 4] Tomsk Polytechnic University, 30 Lenin Avenue, Tomsk, 634050, Russia

E-mail:[1] yankovich@tpu.ru, [2] osipova@tpu.ru

Abstract. The article studies the public health potential risk originated from water consumption and estimated on the basis of the groundwater chemical composition. We have processed the results of chemical groundwater analysis in different aquifers of Tomsk district (Tomsk Oblast, Russia). More than 8400 samples of chemical groundwater analyses were taken during long-term observation period. Human health risk assessment of exposure to contaminants in drinking water was performed in accordance with the risk assessment guidance for public health concerning chemical pollution of the environment (Russian reference number: P 2.1.10.1920-04-M, 2004). Identified potential risks were estimated for consuming water of each aquifer. The comparative analysis of water quality of different aquifers was performed on the basis of the risk coefficient of the total non-carcinogenic effects. The non-carcinogenic risk for the health of the Tomsk district population due to groundwater consumption without prior sanitary treatment was admitted acceptable. A rather similar picture is observed for all aquifers, although deeper aquifers show lower hazard coefficients.

1. Introduction

In the world the issue of safe drinking water supply is of specific relevance. Water is one of the key environmental components destined to permanent contact with human body. The major part of different substances, both necessary for life and toxic ones enters human body with drinking water. Patterns of human exposure to the factors contaminating the environment, particularly at low rates, are still poorly studied. They can be both direct and indirect. Method of risk assessment allows estimating the real dose hazard for humans taking into account the exposure factor (exposure duration, human age, and dose) [1], [2], [3].

The current study presents the comparative analysis of aquifer water quality and assessment of its potability using the risk assessment method.

The research is performed for Tomsk district (Tomsk Oblast). The area includes 128 settlements. The district is suburban, its territory comprises the administrative center with large industrial enterprises that are responsible for high anthropogenic impact on the environment including underground hydrosphere. The settlements are supplied with water both from central system with preliminary water treatment and individual wells (approximately 19.7% of the district population have individual water sources) [4], [5]. Over the study area four aquifers are distinguished – Neogene-Quaternary, Paleogene, Cretaceous, Paleozoic, the waters of which are used by population to greater or lesser extent for drinking and living needs. According to the result of long-term observation of chemical content in ground water, it is observed that a number of substances are found in increased concentrations, often exceeding the maximum allowable concentration, which is explained by some subjective and objective factors.

2. Materials and methods

The authors used the data of TOMSKGEOMNITORING Company including the results of ground water analysis sampled from observation wells tapping waters of different stratigraphic units located in the territory of Tomsk district. Stationary monitoring of ground water dynamics of Mesozoic and Cenozoic deposits and Paleozoic formations has been performed since 1962 in the study area.

Taking into account the distribution regularities, the average values of elemental concentration were obtained for water of each unit. In total, more than 8400 chemical analyses of ground water were used [6]. Water analysis was carried out using the standard certified techniques.

The risk assessment of non-carcinogen effect was performed in accordance with the human risk assessment guideline under the exposure of chemicals in the environment – R. 2.1.10. – 1920–04 [7], [8].

The LADD (Life-time Average Daily Dose), mg/kg per day of chemicals consumed through water ingestion was calculated by Eq. (1), hence, we used standard exposure factors

$$LADD = \frac{C \times V \times ED \times EF}{BW \times AT \times 365} \quad (1)$$

where, C - C_{water}, mg/L, means concentration of chemicals in water. The data were averaged for the long-term monitoring.

V – daily ingestion intake of water to human body, to be assumed 2 L/day for adults;

ED – exposure duration, 30 years; EF – exposure frequency – 350 days/year;

BW – body weight (for adults), 70 kg; AT – average time – 30 years; 365 – number of days in a year.

To estimate the chronic health risks, HRIs were calculated by Eq. (2)

$$HRI_{per} = LADD \,/\, RfD_{per} \quad (2)$$

where, RfD_{per} - the oral toxicity reference dose for each element or compound [7], [8]

$$HRI = \sum HRI_{per} \text{ for each element or compound} \quad (3)$$

The HRI value less than one is considered to be safe for the consumers.

3. Results and discussion

Four aquifer systems were distinguished, the waters of which were used by population for their domestic needs to different extent. Table 1 lists the data on average chemical composition of ground water.

Table 1. Average chemical composition of ground water.

Substances	Units	Average value in ground water			
		Aquifers			
		Neogene-Quaternary	Paleogene	Cretaceous	Paleozoic
The number of analyses		3450	3707	508	738
pH		7.4	7.4	7.3	7.6
HCO_3	mg/L	268.1	313.9	247.5	381.6
SO_4	mg/L	5.67	5.0	4.2	9.3
Cl	mg/L	4.8	3.0	17.5	6.1
NO_2	mg/L	0.06	0.046	0.043	0.049
NO_3	mg/L	0.79	0.7	0.4	0.7
NH_4	mg/L	0.82	0.8	0.8	0.6

Ca	mg/L	61.7	72.3	42.7	90.1
Mg	mg/L	12.5	15.7	16.3	20.5
Na	mg/L	7.3	9.5	39.5	13.4
K	mg/L	1.2	1.4	3.9	1.3
Si	mg/L	7.7	9.2	10.8	9.7
Σ of salt	mg/L	414.4	485.1	469.1	593.8
Permanganate oxidation	mgO_2/L	2.7	2.4	1.9	1.9
F	mg/L	0.22	0.20	0.22	0.21
Al	mg/L	0.2	0.1	0.1	0.1
Fe	mg/L	6.5	4.0	4.8	2.3
Mn	mg/L	0.34	0.2	0.1	0.2
Co	mg/L	0.0019	0.0010	0.0005	0.0019
Ni	mg/L	0.011	0.0083	0,0116	0.0072
Sr	mg/L	0.34	0.4	0.5	0.5
Mo	mg/L	0.0115	0.0044	0.0191	0.0020
Zn	mg/L	0.042	0.0192	0.0115	0.0111
Cu	mg/L	0.027	0.0117	0.0046	0.0055
Hg	mg/L	0.0004	0.0002	0.0005	0.0002

Water of Neogene-Quaternary deposits is fresh, mainly hydrocarbonate calcium, sometimes magnesium-calcium or of blend cation composition. The total dissolved solids in ground water range widely from 136 to 1191 mg/L that can be explained by both natural conditions and anthropogenic effect. Concentrations of sulphate, chloride, sodium, and potassium do not exceed 10 mg/L on the average. The water contains significant amount of silicic acid (up to 39 mg/L). Ammonium ion content in water amounts 0.82 mg/L on average, changing in the range from 0.2 to 3.7 mg/L.

A typical feature of the given water is high contents of iron and manganese, concentrations of which ranges widely (from 0.8 to 29 and from 0.02 to 13 mg/L, respectively). Within the boundaries of Tomsk and Seversk as well as adjacent areas, the deposit waters have changed their chemical composition. They are often chloride-hydrocarbonate, sodium-calcium, very hard, weakly alkaline and alkaline waters with the mineralization value from 0.3 to 1.0 mg/L or more. It is waters of this aquifer that are used by the population as a decentralized water supply.

Water of Paleogene deposits is fresh, with the mineralization value 250-650 mg/L, water medium is neutral and weakly alkaline. By its chemical composition, water is hydrocarbonate, magnesium-calcium, often of blend cation composition. There are ammonium ions everywhere in waters of Paleogene deposits. Their concentrations amount 0.1 – 2.8 mg/L. As is the case with the overlying aquifer system, high concentrations of iron and manganese are typical for waters of Paleogene deposits. Concentration of iron in ground waters of the given deposits ranges widely (from 0.1 to 15 mg/L). Concentration of manganese amounts from 0.02 to 0.6 mg/L. Waters of Paleogene deposits are used for centralized water supply.

Ground water of Cretaceous deposits is generally fresh with the mineralization value from 160 to 810 mg/L, but in some sites there is brackish water with mineralization reaching 1450-4087 mg/L. By its chemical composition waters of Cretaceous deposit are diverse. In the south of the area they are often hydrocarbonate or chloride-hydrocarbonate, calcium or sodium-calcium neutral and weakly alkaline. With the increase in mineralization of Cretaceous deposit waters, there is an increase in chloride and sodium ion concentration in them, anion water composition has become hydrocarbonate-chloride or chloride, cation composition – calcium-sodium or sodium. Low amounts of sulfates (0-14.6 mg/L) and nitrates (1-2 mg/L) are established in the waters of Cretaceous deposit. Ammonium concentrations are also low and change from 0.1 to 2.1 mg/L, but in the brackish water area the ammonium concentration is generally higher and sometimes reaches 2.5-3.3 mg/L. Cretaceous deposit waters are characterized by significant amount of iron and manganese concentrations.

Water of Paleozoic deposit is fresh (mineralization from 200 to 700 mg/L), moderately hard and hard, neutral and weakly alkaline in the most part of the study area. It is referred to hydrocarbonate type in terms of chemical composition, mostly calcium and magnesium-calcium. Sulfate concentration in waters of Paleozoic deposit often exceeds 10 mg/L. As is the case of overlying aquifer system, water of Paleozoic deposit is saturated with total iron, the average concentration being 2.3 mg/L.

The share of this water in domestic water supply is not great; it is mostly used in the south-eastern part of Tomsk district.

In accordance with the methods of human risk assessment using equations 1-3, the quantitative risk indicators are defined: average daily intake and the chronic health risks, HRI as an indicator of toxic effect under the exposure of chemical elements when consuming ground water without preliminary water treatment (Table 2).

Table 2. The life-time average daily dose of element (LADD) in human organism with ground water consumption and the chronic health risks, HRI.

Element	**Aquifer system**	**RfD per**	**LADD**	**HRI**	**Aquifer system**	**RfDper**	**LADD**	**HRI**
		mg/kg×day				mg/kg×day		
Nitrites	Neogene-Quaternary	0.1	0.002	0.02	Paleogene	0.1	0.001	0.01
Nitrates		1.6	0.02	0.01		1.6	0.019	0.01
Calcium		41.4	1.7	0.04		41.4	2.0	0.05
Magnesium		11.0	0.34	0.03		11.0	0.43	0.04
Fluoride		0.06	0.006	0.1		0.06	0.005	0.09
Aluminum		1.0	0.005	0.005		1.0	0.003	0.002
Iron		0.3	0.18	0.59		0.3	0.1	0.37
Manganese		0.14	0.009	0.07		0.14	0.005	0.04
Cobalt		0.02	0.00005	0.003		0.02	0.00003	0.001
Nickel		0.02	0.0003	0.02		0.02	0.0002	0.01
Strontium		0.6	0.009	0.02		0.6	0.01	0.02
Molybdenum		0.005	0.0003	0.06		0.005	0.0001	0.02
Zinc		0.3	0.001	0.004		0.3	0.0005	0.002
Copper		0.02	0.0007	0.04		0.02	0.0003	0.02
Mercury		0.0003	0.00001	0.04		0.0003	0.000005	0.02
Nitrites	Cretaceous	0.1	0.001	0.01	Paleozoic	0.1	0.001	0.01
Nitrates		1.6	0.01	0.007		1.6	0.02	0.01
Calcium		41.4	1.2	0.03		41.4	2.5	0.06
Magnesium		11.0	0.45	0.04		11.0	0.56	0.05
Fluoride		0.06	0.006	0.1		0.06	0.006	0.09
Aluminum		1.0	0.0027	0.003		1.0	0.003	0.003
Iron		0.3	0.13	0.44		0.3	0.06	0.21
Manganese		0.14	0.0027	0.02		0.14	0.005	0.04
Cobalt		0.02	0.00001	0.0007		0.02	0.00005	0.003
Nickel		0.02	0.0003	0.02		0.02	0.0002	0.01
Strontium		0.6	0.014	0.02		0.6	0.014	0.02
Molybdenum		0.005	0.0005	0.1		0.005	0.00005	0.01
Zinc		0.3	0.0003	0.001		0.3	0.0003	0.001

Copper	0.02	0.0001	0.007	0.02	0.0002	0.008
Mercury	0.0003	0.00001	0.045	0.0003	0.000005	0.018

Notes: RfD per - the oral toxicity reference dose, mg/kg×day [7], [8].
LADD – life-time average daily dose, mg/(kg × day);
HRI - the chronic health risks (risk coefficient).
The priority substances are distinguished.

The priority substances of all aquifers are total iron and fluoride with the risk coefficient for human health being more than 0.05. The priority substance for Neogene-Quaternary deposit water is manganese; for Neogene-Quaternary and Cretaceous deposit water – molybdenum; for Paleozoic deposit water and Paleogene deposit water – calcium and magnesium.

The total risk of non-carcinogen effect with regular water consumption from Neogene-Quaternary aquifer system amounts 1.03; Paleogene – 0.70; Cretaceous – 0.84; Paleozoic – 0.55. The share of different substances in the total value of risk coefficient with regular intake of water to human organism is shown in Figure. The similar pattern is observed in all aquifers. Groundwater from deeper aquifers is characterized by less risk coefficients.

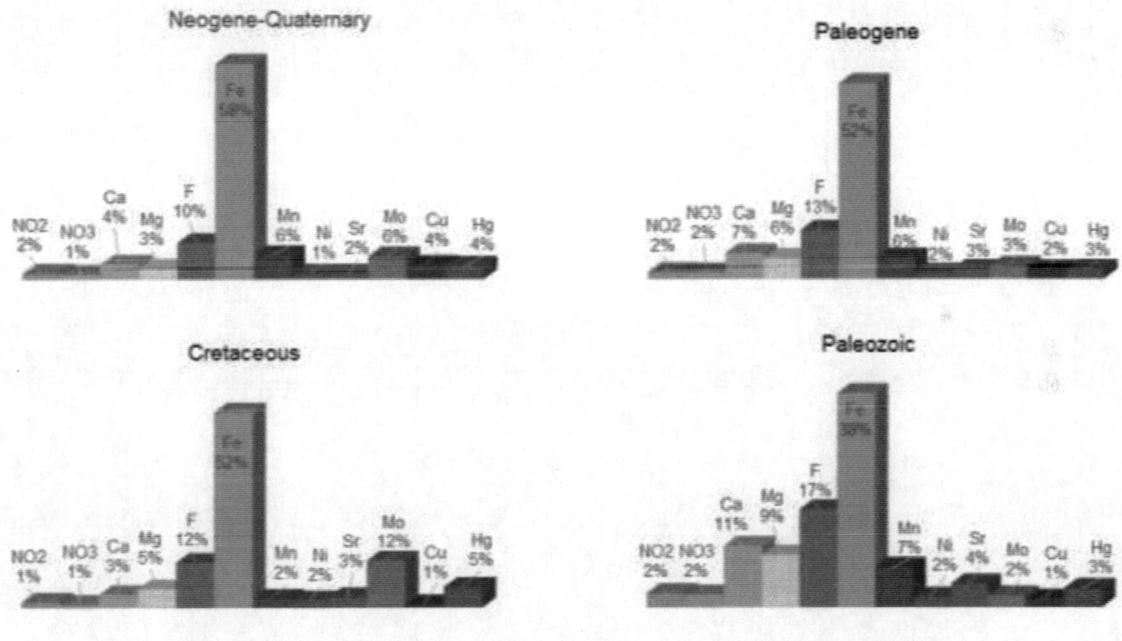

Figure. The share of different substances in total risk coefficient.

4. Conclusion

The non-carcinogen risk for human health in Tomsk region conditioned by ground water consumption from different aquifer systems without preliminary treatment is admissible. However, to produce good drinking water, it is necessary to treat it for removal of elevated chemical concentrations including excess of iron and manganese.

References

[1] Yabel A, Fabro R and Guadalupe J 2015 Spatial distribution of nitrate health risk associated with groundwater use as drinking water in Merida, Mexico *Appl. Geog.* **65** 49–57
[2] Khan K, Lu Y and Khan H 2013 Health risks associated with heavy metals in the drinking water of Swat, northern Pakistan *J. Environ. Sci.* **25**(10) 2003–13
[3] *Environment and health: approaches to risk assessment* Ed by A P Shcherbo 2002 (Sant-

IOP Conf. Series: Earth and Environmental Science **33** (2016) 012027 doi:10.1088/1755-1315/33/1/012027

Petersburg: MAPO) p 370

[4] *Report on conditions and protection of the environment in Tomsk Oblast in 2012* 2013 Ed by A M Adam (Tomsk: Deltaplan) p 172

[5] Lukashevich O D 2007 Ecological risk in using sources of decentralized drinking water supply. *Bezopasnost' zhiznedeyatel'nosti* **3** 15-21

[6] Yankovich E P, Yankovich K S, L'gotin V A and Gulmina G A 2015 Ecogeochemical Characteristics of Underground Water of Ob-Tom Interfluve (RUSSIA) *In: 15-th Int. Multidiscip. Sci. GeoConference SGEM 2015* (Albena, Bulgaria, June 18-24 2015) Book1 **2** 269-276

[7] Rukovodstvo R 2.1.10.1920-04 2004 *Guideline for human health risk assessment under the exposure of chemical substances polluting the environment* (Moscow: Federal Center of Gossanepidnadzor Minzdrav RF) p 273

[8] U.S. Environmental Protection Agency 1998 *Guidelines for ecological assessment EPA/630/R-95/002F* (Washington, DC)

Impact of petroleum products on soil composition and physical-chemical properties

N N Brakorenko[1] **and T V Korotchenko**[2]

[1,2] Institute of Natural Resources, National Research Tomsk Polytechnic University, 30 Lenin Av., Tomsk, 634050, Russia

E-mail: [1]brakorenko@sibmail.com

Abstract. The article describes the grain-size distribution, physical and mechanical properties, swelling and specific electrical resistivity of soils before and after the contact with petroleum products. The changes in mechanical properties of soils contaminated with petroleum products have been stated. It leads to the increase in compressibility values, decline in internal friction angle and cohesion.

1. Introduction

Today, the study of physical-chemical properties of soils within Tomsk region is of particular scientific interest [1-3]. However, the number of the research focused on the analysis of petroleum product (PP) impact on soil composition and properties is not great. This issue has been recently addressed both in Russia and abroad [4-9]. In addition, there are no data on the impact of petroleum products on physical-chemical properties of Tomsk region soils.

Taking into account all the above, the current study that examines the impact of petroleum products on composition and physical-chemical properties of soils in Tomsk region is rather relevant and of practical and scientific value.

2. Materials and methods

The impact of petroleum products on rock properties is basically studied by two methods. According to the first method, the properties are examined by adding oil, then water into the dry soil. Within the second method, dry soil is watered and after this oil is added. As the result, different data can be obtained [10].

For laboratory experiments, natural soil types characterized by standard properties were used. First of all, physical-chemical properties of the soil samples were analyzed under natural conditions. Then, the same soil samples were saturated with benzene 95 RON for different periods: 2 days, 2 weeks, 1 month, and 6 months. The amount was 500 mg/kg.

3. Results and discussion

The long-term effect of petroleum products on soil leads to the changes in soil composition, precisely, it increases the content of fine particulate matter (Table 1).

The increase in fine particulate matter content is explained by the fact that the presence of organic matter, i.e. petroleum products, in soils contributes to developing reducing environment. Under these conditions, the cementing bonds are destroyed due to reduction of ferric iron compounds in clay soils

($Fe^{3+} \rightarrow Fe^{2+}$). The presence of Fe^{2+} in soils stipulates destruction of aggregates and increases the content of fine-grained fractions, which, in turn, leads to water repellency of clay deposits [11].

Table 1. Soil grain-size distribution (based on the experiment data).

№ well	Sampling depth, m	Duration of PP-soil contact	State of soil	Grain-size distribution, % fraction size, mm.									
				1-0.5	0.5-0.25	0.25-0.1	0.1-0.05	sand amount	0.05-0.01	0.01-0.005	dust amount	<0.005	clay amount
726	10.0	1 months	natural	0.03	0.03	2.00	22.93	24.99	39.62	19.01	58.63	16.38	16.38
726	10.0	1 months	Contaminated with PP	0.03	0.03	2.00	22.90	24.96	35.5	17.23	52.73	22.31	22.31
119	4.1	6 months	natural			0.37	5.92	6.29	50.30	8.53	68.83	24.88	24.88
119	4.1	6 months	Contaminated with PP			0.23	3.63	3.86	45.13	14.32	59.45	36.69	36.69

Unlike sandy loam, loam is characterized by more significant changes in composition. It can be explained by the fact that sandy loam generally contains more dust fractions (aggregates of clay fraction) which can be easily broken when coming into contact with petroleum products.

Table 2 presents the basic physical properties of soils both in its natural state and after contact with petroleum products (density, moisture content, and porosity coefficient). Thus, loam porosity coefficient is found to increase with the increase in the level of petroleum product contamination, while porosity coefficient of sandy loam changes insignificantly. It also can be explained by the fact that loam contains more dust fractions (aggregates of clay fraction) that can be broken when coming into contact with petroleum products. As is known, unlike sand soils, clay soils are characterized by higher porosity coefficient [11].

The density of soil alters insignificantly when contaminated with petroleum products (Table 2).

Table 2. Physical properties of soils.

Type of soil	State of soil	Physical properties		
		Density, g/cm³	Porosity coefficient, units	Water content, %
Loam	natural	1.93-1.97; 1.98-2.0	0.63-0.80	20-24; 27-31
	contaminated with PP	1.9-2.01; 2.04-2.1	0.65-0.9	18-26; 29-31
Sandy loam	natural	1.9-1.96; 2.0-2.04	0.41-0.60	6-8; 15-18
	contaminated with PP	1.9-2.0; 2.04-2.1	0.46-0.58	8-13; 17-25

Note: the prevailing values of soil physical properties depend on soil consistency. Two ranges of density and water contact values are given for soils of different consistency.

Mechanical properties of soils are the most important parameters in soil geo-engineering estimates. Therefore, the current study is focused on the analysis of petroleum product impact on the following properties: deformation modulus under load 0.25-0.3 MPa, soil settlement under load ranging from 0.05 to 0.3 MPa, and internal friction angle and cohesion.

As shown in Table 3, soils contaminated with petroleum products demonstrate decline in specific cohesion (loam – by 12.5 %, sandy loam – by 30 %) and internal friction angle (loam – by 10 %, sandy loam – by 20 %). Such changes are stipulated by a number of facts: 1) changes in physical properties of contaminated soils (consistency), as, when contaminated, soils are additionally moistened; 2) adsorption of petroleum products on the particle mineral surface, which, in turn, leads to formation of so-called "lubricant" covering soil particles.

Table 3. Changes in soil mechanical properties induced by petroleum products.

Soil in accordance with GOST 25100-2011	Specific cohesion of soil in natural state, kPa, min–max / average	Specific cohesion of soil contaminated with PP, kPa min–max / average	Internal friction angle of soil in natural state, degree min–max / average	Internal friction angle of soil contaminated with PP, degree min–max / average	Content of PP, mg/kg
Stiff loam (8)	23.0 – 31.7 / 27.2	22.0 – 25.0 / 23.8	17.0 – 26.0 / 22.6	17.0 – 24.0 / 20.1	500
Solid loam* (6)	14.0 – 20.0 / 18.0	10.0 – 15.0 / 13.0	23.0 – 31.0 / 29.5	22.0 –26.0 / 23.5	400 - 1700
Sandy clay (15)	14 – 20 / 17.6	8 – 16 / 11.86	23 – 33 / 29.8	22 – 26 / 23.8	500

* data obtained by Tomsk geological engineering companies when conducting geo-studies within the petrol stations and compared with the results of the similar studies carried out within the territories unaffected by petroleum products; the number of studied objects is given in brackets.

The analysis of petroleum product impact on soil compressive strength has revealed the following regularity:

- soil (high-plastic loam) with water content up to 22% compresses two times less in natural state (soil settlement varies from 38 mm/m for uncontaminated soils up to 59 mm/m for soils contaminated with petroleum products, with load being 0.3 MPa) in comparison with the case when it is contaminated with petroleum products (Table 4);
- soil (high-plastic loam) with water content up to 25.4% is less compressed than natural one under insignificant loads; compressibility increases at loads from 0.2 MPa (Table 4);
- after the contact with gasoline during 2 weeks and under the load 0.4 MPa, compressibility of soil with water content up to 37% (very soft loam) is significantly lower than that of the soil in natural state (soil settlement 80.4 mm/m, load – 0.3 MPa) (Table 4).

Thus, the compressibility of contaminated soils directly depends on the amount of petroleum products, which, in its turn, is defined by water content. In general, contaminated soils are characterized by decreased deformation modulus (compressibility increase).

Table 4. Settlement of soils with different initial water content.

Vertical pressure, MPa	Soil settlement in natural state, mm/m			Settlement of soils contaminated with PP, mm/m		
	Water content, 22%	*Water content 25,4%*	*Water content, 37 %*	*Water content, 22%*	*Water content, 25,4%*	*Water content, 37 %*
0.05	16	31.6	18.4	23.2	19.2	20.4
0.10	21.6	43.2	34.4	29.2	38.8	33.2
0.15	26.8	50.8	43.2	38.8	50	37.6
0.20	30	56	51.6	42	61.6	42
0.30	34	65.6	68.4	48.4	76	48.8
0.40	38	71.6	80.4	59.2	91.6	56.4

It has been revealed that swelling of soils in water significantly exceeds soils swell in gasoline (Figure 1). It is due to the fact that petroleum product films can easily envelope the particles and aggregates of soil clay particles leading to soil water repellency.

As is known, waters trapped in soil pores serve as natural electrolyte characterized by different composition and ionic conductivity. Depending on chemical composition and concentration, specific resistivity of natural waters varies over a wide range (from 10^{-2} to 10^3 ohm • m and more) and reduces with increasing humidity and salinity. The soils contaminated with petroleum products demonstrate the decrease in specific resistivity (Figure 2), which can be explained by alteration of pore water salinity due to the contamination with petroleum products. Therefore, there is a need to characterize macro- and microcomponent composition of ground waters contaminated with petroleum products for better assessment of environmental impact associated with petroleum products.

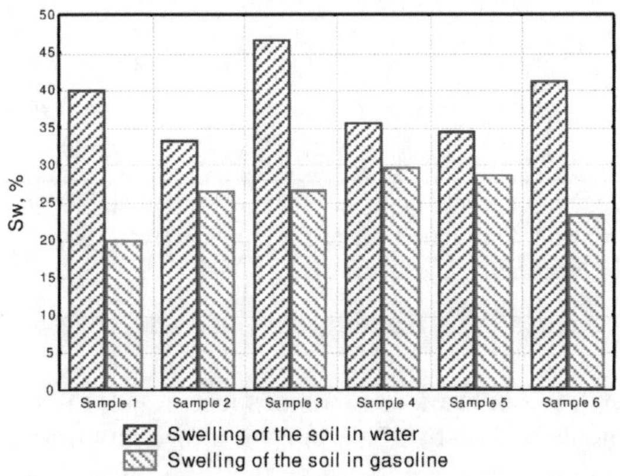

Figure 1. Free soil swell.

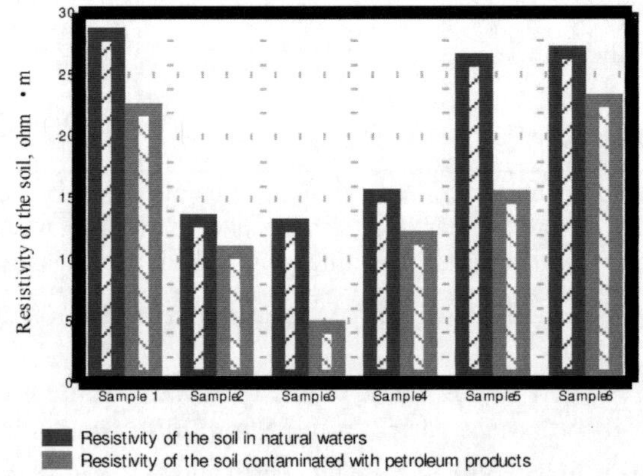

Figure 2. Specific electrical resistivity of soil.

4. Conclusion

On the basis of the above, it is possible to come to the following conclusions:

1) depending on duration and iron content in soil, interaction of soils with petroleum products stipulates alteration of grain-size distribution of sandy-clayey soils, precisely, it increases the content of fine particulate matter;

2) depending on initial water content, contamination of soils with petroleum products leads to the changes in soil compressibility. Excessive water content and changes in soil composition (soil plasticity increase) can contribute to soil settlement near the building basements;

3) contamination of clay soils with petroleum products leads to the decrease in internal friction angle and cohesion.

All the revealed changes basically have adverse effects on soil stability and corresponding buildings and facilities.

References

[1] Kramarenko V V, Nikitenkov A N, Molokov V Y and Matveenko I A 2015 The Relationship of Physical Property Indicators and Clay Soil Structural Strength of Tomsk Oblast Territory *IOP Conf. Ser.: Earth . Environ. Sci.* **27**

[2] Pokrovsky V D, Dutova E M, Kuzevanov K I, Pokrovsky D S and Nalivaiko N G 2015 Hydrogeological Conditions Changes of Tomsk, Russia *IOP Conf. Ser.: Earth Environ. Sci.* **27**

[3] Strokova L A and Yemelyanov T V 2006 Geotechnical criteria for assessing the sustainability of the regional geological environment to technogenic impacts (for example, the Tomsk Ob River) *Obskoy Vestnik* **4** 349-359

[4] Seredin V V 2000 Rehabilitation of territories polluted with oil and petroleum products *Geoekologiya* **6** 525-540

[5] Zhuravlev I I, Motenko R G, Ershov E D 2005 Formation thermal properties of frozen dispersed soils at pollution by oil and petroleum products *Geoecology. Engineering geology. Hydrogeology. Geocryology* **1** 50-60

[6] Kazakov A I, Arbuzov Y V, Kovalevsky Y V 1998 Impact of petroleum product facilities on the geological environment *Geoecology* **1** 54-74
[7] *Studies on the Effects of Certain Soil Properties on the Biodegradation of Oils Determined by the Manometric Respirometric Method Juhani Kaakinen, Pekka Vähäoja, Toivo Kuokkanen, Katri Roppola* URL: http://www.sciencedirect.com
[8] *Effects of crude oil contamination on geotechnical properties of clayey and sandy soils Mashalah Khamehchiyan Amir Hossein Charkhabi and Majid Tajik* URL: http://www.sciencedirect.com
[9] Effect of hydrocarbon pollution on the microbial properties of a sandy and a clay soil *Valeria Labud, Carlos Garcia and Teresa Hernandez* URL http://www.sciencedirect.com
[10] Shevchenko L V and Shirshova I A 2008 Mechanical properties of frozen clay soil contaminated with oil *Geoecology. Engineering geology. Hydrogeology. Geocryology* **1** 78-84
[11] Sergeeva E M 1982 *Soil* (Moscow: Moscow State University) p 392
[12] Trofimov V T, Korolev V A, Voznesensky E A et al 2005 *Soil: a textbook for high schools* (Moscow: Moscow State University) p 1023

Contemporary approaches to studying and mapping of active water exchange zone of ground water

C Ye Moraru

Institute of Geology and Seismology of Academy of Science, Kishinev, Moldova

E-mail: cmoraru@yahoo.com

Abstract. The article deals with a zone of ground water active exchange. New principles of the zone study and mapping under the platform hydrogeological condition are discussed. The assessment and distribution techniques are suggested for the active water exchange zone under the condition of hydrogeological parameterization uncertainty. The efficiency and significance of the suggested techniques are proved using the example of ground water in the southwest of Black Sea artesian basin.

1. Introduction

Vertical hydrodynamic zonality is referred to the basic regularities of ground water distribution in the upper (water-saturated) part of lithosphere. The first who wrote about this regularity was V.I. Vernadsky (1933). B.L. Lichkov was the first who described hydrogeodynamic zonality (1933). The three hydrogeodynamic zones were distinguished by F.A. Makarenko (1939).

N.K. Ignatovich [2] proved the concept of three hydrogeological zones in practice: active water exchange, impeded circulation (migration), and non-moving water. According to Ignatovich (1944), active water exchange zone is an upper part of hydrogeological section including ground water and upper level of artesian water involved in water exchange with atmosphere and river inflow.

Almost simultaneously with Ignatovich's research [2], the works by K.V. Filatov (1947), Z.A. Makeev (1948), and M.A. Gatal'skiy (1954) were published. The works by Zavarzin L.G. (1969), Yezhov Yu.A., Vdovin Yu.P. (1970), Zaitsev I.K. (1979), Valukonis G.Yu., Khod'kova A.Ye. (1968,1973) analysed the theoretical and practical issues of hydrogeodynamic zonality up to 1973. Ye.V. Pinneker (1983) also paid much attention to this question.

In 1978 S.L. Shvartsev [7] was the first who generalized the data on hydrochemistry of hypergenesis zone. It was stated that there are few works in the sphere of hydrogeochemistry, whereas those about intensive (active) exchange water zone are virtually absent. Shvartsev's ideas were developed and specified by many researchers. Among them one can mention an extensive work by Ye.M. Dutova on hydrogeochemistry of hypergenesis zone in the Altay-Sayan folded belt [1].

The foreign concept of low temperature groundwater geochemistry [8-14] is similar to that of active water exchange zone.

The most practically applicable and well-grounded classification is Ignatovich's one [2] (or Ignatovich – Lichkov – Makarenko's). The contemporary ideas of active water exchange zone are not clear enough (Kurennoy V.V., 2010). Vsevolozhsky V.A. (2007) noted that delineation methods for this zone were far from being perfect. The number of works on active water exchange is few, though ground water of this zone is of great significance (Shvartsev S.L [7]).

The author of the article suggests a set of techniques to delineate and map the active water exchange zone of platform ground water.

2. Materials and methods

The research is based on long-term results of complex hydrogeological investigation in the southwest of the Black Sea artesian basin including the territory of Moldova, south of Ukraine, and Subcarpathia part of Rumania. To collect the data, the field expedition was organized to sample ground water and rocks, to measure hydrogeological, hydrogeochemical parameters, and in some cases – to obtain ecological, meteorological, engineering and other information.

The analysis includes laboratory and field (express) identification of chemical elements and physic-chemical parameters. In the laboratory the macro- and microelements (atomic spectroscopy laboratory of Chemistry Institute of Moldova Academy of Science), helium (device INGEM-1, operator is the author of the article) are determined. The data on tritium content in natural water and water temperature of deep wells are taken from the archive of the Hydrogeology Laboratory of Geology and Seismology Institute (Moldova Academy of Science).

The number of author's chemical analyses (element determination) is about 2500 for ground water, more than 3000 helium samples are determined in water and natural gases. Besides, geologic-hydrogeological data are collected and generalized for more than 1000 wells (stratigraphy, water level, flow rate etc.).

Data processing is performed using the modern quantitative methods of geologic information processing, techniques of physical-chemical modeling and mapping. The following software Microsoft 2013, SPSS10-14, AquaChem 4.0, Surfer 7.0-8.0-9.0-11.0, ModFlow (PMWin 4.1, Visual ModFlow), RockWorks 14.0, CorelDraw 12.0 and GWW, as well as GIS software are used.

3. Results and discussion

At the moment there is no common method of hydrogeodynamic zone delineation, zone of active water exchange, in particular. The following set of techniques is suggested to be applied collectively [5].

A) Hydrogeochemical method. Mineralization value is chosen as a form of generalized factor of ground water hydrogeochemical condition. At present, there are a number of ground water classifications in terms of mineralization value. It has been suggested to use Vernadsky's classification (1933 – 1936) with changes by V.A. Priklonsky (1933, 1949), according to which water is divided into 4 groups: fresh – with mineralization value of up to 1.0 g/L; brackish - 1.0 – 10.0 g/L (1.0 - 3.0 g/L and 3.0 - 10.0 g/L according to Priklonsky); saline - 10.0 – 50.0 g/L; brine - > 50.0 g/L. The active water exchange zone contains mostly fresh water with mineralization up to 1.0 g/L. The data on ground water composition (water type, element composition etc.) are also used in this method.

B) Geothermal method. Ground water is involved in formation of physical-chemical equilibrium systems with the environment, the basis of which is presented in pairs: "water–rock", "water–gas", "water–dissolved substance" and "water–living world" (Vernadsky V.I., 1933; Samarina V.S., 1977). Water as a chemical substance increases or decreases functionally its participation in the chemical reactions depending on its temperature (or system) in the equilibrium systems.

Water-bearing rock solution or formation of new minerals occurs mostly in "water - rock" system. The data analysis on solid solubility in water at different temperatures shows that solubility in water occurs in low quality at the ambient temperature of about 20°C, but at temperature of more than 20°C the amount of dissolved substances sharply increases [3]. The analysis of natural gas solubility values (N, H_2, O_2, CH_4, H_2, CO_2, and C_2H_6) in water at the temperature range from 0°C to 100°C reveals that the maximum decline in natural gas solubility in water is observed at the temperature range from 0.0 to 20°C [6]. Hence, the marginal temperature for basic equilibrium systems is 20°C. This temperature value is suggested to be used for low boundary delineation of the active water exchange zone.

B) Tritium in ground water. Tritium distribution in ground water is known to be a direct indicator of its relationship with precipitation and surface water (Zelenin I.V.,1974; Dubinchuk V.T.,1979;

Morkovkina I.K., Romanov V.V., 1975). Tritium falls out with rain water, flows into surface and ground waters. There are different patterns of tritium migration in ground water (Zelenin I.V., 1974). We take a simple and effective assumption: presence of tritium in ground water reveals its connection with the Earth surface and vice versa – its absence is an indicator of the absence of such a connection. The low boundary of active water exchange zone can be determined by zero (tritium absence) or close to zero tritium content.

C) Helium in ground water. Helium distribution in ground water is regulated by the following factors: tectonic fault location, background concentration of the element within the aquifer, crystalline foundation location and downward infiltration vertically to rainfalls in ground water. The latter is closely related to the development of active water exchange zone. Helium concentration deviation from the background concentration towards its decrease clearly shows mixing of aquifer water with water of overlying aquifers or meteoric water that, as a rule, contains low helium concentration or no helium at all. Helium field contrast ratio best demonstrates the given phenomenon [4]. Helium field contrast can express the spatial location of active water exchange zone.

D) Hydrogeological stratigraphy. In the platform condition, aquifers, as a rule, correspond to the definite stratigraphic units. Structurally, aquifers are separated by impermeable (or relatively impermeable) strata. At low bedding angle of aquifer (or closer to horizontal location) similar hydrogeochemical properties are typical for large territories. Memphian layers, Mississippian depression (the USA) can serve as examples [10].

Stretching and location of active water exchange zone is structurally connected with and confined to the definite aquifers or their parts. Therefore, hydrogeological stratigraphy characterizes indirectly the spatial location of the active water exchange zone.

Ground water that is involved into the first from the surface aquifers and has complex hydrogeodynamic pattern is, by its definition, a part of active water exchange zone.

The suggested set of techniques for assessment and location of active water exchange zone is used for ground water of the southwest Black Sea artesian basin. Tectonically, in terms of folded basement age, the whole study area is divided into the area of Precambrian folded basement (margin of the East-European Platform) and the area of Hercynian folded basement (Scythian Plate). Hydrogeologically, the study area is southwest part of the Black Sea artesian basin of the first order.

As a result of the suggested technique, the geometrical parameters of the active water exchange zone in the southwest of the Black Sea artesian basin (by the example of Moldova territory) are characterized by the following features: a) the thickness of active water exchange zone ranges unevenly within 10.0 – 550.0 m from the Earth surface. The active water exchange zone includes: in the north – Quaternary, Neogene, and Cretaceous-Silurian aquifers; in the center – Quaternary and Neogene aquifers; in the south – Quaternary and Neogene aquifers (to lower Sarmatia); b) the thickness of active water exchange zone increases towards southwest and southern directions and is regulated by geologic structure and hydrogeological conditions of interstratal aquifers; c) according to the plan, active water exchange zone location coincides with the location of ground water recharge area; d) areas where active water exchange zone is located are characterized by maximum density of water supply wells; e) as for a percentage of the total aquifer area location, the active water exchange zone location has the following values: Cretaceous aquifer – 55%, lower Sarmatia – 64%, middle Sarmatia – 71%, upper Sarmatia - Pontian – 91% and Quaternary aquifers – 100%.

4. Conclusion

For platform areas, in the condition of hydrogeological parameterization uncertainty a set of techniques is suggested to delineate the boundaries of active water exchange zone vertically and according to the plan. It consists of the sequence of the following methods: (a) hydrogeological stratification, (b) hydrogeochemical method, (c) geothermal method, (d) tritium in ground water, and (e) helium technique.

Active water exchange zone of ground water in the southwest of the Black Sea artesian basin is delineated both vertically and in plan using a set of suggested techniques. It comprises ground and intermediate waters of Quaternary, Pontian, Meotian, upper-, middle-, lower- Sarmatia, Cretaceous, and Silurian periods.

Acknowledgments
When making the research the author used a number of valuable comments and recommendations by Prof., Doc. of Geol.-Mineral. Sciences Shvets V.M., Prof., Doc. of Geol.-Mineral. Sciences Shvartsev S.L., Prof., Doc. of Chemcal Sciences Ryhenko V.N., Prof., Doc. of Geol.-Mineral. Sciences Khausov A.P., Doc. of Geol.-Mineral. Sciences Galtskaya I.V., and Prof., Doc. of Geol.-Mineral. Sciences Pozdnyakov S.P., to whom the author expresses his sincere gratitude.

References

[1] Dutova Ye M 2005 *Hydrogeochemistry of hypergenesis zone in Altay-Sayan folded belt*: abstract of dissertation … Doctor of Geol.-Mineral. Sciences: / Ye.M. Dutova – Tomsk p 46
[2] Ignatovich N K 1948 *Hydrogeology of Russian platform* (Moscow: Gosgeolizdat) p 333
[3] Lurie Yu Yu 1947 Calculation and reference tables for chemists (Moscow: State Scientific-engineering Publishing House of Chemical Literature) p 331
[4] Moraru C Ye 1987 *Hydrogeochemistry of helium and its role as an indicator of aquifer interconnection (by the example of Moldova artesian slope):* abstract of dissertation … Cand. of Geol.-Mineral. Sciences: / C.Ye. Moraru (Moscow: Moscow Geological Exploration Institute) p 16
[5] Moraru C Ye 2013 *Hydrogeochemistry of active water exchange zone in the southwest of the Black Sea artesian basin (geoecological aspects):* abstract of dissertation … Doctor of Geol.-Mineral. Sciences: / C.Ye. Moraru (Moscow: Institute of Environmental Geoscience of Russian Academy of Science) p 52
[6] *The reference book of hydrogeologist* / Ed. by Altovskiy M Ye 1962 (Moscow: State Scientific-engineering Publishing House of Literature on Geology and Subsurface Resource Conservation) p 616
[7] Shvartsev S L 1998 *Hydrogeochemistry of Hypergenesis zone* (Moscow: Nedra) p 365
[8] Guangzhi Tu 1996 *Low-Temperature Geochemistry* (Science Pr.) p 202
[9] Landmir D 1997 *Aqueous Environmental Geochemistry* (Prentice Hall) p 600
[10] Moraru C and Anderson J A 2005 *Comparative Assessment of the Ground Water Quality of the Republic of Moldova and the Memphis, TN area of the United States of America* (Memphis, the University of Memphis) p 187
[11] Moraru C 2014 Groundwater quality in the Republic of Moldova *In: Management of water quality in Moldova* (Springer) 177-197
[12] Melian R, Myrlean N, Gurev A, Moraru C and Radstake F 1999 Groundwater quality and Rural drinking water supplies in the Republic of Moldova *J. Hydrogeology* 7 188-196
[13] Gumm L P et all. 2015 Dissolved noble gases and stable isotopes as tracers of preferential fluid flow along faults in the Lower Rhine Embayment, Germany *J. Hydrogeology* (DOI: 10.1007/s10040-015-1321-7)
[14] Williams A J, Crossey L J, Karlstrom K E, Newell D L, Person M and Woolsey E 2013 Hydrogeochemistry of the Middle Rio Grande aquifer system - fluid mixing and salinization of the Rio Grande due to fault inputs *J. Chemical Geology* **351** 281-298

Modeling pellet impact drilling process

A V Kovalyov[1], S Ya Ryabchikov[2], Ye D Isaev[3] and O S Ulyanova[4]

[1,2,3,4] National Research Tomsk Polytechnic University, 30 Lenina Ave., Tomsk, 634050, Russia

E-mail: [1]arteka011287@mail.ru

Abstract. The paper describes pellet impact drilling which could be used to increase the drilling speed and the rate of penetration when drilling hard rocks. Pellet impact drilling implies rock destruction by metal pellets with high kinetic energy in the immediate vicinity of the earth formation encountered. The pellets are circulated in the bottom hole by a high velocity fluid jet, which is the principle component of the ejector pellet impact drill bit. The experiments conducted has allowed modeling the process of pellet impact drilling, which creates the scientific and methodological basis for engineering design of drilling operations under different geo-technical conditions.

1. Introduction

Currently, there is a global trend to drill hard rock for various purposes. Hard rock drilling is characterized by low drilling rate and low rate of penetration. New materials and new design of drill bits for mechanical drilling can improve drilling efficiency. Despite the continuous improvement, the enhanced hard rock drilling is still a challenge.

Therefore, new alternative methods for rock drilling are urgent now. According to some researchers [1, 4, 6], one of the most promising drilling methods is hydro-dynamic drilling by high-speed fluid. This method ensures transmission of significant energy to the bottom hole with the multifold increase in the drilling rate, as well as the rate of penetration. Furthermore, this method is easy to apply within the current mechanical drilling technology where mud is used for cutting transport. However, hydro-dynamic drilling in its traditional version is not enough efficient for hard rock drilling.

Pellet impact drilling implies rock destruction by metal pellets circulating constantly in the bottom hole, which will solve a number of engineering problems connected with hydro-dynamic drilling. This method is characterized by a number of significant advantages: simplicity of drill bit design, no need for drill bit rotation and axial load. According to the data on drilling performance [3, 9] pellet impact drilling is associated with the increase in the drilling rate and the rate of penetration in comparison with the traditional drill bits.

We have proved that the most effective method to drill hard rock is to use ejector pellet impact drill bit with the nozzle and the tubular mixing chamber arranged consequently in line [5]. The efficiency of this drill bit is connected with the accuracy of geometric bit design and drilling parameters. The existing methods of pellet impact drill bit engineering design are limited in application and appropriate to design a particular drill bit only. Therefore, designing a specified model of pellet impact drilling process is currently urgent since it would create a scientific and methodological basis for engineering design under different geological and technical conditions.

2. Designing a pellet impact drilling process

Designing a pellet impact drilling process was made on the basis of experimental data, in a particular, high-speed photography. A hydrodynamic layout of drill bit performance is shown in Figure 1.

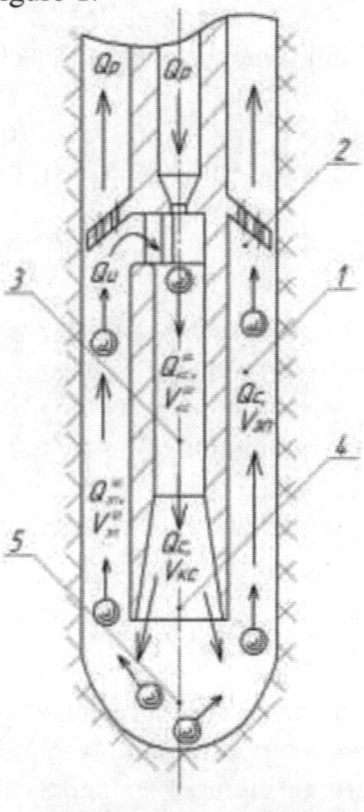

Figure 1 A hydrodynamic layout of drill bit performance:
1 – annular space; 2 – suction zone; 3 – mixing chamber; 4 – diffuser; 5 – a space between drill bit and bottom hole;
Q_{dm} – consumption of drilling mud supplying to the drill bit;
Q_i – injected drilling mud consumption;
Q_{mc} – mixed mud consumption;
V_{as}, V_{mch} – mud velocity in annulus space and drill bit;
$Q^p{}_{mch}$, $V^p{}_{mch}$ – pellet consumption and pellet velocity in mixing chamber;
$Q^p{}_{as}$, $V^p{}_{as}$ – pellet consumption and pellet velocity in annulus space.

Pellets are used in the following operations:
1) pellets are lifted in annular space 1 up to suction zone 2;
2) pellet are sent to mixing chamber 3 from suction zone 2;
3) pellets move in mixing chamber 3 and diffuser 4;
4) pellets move between the drill bit and bottom hole 5

A distinctive feature of a fluid jet is circulation of drilling mud and pellets. [3]. Thus, the circulation of drilling mud (Q_{dm}) and pellets ($Q^p{}_{db}$, $Q^p{}_{as}$) is observed in the process of drilling by the drill bit. It is necessary to note that operation mode of a fluid jet is changed by adding pellets in drilling mud.

3. Hoisting pellets in annulus space

The analysis of pellet movement in annulus space (Figure 2) obviously shows its unequal character.

Figure 2. Stroboscopic image of pellet movement in annulus space.

Average velocity of pellets hoisting (m/s) in the case of even movement can be defined by:

$$V^p_{as} = V_{as} - U \quad , \qquad (1)$$

where V_{as} – velocity of drilling mud circulation in annulus space, defined by the equality of mixed mud consumption in mixing chamber and annulus space

(2)

where d_{mch} – diameter of a mixing chamber; D_{well} – well diameter; d_{db} – outer diameter of the drill bit.

U – free-falling velocity of pellets in drilling mud for transient and turbulent regime of flow (m/s), defined by Rittinger's formula

(3)

where ω - Rittinger's constant for pellet in average 5,11 , d_p – pellet diameter, ρ_p – density of pellets, ρ_{dm} – density of drilling mud.

The maximum number of pellets in annulus space (N^p_{asmax}) is defined using our derived formula, based on the continuous motion of pellets. It is assumed that pellets are arranged in several rows.

(4)

where l_{db} – length of the drill bit.

According to the results of high-speed photography, it is possible to observe that within a pellet portion supply, there are gaps between pellets within a row, as well as between rows, even if the number of pellets is optimal. Therefore, the number of pellets can be defined by the following formula:

, (5)

where C_1 – coefficient of a gap between rows; C_2 – coefficient of a gap between pellets in a row.

Formula (5) is appropriate for calculation of an optimal portion of pellets supply. Validity of the formula is testified by the results of experimental research of the influence of the length of the mixing chamber on pellet impact drilling efficiency. The increase in length of the mixing chamber necessitates a greater number of pellets, which is shown in the mentioned-above formula.

It should be noted that efficient cutting transport may be ensured through:

- increasing the capacity of an arrestor;
- choosing the optimal coefficient of ejection that ensures the appropriate velocity of pellets in the mixing chamber.

4. Pellet supply to the drill bit

In upper part of the drill bit, pellets lifted up by rising current interact with the arrestor and then are sent to the mixing chamber through the process windows.

To prevent pellets jamming under the arrestor the following condition should be satisfied:

(6)

According to the tests, formula (6) is valid if the coefficients of ejection are more than 2.

5. Pellet movement in the mixing chamber and the diffuser

Pellet movement trajectory from the mixing chamber to the diffuser (figure 3) has been studied with high-speed photography. It was revealed that pellets in the mixing chamber do not move parallel to its axis due to the fact that velocity vector is directed at the angle to an axis of a well at the point of pellets' entrance in a mixing chamber.

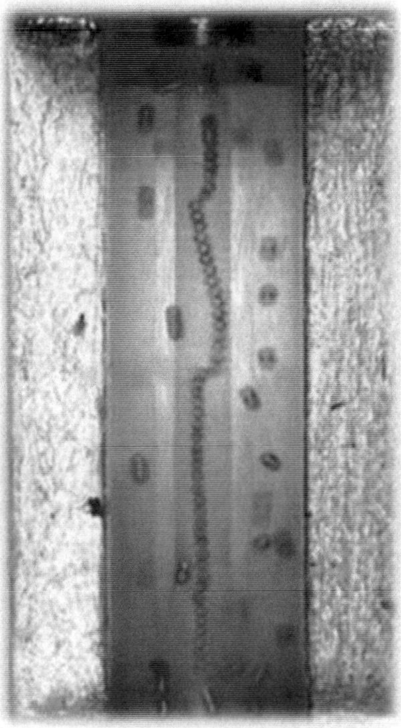

Figure 3. Stroboscopic image of pellet movement in the drill bit.

Therefore, the ricochet of pellets after hitting a wall is observed in the mixing chamber. At the same time, the trajectory deviation of pellets is not detected in the diffuser.

Furthermore, the analysis of video sequences shows when the pellets enter the mixing chamber; they are accelerated by the drilling mud flowing from a nozzle. At a distance from the inlet section of the mixing chamber the velocity profile of the fluid jet becomes equal; pellet velocity reduces to velocity V^{p}_{mch} and remains constant.

Pellet velocity in the mixing chamber is defined as:

(7)

where θ – coefficient depending on the ratio of the mixing chamber diameter to the pellet diameter, pellet consumption in a mixing chamber and drilling mud properties [7].

V_{mch} – velocity of drilling mud in the mixing chamber defined as:

——— (8)

where n_e – coefficient of ejection of the fluid jet.

The maximum number of pellets in the drill bit is defined as:

— (9)

Pellets consumption is equal to

, 10)

where t – the time of drilling.

Substituting (9) into (10) in case of even movement, we will obtain:

—— ——, (11)

In fact, not all pellet in the drill bit move in a row and there is a certain distance between the rows. In this regard, formula (11) should be corrected by adding coefficient C_3 taking into account the gaps between pellets in a row and the rows.

———, (12)

It should be noted that the empirical dependence of pellets consumption in the mixing chamber on pellets velocity is confirmed by the American researchers as well [2].

6. Pellet movement in bottom-hole zone

After leaving the diffuser pellets hit the rock and crush it. Figure 2 shows that the distance between the drill bit and bottom hole is gradually filled up with pellets because of considerable flow turbulence of drilling mud after its interaction with the bottom hole and pellets encounter as well. In addition, a comparison of the actual geometric dimensions of the bit, pellets and the distance between the drill bit and the bottom hole, as well as trajectory analysis show that the marginal zone of bottom hole is crashed mostly by ricocheted pellets and supplied pellets. This once again confirms the justification of a minimum distance between the drill bit and bottom-hole in drilling of a well.

IOP Conf. Series: Earth and Environmental Science **33** (2016) 012030 doi:10.1088/1755-1315/33/1/012030

7. Conclusion

The suggested model of pellet impact drilling process creates a scientific and methodological basis for engineering design of drilling operations under different geo-technical conditions, and ensures through:

- to calculate pellet consumption and pellet movement velocity in annulus space and the drill bit;
- to ensure the efficient cutting transport from bottom hole;
- to prevent pellets jamming under the arrestor of the drill bit;
- to calculate the optimal portion of pellet supply for the drill bit.

References

[1] Davidenko A N 2013 *Abrazivno-mekhanicheskoe udarnoe burenie skvazhin: monografiya* p 110

[2] Eckel I E, Deily F H, Ledgerwood L W 1956 Development and testing of jet pump pellet impact drill bits *Transaction AIME* **207** 15

[3] Zaurbekov S A 1995 *Povyshenie effektivnosti prizaboynykh gidrodinamicheskikh protsessov pri sharostruynom burenii skvazhin Author's abstract* p 18

[4] Kovalyov A V, Ryabchikov S Ya, Aliev F R, Yakushev D A and Gorbenko V M 2015 Methodical issues of evaluating methane capacity in small Problems of hydrodynamic methods of wells drilling and the main directions in their solution *Bulletin of the Tomsk Polytechnic University* **326** (36) 12

[5] Kovalyov A V, Ryabchikov S Ya, Isaev Ye D, Aliev F R, Gorbenko M V and Strelnikova A B 2015 Designing the ejector pellet impact drill bit for hard and tough rock drilling *IOP Conf. Ser.: Earth Environ. Sci.* **24** URL: http://dx.doi.org/10.1088/1755-1315/24/1/012016

[6] Kozhevnikov A A and Davidenko A N 1987 *Gidromekhanicheskij i ehrozionnyj sposoby razrusheniya gornyh porod pri burenii skvazhin* p 45

[7] Sokolov E Ya and Zinger N M 1989 *Strujnye apparaty* p 350

[8] Spivak A I, Popov A N, et al. 2003 *Tekhnologiya bureniya neftyanyh i gazovyh skvazhin*: uchebnik dlya vuzov p 509

[9] Uvakov A B 1969 *Sharosruynoyue bureniye* p 207

Selection of acid compositions in well construction in difficult geological conditions

M V Mishchenko[1] and M R Kamartdinov[2]

[1, 2] Tomsk Polytechnic University, 30 Lenin Avenue, Tomsk, 634050, Russia

E-mail: [1] MishenkoMV@hw.tpu.ru, [2] KamartdinovMR@hw.tpu.ru

Abstract. Within the scope of the current work we have presented an approach towards selecting and substantiation of acid composition in accordance with petrophysical characteristics of formations for acid treatment of bottom-hole formation zones. Article presents the results of lab tests of selected acid compositions, in conditions, which model thermobaric conditions of a payzone, combined with an evaluation of hydraulic permeability change, and this, in its turn, should allow us to evaluate the quality of impact of the acid composition recipe on the reservoir formation.

1. Introduction

Processing bottom-hole formation zone with acid compositions is a relatively inexpensive, effective and commonly used methods to increase the permeability of the bottom-hole zone [5]. The method is based on the effects of water-based solution of acid reacting with minerals, forming the reservoir rock, and introduced solid minerals, blocking the bottom-hole zone. Therefore, one must use acid compositions containing chemical additives that meet specific geological and physical conditions and provide optimum depth of penetration of the active reagent. This approach can increase the effectiveness of the acid treatment and avoid the negative consequences, in particular, the formation of stable emulsions, insoluble precipitates causing secondary mudding of bottom-hole formation zone [3, 6, 7, 8]. Well acidizing with sandstone reservoirs, in most cases does not require influence on the reservoir matrix. It is enough to remove colmatants of technogenic origin or carbonate component of reservoir. One of the common ways to treat well bottom-hole zone is the use of hydrochloric acid [4, 7].

2. Preparation and laboratory work

To select the most effective acid composition, the following laboratory tests were performed: samples preparation, study of porosity and permeability coefficients, determination of their mineralogical composition, definition of the impact of acidic compounds on the rock.

Preparation of samples for the study was conducted in accordance with GOST 26450.0-85 whereby core material samples were subjected to extraction with alcohol-benzene mixture (in the ratio 3:1), washed in distilled water and dried to constant weight [1]. Upon preparatory phase completion, porosity and permeability coefficients were studied [2].

To determine the mineralogical composition and quantitative relations of minerals in the rock, X-ray diffraction (XRD) method was used. Of all methods of determining the mineral composition of rocks, X-ray analysis is the most informative. For clay minerals XRD is the only reliable diagnostic tool. The quantitative mineral content in the sample was established against basal diagnostic reflections in the diffraction patterns obtained under strictly constant conditions of sample survey [9]. Based on this analysis, compositions for acidizing were proposed.

For greater effect of acid treatment, it is proposed to use mud acid (a mixture of hydrochloric and hydrofluoric acids in different proportions), which, in turn, will optimally recover reservoir permeability and porosity properties (PPP).

Laboratory testing of acid compositions. For laboratory analysis, core material composition was taken, the main parameters of which are presented in Table 1.

Table 1. Core material for laboratory analysis.

Suite/Formation	Interval	Lithological composition
Dolganskaya (K_{1-2} dl)	1090.0–1096.9	Uneven interbedding of siltstones sandy, clayey, quartz-feldspar, micaceous and clays dark grey, silty, micaceous.
Malokhetskaya (K1mh)	2269.8–2266.2	Sandstone, silty clayey
	2270.7–2267.1	Sandstone, silty clayey
	2275.4–2273.5	Sandstone, silty clayey
Suhodudinskaya (K1cd)	2504.0–2509.6	Sandstone, silty clayey, weakly micaceous
	2815.0–2826.2	Sandstone, silty clayey
	2816.4–2815.1	Sandstone, silty clayey
	2957.7–2970.7	Siltstones sandy, clayey, micaceous; Sandstone, silty clayey

In the samples, general mineralogical composition was determined. The results are shown in Table 2. In determining the mineral composition of the rocks the total content of all minerals and cement was assumed to be 100%.

Table 2. Mineral composition of rocks according to the XRD.

Suite/Formation	Mineral composition							Total, %
	Calcite	Kaolinite	Quartz	Quartz-feldspar	Muscovite / Illit	Plagioclase	Chlorite	
Dolganskaya	0.0	3.0	28.8	5.6	21.3	31.6	9.7	100
Malokhetskaya	0.0	3.0	30.0	7.6	18.3	30.6	10.5	100
Suhodudinskaya	0.0	1.1	44.9	8.4	4.2	37.5	3.9	100

X-ray diffractograms of core samples are shown in Figure 1-3. On disintegrated rock samples, interaction of acidic compounds (HCl 10%) with different groups of minerals that make up the sample was studied. For each sample six batches of five different test compounds were prepared. Batches were filled with 30 ml of acid solution, mixed and placed in an oven for 4 hours at reservoir temperature. After a predetermined time, precipitates were transferred onto pre-dried filters, washed with distilled water until neutral state, dried to a constant weight. Weight loss was determined by the difference in the sample weights after interaction with the test compositions. The results are shown in Table 3.

IOP Conf. Series: Earth and Environmental Science **33** (2016) 012031 doi:10.1088/1755-1315/33/1/012031

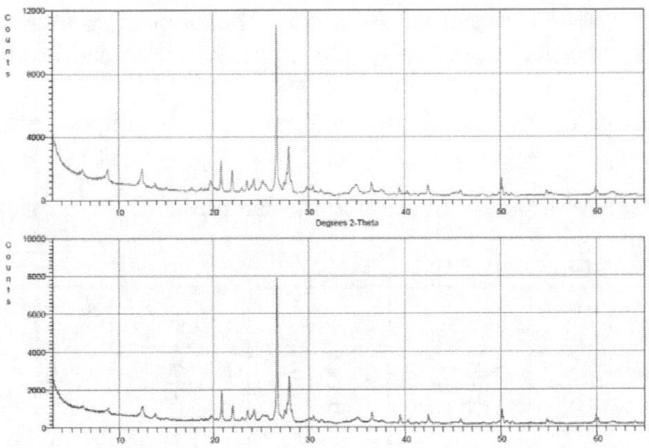

Figure 1. X-ray diffraction patterns before treatment with acid compositions, Dolganskay formation.

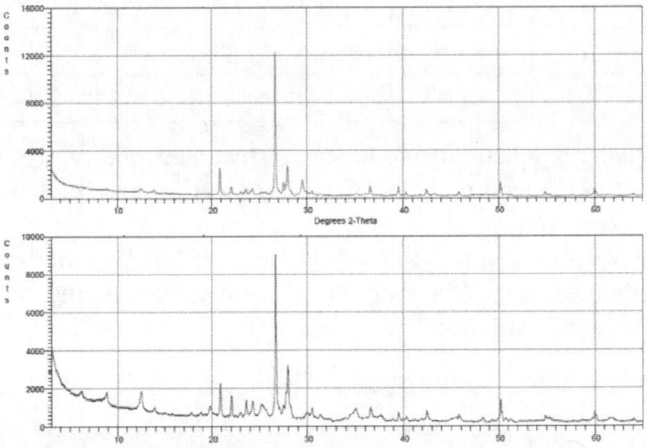

Figure 2. X-ray diffraction patterns before treatment with acid compositions, Malohetskaya formation

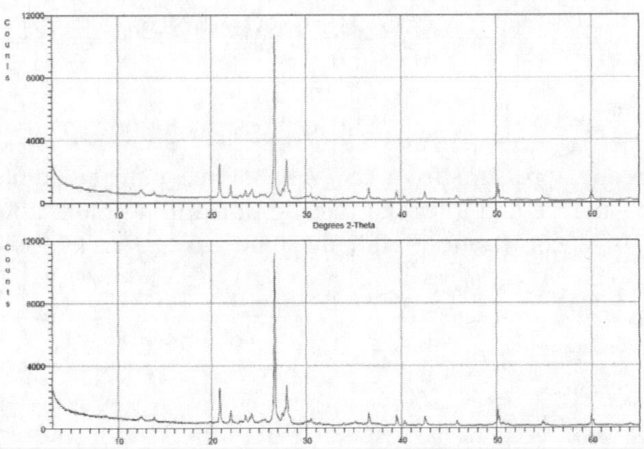

Figure 3. X-ray diffraction before treatment with acid compositions, Suhodudinskaya formation

Table 3. Weight loss after crushed rock samples acidizing.

Suit	Weight before acidizing, gr.	Weight after acidizing, gr.	Weight change, %
Dolganskaya	7.221	7.215	1.05
Malokhetskaya	9.510	9.485	0.58
Suhodudinskaya	5.413	5.330	1.53

For sand reservoirs, this weight change after acid treatment was minor. The biggest change in weight (approximately 1.5%) showed the sample of Suhodudinskaya suite. It can be explained by the highest clay content among selected test samples.

The study results of the mineral composition change of pelitic fraction in the samples after exposure to 10% HCl acid solution are shown in Table 4.

Table 4 .Mineral composition change of pelitic fraction in the samples.

Suit	Mineral composition change, %						
	Calcite	Kaolinite	Quartz	Quartz-feldspar	Muscovite / Illit	Plagioclase	Chlorite
Dolganskaya	+0.1	+0.2	+1.2	-2.6	-6.8	-4.2	+0.5
Malokhetskaya	-0.2	+0.0	-2.3	+0.6	-8.5	-7.4	+1.9
Suhodudinskaya	-0.1	-0.4	+1.9	+1.4	-7.0	-6.2	-2.1

As the table shows, the major quantitative changes when exposed to acid treatment take place in illite and plagioclase. Changes in the content of other components of pelitic fraction are logically associated with statistical error in measurement.

In accordance with the method, filtration experiments simulating acid effect on the bottom-hole zone were performed. The compositions most often used in Western Siberia were chosen as the acid ones (see Table 5). For basic acid composition, 10% HCl solution was used.

Table 5. Acid compositions used in filtration experiments.

No.	Acid composition	Producer
1	HCl (10% solution)	
2	Himeko TK-2 KM	ZAO «HIMEKO-GANG»:
3	Modified composition of hydrochloric acid	AO «POLIEKS»
4	FLAKSOKOR 210	ZAO "Petrohim"
5	Petrohim-KGS	GC "Mirrico" basic solution

Figure 4 shows the behavior of the differential pressure from the flow of the pumped liquid (kerosene) at various stages of the experiment - before and after the injection of drilling mud, after acid exposure. Figure 5 shows a photograph of a core sample with the mud cake after the acid treatment.

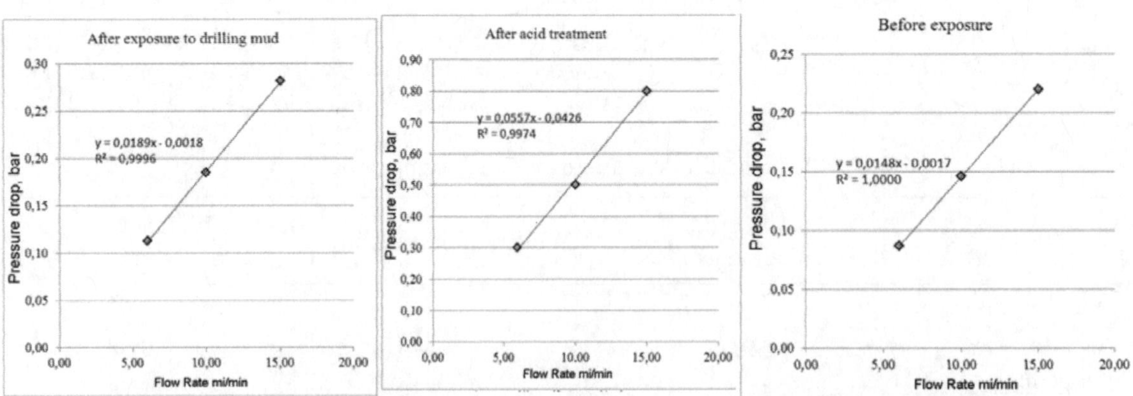

Figure 4. Dependence of the differential pressure across the sample from fluid flow.

Results of filtration experiments, which determine the permeability recovery coefficient (K_{recov}) after exposure to drilling mud and subsequent treatment with acid compositions, are shown in Table. 6.

As seen from Table, 10% HCl effect does not improve the recovery coefficient. In general deterioration of permeability by oil after 10% HCl acid impact on the core can be explained by two different processes: a positive effect on the permeability of the core by removing residues of colmatant is offset by negative influence of the water-based phase of the acid solution entering the pore space. The sharp decrease in permeability of the core sampled from producing horizons after the impact of technological water-based fluids has repeatedly been noted earlier by other researchers.

Figure 5. Photograph of a core sample with the mud cake after the acid treatment.

Table 6. Results of filtration experiments to determine the effectiveness of acid treatments.

Suit	Drilling mud type	Acid composition	K_{recov}, after mud cake collapse, %	K_{recov}, after acid treatment, %
Malohetskaya	Basic	HCl (10 % solution)	48.26	26.57
Malohetskaya	Polymer+KCl	Himeko TK-2 KM	92.60	86.67
Nizhnehetskaya	OBM	Modified composition of hydrochloric acid	82.61	96.70
Nizhnehetskaya	Basic	HCl (10 % solution)	46.24	24.06
Suhodudinskaya	OBM	FLAKSOKOR 210	53.15	25.81
Nizhnehetskaya	Basic	Petrohim-KGS	49.26	93.32

4. Conclusion

The conducted tests on acid treatment of various compositions based on HCl showed little impact on the rock of formations under consideration (the change in weight of the sample after the acid treatment is not more than 1.5%, minor changes in the mineral composition based on results of XRD). Meanwhile, the acidic compounds actively interact with colmatant of the considered drilling muds.

Conducted filtration experiments, simulating acid treatment of bottom-hole zone after exposure to drilling mud, showed mixed effects from exposure to different acidic compounds in the mud cake. Direct measurements of permeability before and after acid treatment, as well as photos of the surface of the sample with a mud cake, show high activity of acid composition and destruction of the mud cake. Meanwhile, the experimental results show that in most cases washing off the mud cake is more efficient than the acid treatment without washing off. Furthermore, the use of acid solutions (10% HCl, water-based) leads to a significant reduction of permeability due to increased saturation of the pore space with water-based acid compositions, preventing filtration of oil (kerosene).

References

[1] GOST 26450.0-85. *Porody gornyye. Obshchiye trebovaniya k otboru i podgotovke prob dlya opredeleniya kollektorskikh svoystv* (in Russian)]

[2] GOST 26450.2-85. *Porody gornyye. Metod opredeleniya absolyutnoy gazopronitsayemosti pri statsionarnoy i nestatsionarnoy fil'tratsii* (in Russian)

[3] Mavliev A R, Almukhametov M A, Petrov I A, Shishlov A S, Fetisova V R and Rogachev M K 2012 Substantiation and development of acid deflector compositions based on water-in-oil emulsions to improvement of the effectiveness of the bottom-hole treatment *Neftyanoye*

khozyaystvo **12** 56-58 (in Russian)
[4] Nikolayev N I, Shipulin A V and Kupavykh K S 2015 Rezul'taty issledovaniy i effektivnost' primeneniya kompleksnoy tekhnologii khimicheskoy obrabotki prizaboynoy zony plasta *Territoriya Neftegaz* **4** 79-83 (in Russian)
[5] Petrov I A, Azamatov M A and Drofa P M 2010 Kompleksnyy podkhod k obrabotke prizaboynoy zony plasta kak sposob intensifikatsii dobychi *Georesursy* **1** 7-10 (in Russian)
[6] Fedorenko V Yu, Nig'matullin M M, Petuhov A S, Gavrilov V V, Bespalov M V and Afanas'ev O L 2015 Additives for acid compositions *Neftyanoye khozyaystvo* **2** 88-90 (in Russian)
[7] Filin V V and Khaliullin A A 2012 Povysheniye effektivnosti promyvki skvazhin posle obrabotki prizaboynoy zony plasta kislotnymi kompozitsiyami na mestorozhdeniyakh Zapadnoy Sibiri *Prospects for the development of science, engineering and technology: Materials of the II International Scientific and Practical Conference* 141-146 (in Russian)
[8] Khismetov T V, Bernshteyn A M, Magadova L A, Davletov Z R and Nikitin A N 2012 Podbor kislotnykh kompozitsiy dlya obrabotok prizaboynoy zony po dannym yaderno-fizicheskikh metodov *Neftyanoye khozyaystvo* **5** 78-81(in Russian)
[9] Shlykov V G 1991 *Rentgenovskiye issledovaniya gruntov* (Moscow: MGU) p 184 (in Russian)

Oil and Gas Exploration Planning using VOI Technique

D N Peskova[1], A V Sizykh[2] and V S Rukavishnikov[3]

[1,2]Gazpromneft-ntc, Saint Petersburg 190000, Russia
[3]Tomsk Polytechnic University, Tomsk 634034, Russia

E-mail: [3]RukavishnikovVS@hw.tpu.ru

Abstract. Paper deals with actual problem about making decisions during field development. The main aim was to apply method "Value of information" in order to estimate the necessity of field exploration works and show the effectiveness of this method. The object of analysis - field X, which is located in the Eastern Siberia. The reservoir is B13 formation of Vend age. The Field has complex structure, and divided into blocks by faults. During evaluation of the project, main uncertainties and oil in place were obtained for three blocks of the field. According to uncertainty analysis, it was suggested to drill a new exploration well, and value of information method was applied to estimate results from this exploration works. Economic evaluation of the value of information method was made by choosing optimal development strategy. According to the obtained results, drilling of the exploration wells for blocks 1 and 3 of the field X is a good decision, while drilling a well in the second block is risky and not recommended. Also using the value of information, optimal well locations were advised - well 1_1e for the first block, and well 3_3 for the third block.

1. Introduction

The paper deals with actual problem about making decisions during field development. Nowadays Gazpromneft spends a lot of money for exploration of the field while not all getting information is useful. So, the main idea was to show how method "Value of information" can be used for proving the necessity of exploration works and what potential economic benefits it can bring.

The object of analysis - field X, which is located in the Eastern Siberia. The reservoir is B13 formation of Vend age. According to seismic surveys, infill drilling data and results of well logging, the field is divided into several blocks, boundaries of which are zones of the deep faults. During interpretation of seismic data it was found and traced a lot of different faults. Reservoir properties are very heterogeneous. All calculations were done for 3 blocks of the field.

The value-of information analysis is principal estimation of expected monetary value (EMV) from obtained information. It obtains benefits that additional information can give in comparison if no information available:

VOI= [EMV with additional information] – [EMV without additional information].

If the difference in expected monetary values between scenario with and without data acquisition is not enough (or even negative) to cover the costs of exploration work, it can be a good reason not to acquire the data.

Method of estimation the value of information can help to obtain the potential value of additional information and can be used as powerful technique for making decisions in the oil and gas industry. In general, VOI analysis focused on the main uncertainties which can affect the amount of STOOIP and NPV value, and how additional information can change the decisions about development of the field and further investment decisions. This method leads to effective usage of data. As a result, the main idea of VOI analysis is to estimate how much information obtained during exploration work is worth, which, in its turn, enables better evaluation of profitability of the project as a whole [1, 2, 3, 4, 5, 6].

2. Method

The first step of the analysis was to estimate STOOIP and associated uncertainties. For doing this, geological model for the field X was created based on seismic surveys, interpreted well logs and variograms. Distribution of STOOIP was obtained using Monte Carlo method (figure 1). And the analysis of influenced factors was done.

First of all, structural uncertainties are related to seismic survey and horizon modeling process. Depending on seismic surveys resolution, the standard deviations were obtained (for 3D seismic – 8m, 2D seismic – 28m).

The second uncertainty type is size and distribution of sandbodies in 3D space. Dimension of reservoir bodies depends on the variogram ranges. Because the field is only on exploration stage and there is a large distance between wells, statistical analysis using well data is not quite accurate. That is why ranges of variogram were taken as triangular distribution. Changes in ranges were estimated through statistical analysis and literature information [7, 8, 9, 10] according to the depositional environment. Uncertainty in distribution of the reservoir was taken into account by stochastic element of the model.

Thirdly, there is uncertainty connected with OWC level. Its estimation was made on the base of having well data and structural map. As a result, the interval of OWC level change was estimated for each block (lower OWC level – the bottom of sandbodies for each block, the upper OWC – lower interval of perforation) and triangular distributions of OWC were used during STOOIP estimation.

Porosity and oil saturation were defined as normal distribution with associated mean value and standard deviation (for porosity: mean – 0.11, standard deviation – 0.01; for oil saturation: mean – 0.66, standard deviation – 0.036).

As a result, P50, P10 and P90 probabilities of STOOIP and associated with them uncertainties were obtained (figures 1). Sensitivity analysis is the first step in VOI analysis. It shows key uncertainties, which are affected on STOOIP, and expected exploration works according to them.

As it can be seen from figure 1, level of OWC is one of the most important factors, so it was planned to drill exploration wells for clarifying these values.

The next step of the VOI analysis is estimation of the probable results, which can be obtained after drilling of the exploration wells. For each block the possible positions of wells were suggested (Figure 2), for example, for the first block – 1_1e, 1_2e, 1_3e and 1_4e wells.

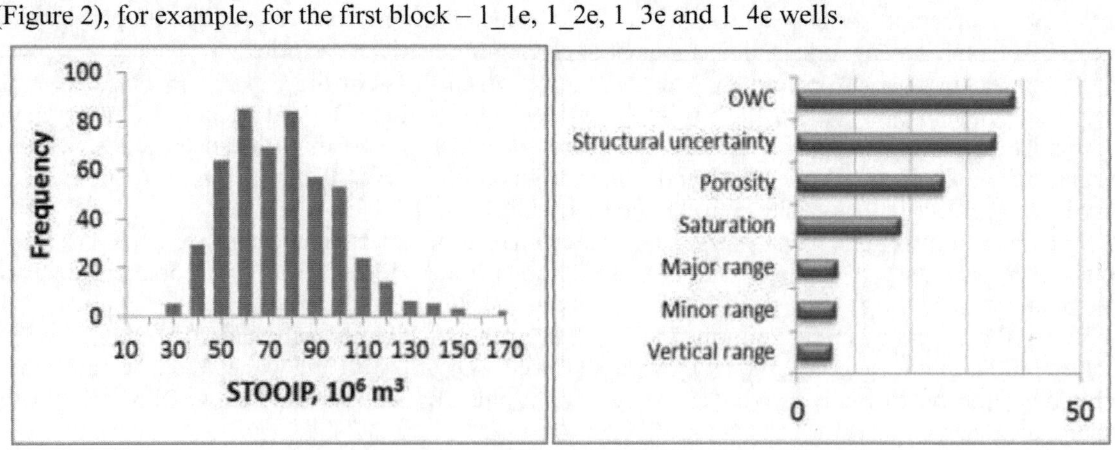

Figure 1. STOOIP distribution and sensitivity analysis.

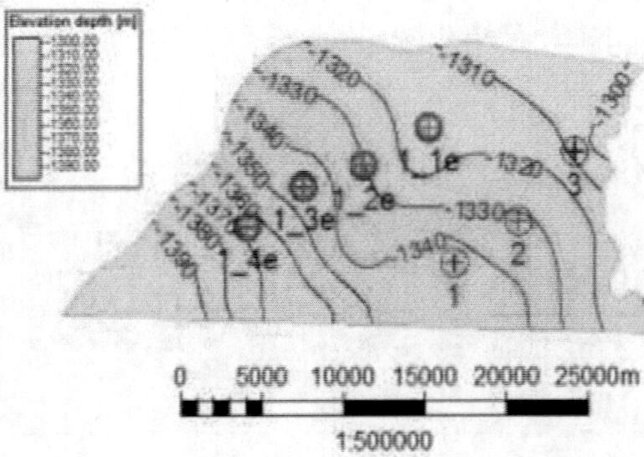

Figure 2. Structural map of block (red – modelled wells, black – real wells).

New exploration well can perforate three zones – oil zone, water zone or oil water zone. Because of structural uncertainty and uncertainty in OWC level, opened-up zone is different. In order to determine the probability of each zone occurrence, workflow was created for 500 realizations for the definite version of well location. The algorithm of identification of a zone is the following:

In the modelled well, property "above contact" is estimated for the reservoir facies. It is equal to 0 if sandbodies are located below oil water contact, and more then 0, if they are situated above OWC.

According to the value of "above contact", it is possible to estimate in which zone this well broachs: above contact>0 – oil zone, above contact=0 – water zone, above contact≥0 – oil water zone.

As a result, modelled STOOIP is assigned to the zone which was obtained.

Based on the described algorithm, workflow program was created. After modelling of 500 different realization of STOOIP, which depend on distributions of analyzed uncertainties, the probability of occurrence of each zone was estimated. This method allows not only estimating the result, which can give exploration well, but also estimating STOOIP for the each option. Distribution of STOOIP in each zone for different well is shown in Figures 3, 4.

The information that minimizes the existing uncertainties has the highest value. Uncertainties in OWC level can be decreased or delayed if well opens-up oil water zone. From this point of view, the well which has more probability is the best option.

The important step of VOI analysis is economic evaluation. Decrease in the uncertainties leads to more accurate STOOIP estimation and also allows taking optimal decision on further development of the field.

Decision on optimal development strategy was chosen according to the net pay value, which was calculated for given distribution of STOOIP. The development scenarios were altered by changing such parameters as number of wells and pipeline capacity. As a result, the option with the maximum mean NPV for a definite STOOIP distribution was taken as optimal.

Because the distribution of STOOIP for each case differs from each other, development strategy and, as a consequence, NPV value will change. Therefore, NPV with and without information will be different.

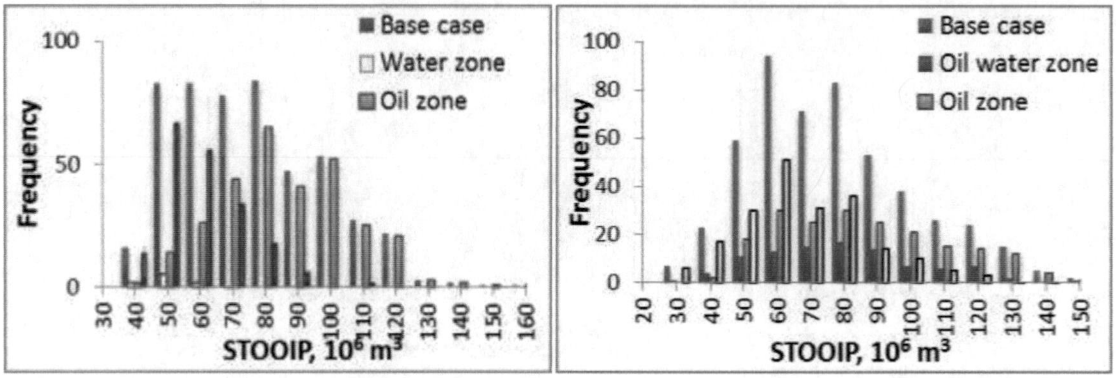

Figure 3. Distribution of STOOIP for different zones for wells 1_1e (left) and 1_2e (right).

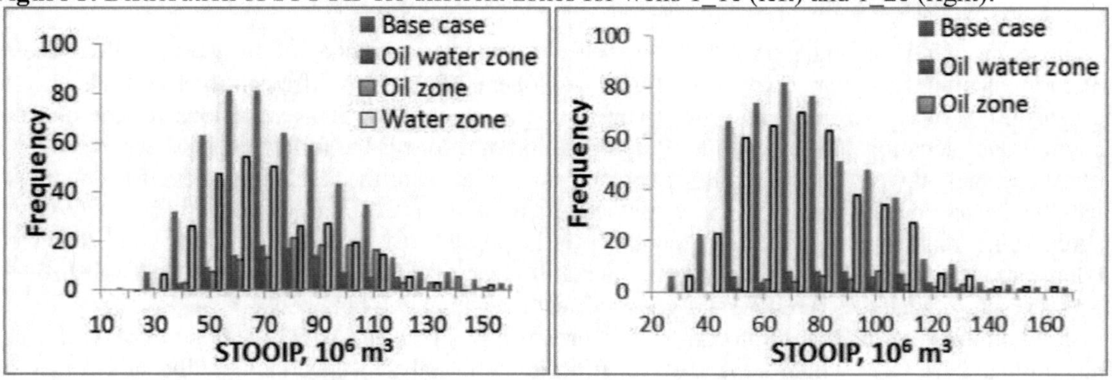

Figure 4. Distribution of STOOIP for different zones for wells 1_3e (left) and 1_4e (right).

Taking into account the probability of occurrence of each case, and calculated NPV, the value of information can be estimated as difference between the net pay value, estimated according to the modelled variants, NPV for the base strategy of development without information and the costs, which are spent on drilling of exploration well:

According to the results, drilling of the exploration wells for blocks 1 and 3 of the field X is a good decision (VOI for first block – 1.5 billion rub, for third block – 3 billion rub), while drilling of a well in the second block is risky (VOI=-170 mln.rub) and not recommended, which can lead to the decrease in profits from this block. Optimal well location according to the value of information – well 1_1e for the first block, and well 3_3 for the third block.

In order to estimate the accuracy of the method, real and modelled data must be compared. During 2014 and 2015 years, there were drilled exploration wells for the field X, corresponding to model wells. After it, STOOIP was estimated more correctly, and associated with them uncertainties change their intervals. Change in STOOIP value is affected on development decision as it was described above. To obtain the real value of information, possible profit was calculated if no exploration work was done, and the development strategy was not changed. STOOIP values were assigned from analysis after drilling exploration work that allows not overestimating the value of information.

The value of information is estimated as difference between NPV with and without information, taking into account the cost of drilling the exploration well (170 mln.rub). The final results of VOI values: for first block: 4.9 billion rub, second block - -170 mln.rub, third block – 10.7 billion rub.

Comparing modelled and real values of the information (VOI), it can be concluded that this method shows good result during making decisions on drilling of the exploration wells. For the first block, as it can be seen from Figure 5, modelling of exploration well was recognized because the change in STOOIP distribution may change the decision on the development of the field. In fact we can see, the

STOOIP was increased and development strategy was changed leading to big profits due to new information. Block 2 was not recognized for drilling of exploration well, so in reality the well brings profit losses. In the third block VOI analysis shows the justifiability of the exploration work. Despite the fact that STOOIP was decreased, the value of information is positive, because information about more accurate STOOIP distribution allowed us to revise the development strategy, which, in turn, brings profits in comparison with the previous strategy.

Block	VOI, mln.r
1	4 892.93
2	-170
3	10 687

Figure 5. VOI results.

3. Conclusion

During analysis, it was obtained that this method shows good result during making decisions on the drilling of the exploration wells. Economic effectiveness of applied method was proved by real data from the field X.

Also, it can be added that VOI analysis can be used as the most powerful tool for making decisions not only about the necessity of the exploration work, but it can be used for selecting different works and ranking the exploration work inside the field.

Taking into account exploration and prospect well data, the resulting map shows considerable detalization improvement for HC generation potential density of Bazhen suite in Tomsk region, and allows zoning the area according to the volumes of raw HC in bituminous formation. The comparative data analysis against specific geochemical core analysis results in previously and newly drilled wells showed good correlation, so it is possible to consider the presented map as a basis for commercial development of Bazhen suite reserves.

In order to differentiate the loss of bituminous formation generation potential resulting from hydrocarbon migration to underlying deposits, three types of deposit zones were distinguished: 1 - sandstones, 2 – alternating mudstones, sandstones and siltstones, 3 - shales. These types of lithological heterogeneity of deposits underlying the Bazhen suite allow estimating its residual generation potential at both qualitative (type 2) and quantitative (types 1, 3) levels. Taking into account the specific character of shale oil development, each type requires its own set of operating conditions.

References

[1] Sizykh A V and Belozerov B V 2015 Complex analysis of geological uncertainties and estimation of the exploration work efficiency with the help of 3D model *EAGE Conference* (Tumen).

[2] Rexilius J P, Prochnow S J and Morgan M 2014 Utilizing Fit-For-Purpose Decision Analysis Techniques to Maximize Value From Unconventionals: Case Studies from the Wolfcamp Play, Permian *SPE Unconventional Recources Conference* (Canada)

[3] Brashear J P 1999 Interdependencies Among E&P Projects and Portfolio Risk Management *SPE Annual Technical Conference and Exhibition* (Houston)

[4] Bratvord R B, J Bickel E and Lohne H P 2007 Value of information in the Oil and Gas Industry: Pasr, Present, and Future *SPE Annual Technical Conference and Exhibition*, (Anahaim)

[5] Koninx J P M 2000 Value-of-Information – from Cost-Cutting to Value-Creation *SPE Asia Pacific Oil and Gas Conference and Exhibition* (Australia)

[6] Beggs S and Bratvord R B 1992 The value of Flexibility in Managing Uncertainty in Oil and Gas Investments *SPE Annual Technical Conference and Exhibition* (San Antonio)

[7] Walker R G 1992 *A Facies models response to sea level change* (Canada)
[8] Einsele G 1992 *Sedimentary basins: evaluation, facies, and sediment budget* (Spriger, Berlin Heidelberg)
[9] Wang D and Helmy M 2010 Enhancement of Glacial Reservoir Modeling by Microresistivity Image: A Case Study from Oman *SPWLA 51st Annual Logging Symposium* (Perth, Australia)
[10] Easterbrook D J 1982 Characteristic Features of Glacial Sediments *AAPG Memoir* **31**

Bazhenov fm unconventional reservoir 3D geological modeling methodology

A Telnova[1], V Baranov[2] and N Bukhanov[3]

[1]Gazpromneft-ntc, Saint Petersburg 190000, Russia
[2,3]Tomsk Polytechnic University, Tomsk 634034, Russia

E-mail: [2]BaranovVE@hw.tpu.ru

Abstract. The Bazhenov Formation has been studied for more than 50 years, but its petroleum potential, optimal STOIIP or resource estimation approaches, the methodology used to select a reservoir, determine its properties are still unclear. The distinctive features of bituminous shale are specific geochemical properties chosen as basic parameters to perform the geological modeling of the Bazhenov deposits and determine the key areas. The main objective of this paper is to choose an optimal 3D geological modeling algorithm and test conventional (petrophysical) and specific (geochemical) properties.

1. Introduction

The Bazhenov Formation has been studied for more than 50 years, but its petroleum potential, optimal STOIIP or resource estimation approaches, the methodology used to select a reservoir, determine its properties are still unclear. The distinctive features of bituminous shale are specific geochemical properties chosen as basic parameters to perform the geological modeling of the Bazhenov deposits and determine the key areas. The main objective of this paper is to choose an optimal 3D geological modeling algorithm and test conventional (petrophysical) and specific (geochemical) properties. Moreover, some specialists characterize the reservoir as fractured and therefore the tectonic aspect plays an important role in the process of Bazhenov Formation reservoir development. The Bazhenov deposits of field A located in the south-western part of Tomsk Region are considered in this paper. Exploration operations are currently conducted in this area and one of the goals of this study is to estimate the petroleum potential and prospects of this field.

2. Materials and methods

Geochemical and petrophysical properties were chosen as the most important parameters used to determine oil-in-place and identify key zones for reservoir development. The author concludes that a set of core data, well logging and seismic data should be used for qualitative 3D geological modeling.

Lithotype differentiation was performed to ensure the accurate modeling of properties. One of the wells was selected as a reference well where core data and a complete necessary well logging complex are present. Eight lithotypes were detected and all of them were united into 2 groups: upper – more kerogenic silicite and low – more carbonated lithotypes, using a complex of core mineralogy, thin sections description and well logs (Figure 1).

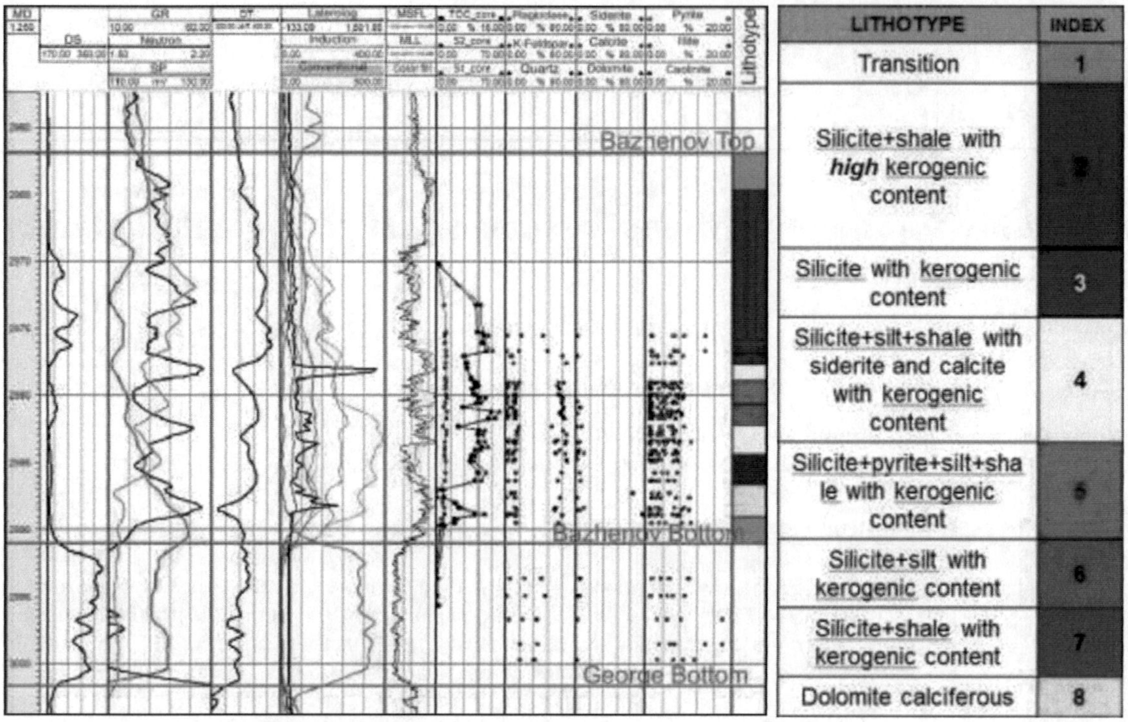

Figure 1. Lithotype differentiation in a reference well.

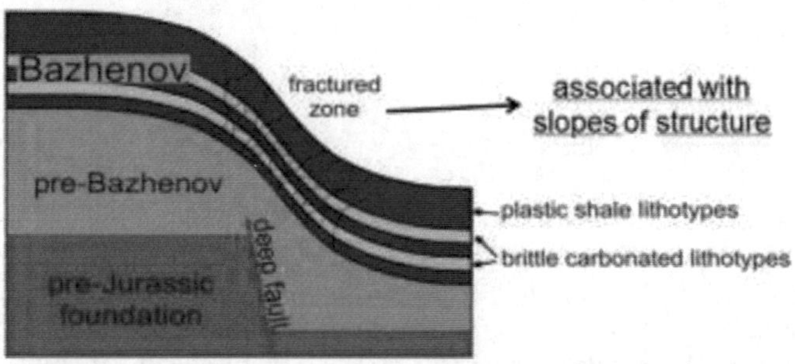

Figure 2. Diagram of fractured zones and deep faults.

It should be mentioned that one can see a difference in the rates of the synthetic model and CM during the initial period of time. It confirms a transient flow mode that cannot be evaluated with the CM method.

An empirical dependence between core properties and well log readings was found separately for each lithotype. In particular: Total Organic Content versus Gamma-Ray and Porosity versus Microlaterolog. The Direct TOC vs. GR dependence is explained by the high content of uranium in organic matter. The standard porosity well log methods (Acoustic and Neutron) are not suitable due to the small values of porosities and the impact of the mineral rock composition and TOC on well log readings. Porosity is determined with the thermogravimetric method that takes into account the matrix and fracture pore volume. The empirical dependence "Hydrogen index and S1 parameter versus TOC" was obtained. Thus, petrophysical and geochemical properties were determined for other wells without core.

Seismic data were used to analyze the distribution of properties in interwell space. This paper presents only 2D profiles of low frequency. This frequency is not sufficient to recognize any inner features of the Bazhenov Formation. The derivative analysis of the seismograms was made. The result of this analysis is an additional boundary between 2 zones of the Bazhenov Formation: upper – more organic rich and shale content (low AI) and low – less organic rich and more carbonate content (high AI). This information allows distributing lithotypes in interwell space more correctly.

A tectonic analysis was made and faults were interpreted in the Bazhenov Formation and deeper in the pre-Bazhenov rocks. According to pre-Jurassic foundation tectonic map, the main faults have an orthogonal direction – north-eastern and north-western. The interpretation of the faults in all 2D profiles shows that the faults have the same directions and the majority of the faults do not cross the Bazhenov Formation, but they create fractured zones and can be a reason for fractured reservoir development. The fractured zones in the Bazhenov Formation are associated with deep faults and with the high dip angle of the Bazhenov structure. In other words, the fractured zones are connected with the slopes of structure folds. This idea is shown in Figure 2. Moreover, brittle carbonated lithotypes are more likely to form fractures. Furthermore, an oil inflow was obtained from the wells located in the areas of high dips. Consequently, the productivity index can be connected with the dip angle.

According to most specialists [2, 3], the Bazhenov Formation has two basic types of fractures: auto fluid fractures and tectonic fractures (Figure 3).

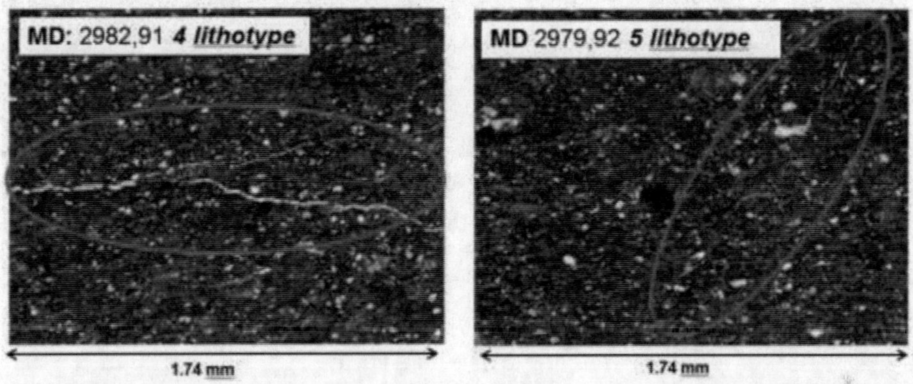

Figure 3. Photos of thin sections in a reference well: Left – Auto fluid fracturing (along horizontal lamination); Right – Tectonic fracturing (vertical fractures due to deep faults).

Organic matter converts into mobile hydrocarbons as it buries and enters an oil window. The growing volume of fluid leads to auto fluid fracturing, and microfractures are mainly formed in less consolidated zones with lateral orientation in conformity with sedimentation lamination. This process is the main reason for the abnormally high formation pressure. Tectonic fractures are mainly formed in fractured zones associated with deep faults.

The models of petrophysical and geochemical properties were based on the lithotype model that was constructed using Indicator Ordinary Kriging, as the Bazhenov Formation sediments are laterally extended at large distances and lithotypes are correlated in different wells.

Dip surface characterizing the fractured zones and, consequently, more voided zones or more porous zones was used as a 2D trend to model porosity with Sequential Gaussian Simulation. Permeability was modeled using an empirical equation with porosity.

Before describing the geochemical modeling process, we should briefly characterize the catagenic maturity of the Bazhenov Formation. According to the tectonic map of the sedimentary cover, Field A is located in the Nyurol Basin underlying surrounding positive structures. According to the tectonic map of the Pre-Jurassic foundation, Field A is above basic intrusive rocks. These 2 conceptual facts lead to high heating of the Bazhenov rocks and, consequently, to their high maturity. The Bazhenov deposits of Field A are in the oil window phase. Moreover, according to the catagenic maturity map developed by Goncharov, field A is located in the zone of a high Tmax value and, according to core data from the reference well, Tmax equals 440 degrees. This information explains the importance of determining geochemical properties.

Geochemical properties such as Total Organic Content, S1 parameter (Figure 4) and Hydrogen Index were modeled with the use of Sequential Gaussian Simulation based on the lithotype model. The

IOP Conf. Series: Earth and Environmental Science **33** (2016) 012033 doi:10.1088/1755-1315/33/1/012033

highest TOC is concentrated in the second kerogenic silicite lithotype while carbonated lithotypes are characterized by low TOC values. Hydrogen Index characterizes the oil-generation potential of source rocks. According to the HI values, the source rocks of the Bazhenov Formation at Field A are in the oil window phase. HI is connected with TOC and therefore carbonated lithotypes are characterized by low HI thanks to which they are less organic-rich. The S1 parameter characterizes already generated mobile oil. The S1 parameter is only low for dolomites. Almost all lithotypes are characterized by high S1.

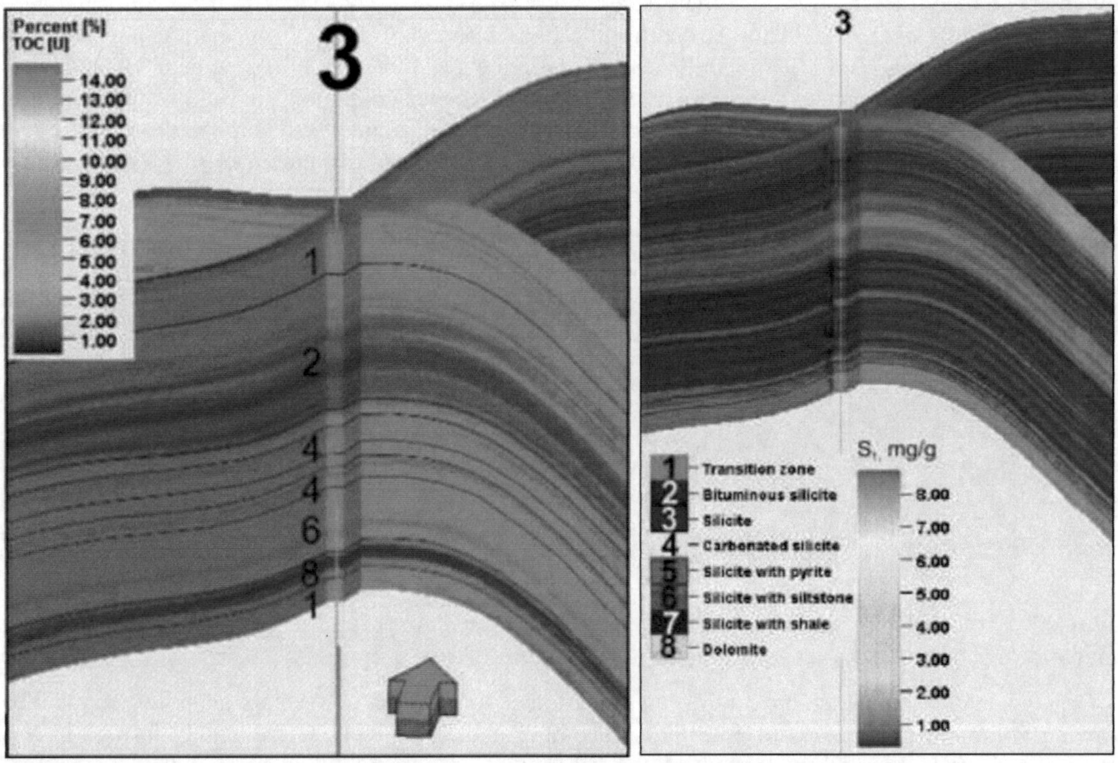

Figure 4. Total Organic Content and S1 parameter 3D model with lithotype differentiation.

Resources were estimated with a conventional method on the basis of the pore volume and also the S1 parameter as the mass fraction of rock [4] and the volume and average density of rocks. The oil-generation potential was estimated with the hydrogen index as a mass of the total organic content. The total organic content was calculated as the mass fraction of rock on the basis of the volume and density of rock. HI was defined as the mass fraction of this TOC. All these parameters were calculated with the help of lithotype differentiation. The following equations were used to estimate resources:

Mobile oil:

Conventional method:

$$\text{Oil-in-place} = \Sigma\, (V_i{*}\text{Poro}){*}S_o/B_o$$

where V_i – lithotype volume, S_o – oil saturation =1, B_o = Volume Factor.

Method using S_1 parameter:

$$\text{Oil-in-place} = \Sigma\, (V_i{*}S_1){*}S_o{*}\,\rho_{rock}$$

where S_1 – mass fraction of already generated mobile oil [mg of HC/g of rock]

Oil-generation potential:

$$\text{Oil mass} = \Sigma\, (V_i{*}\text{HI}){*}\,S_o{*}\,\rho_{rock}$$

where HI – mass fraction of the total organic content [mg of hydrocarbons/g of TOC].

3. Conclusion

Geochemical modeling with lithotype differentiation is an instrument to estimate the oil-potential of a shale formation. Tectonic analysis is necessary to determine promising zones in fractured unconventional reservoirs and to set trends for petrophysical modeling. Modeling geochemical properties is an additional method used to calculate mobile oil resources and oil-generation potential. Also, lithotype identification can be used to choose well intervention. For example, carbonated lithotypes are more prone to hydraulic fracturing while kerogenic silicites can possibly develop with combustion or thermal gas treatment. An additional study is necessary for this analysis.

References

[1] Alekseev A D 2014The Bazhenov Formation: In Search of Big Shale Oil in Upper Salym *Rogtecmagazine*

[2] Braduchan U V, Gurari F G, Zakharov V A and et. al. 1986 *The Bazhenov Horizon of Western Siberia (Stratigraphy, Paleogeography, Ecosystem, Oil Potential)* (Novosibirsk)

[3] Khalimov E M, Melik-Pashaev V S 1980 Research of Big Oil Accumulations in the Bazhenov Formation *Geology of oil and gas* **6** 1 – 10

[4] Bilibin S I, Dyakonova T F, Isakova T G and et al. 2015 The Algorithms of Determining STOIIP Calculation Parameters for the Bazhenov Formation Deposits for the Salym Group *Scientific technical magazine "Rosneft" Geology and Geophysics* 9–17

IOP Conf. Series: Earth and Environmental Science **33** (2016) 012034 doi:10.1088/1755-1315/33/1/012034

Bazhenov Fm Classification Based on Wireline Logs

D A Simonov[1], V Baranov[2] and N Bukhanov[3]

[1, 2, 3] Tomsk Polytechnic University, Tomsk 634034, Russia

E-mail: [2]BaranovVE@hw.tpu.ru

Abstract. This paper considers the main aspects of Bazhenov Formation interpretation and application of machine learning algorithms for the Kolpashev type section of the Bazhenov Formation, application of automatic classification algorithms that would change the scale of research from small to large. Machine learning algorithms help interpret the Bazhenov Formation in a reference well and in other wells. During this study, unsupervised and supervised machine learning algorithms were applied to interpret lithology and reservoir properties. This greatly simplifies the routine problem of manual interpretation and has an economic effect on the cost of laboratory analysis.

1. Introduction

Today one of the critical tasks in oil and gas geology is to study unconventional reservoirs being a source rock for tight reserves. The Bazhenov Formation is exactly a reservoir of that kind. The Bazhenov Formation is the analog of the Bakken field that is successfully developed now [1]. The commercial development of the Bazhenov Formation was proved in the 1960s, but its comprehensive commercial development has started only recently. The Bazhenov Formation is predominantly commercially developed in the central part of Western Siberia. This is a group of fields located in the territory of the Salym megabank. But even this field that is comprehensively studied with drilling has areas where well productivity is very variable. There can be a ten-time difference in oil rates where highly productive wells are located near non-productive ones. This can be explained by the highly heterogeneous structure of the formation and the absence of a unified theory of formation research. Detailed geophysical data, modern core sampling and laboratory analysis allow obtaining a lot of data about the formation. However, very often core data on many wells are absent due to their high cost. Therefore it is difficult to interpret the section of the Bazhenov Formation, especially if it refers to productive intervals consisting of carbonate and siliceous rocks. Formation interpretation is a routine requiring a technique that would help interpret lithology and petrophysical data in wells that are not characterized with core. This task is a machine learning procedure that is gaining momentum now. This procedure deals with mathematical science that can be applied to the task of interpreting formation lithology and reservoir properties in case of the absence of data.

2. Materials and methods

The task of interpreting well logging data on the Bazhenov Formation interval implies determining specific layers in the section, their lithology, a possible association in the group (lithotypes) and predicting petrophysical and geochemical properties through statistical relationships. These relationships between data can be obtained separately for each lithotype at the stage of lithotype interpretation using well logging. It deals with the fact that relationships between data, for example, between the porosity and well logging porosity methods (neutron and sonic logging), represent a cloud of points, the regression equation of which cannot characterize the entire section of the formation. This is primarily due to the extremely heterogeneous composition of the Bazhenov Formation characterized by different proportions of mineral components. Therefore, the interpretation of lithotypes is an important stage of well logging data interpretation, during which it is possible to predict the petrophysical properties and determine promising development zones on the Bazhenov Formation interval.

The first important step was to match core data with well logging diagrams. Core data matching was made on the basis of measurements taken in other areas: Salym. In this area, core was matched with depths on the basis of studying the relationship of radioactive elements such as uranium in the core samples and measuring natural radioactivity in a well, because uranium is the most significant contributor to the radioactivity of rocks in the Bazhenov Formation [2] The second step of Bazhenov Formation interpretation was using accessible log curves. Curves of gamma, neutron, acoustic, electric logs (both conventional logs at different depths of investigation and microresistivity logs) are required to determine specific layers in the formation section. Caliper is also an essential tool for Bazhenov Formation section interpretation, especially when there is a need to distinguish between dense and brittle sections confined to specific lithotypes. Caliper is sensitive to shale intervals due to their ability to demonstrate flaggyness. The methods of GR, LL, NGL and DT allowed clearly determining dense rocks associated with potential reservoir intervals. Also, these layers were determined on the basis of microresistivity logs that indicate the invasion of drilling mud filtrate into the formation. This may be due to both the natural and artificial fracturing of rock (caused by drilling) on these intervals. It is impossible to calculate and predict petrophysical properties at this step due to the absence of relationships between petrophysical values on core and well logging parameters for the entire section of the formation where the relations are a "cloud" of points without an explicit dependence. It is convenient to use relationships obtained from the interpreted groups of rocks – lithotypes (Figure 1, 2).

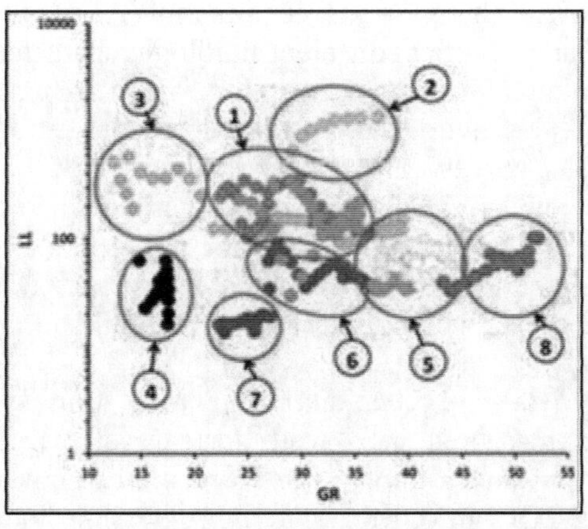

Figure 1. GR-LL plot for lithotype identification.

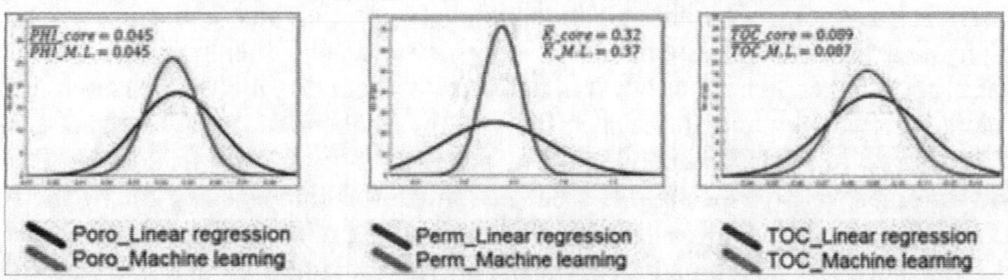

Figure 2. Different algorithm results.

Determining lithotypes at the stage of well logging analysis allows studying the interpreted layers and finding specific features on the basis of which these layers can be grouped. Only three logs of the whole complex of well log curves are essential to determine lithotypes: GR, LL and NGL. The method of specific area identification on GR-LL and GR-NGL was used to interpret the lithotypes. In this case, the NGL units should be converted into NPHI because it is inappropriate to use conventional units for neutron logs due to the different calibration of logging devices. The lithotypes are determined most clearly on the GR-LL plot, but there are areas on this plot where it is impossible to identify any lithotypes. In this case, the GR-NPHI plot should be used. The plot allows making several conclusions. As we can see from the GR-LL plot, there is a small decrease in resistivity vs. the increase in GR values. The upper part of the Bazhenov Formation is characterized by higher values of gamma ray logging and lower values of resistivity logging and vice versa for the lower part of the section. It means the increase in the carbonate content in the lower part of the formation. Higher carbonate content testifies to the greater effective thickness of the lithotypes being a potential reservoir. With lithotype interpretation, it was possible to determine such lithotypes as: 1) Siliceous argillite with low TOC; 2) Siliceous rocks (silicite); 3) Shaly-carboniferous siliceous rock; 4) Transition zone-2 (TZ-2); 5) Interbedding shaly-kerogen and siliceous rocks; 6) Shaly-carboniferous siliceous rocks; 7) Transition zone-1 (TZ-1); 8) Siliceous rocks with high TOC (radiolarite)

Core material was interpreted with the help of the data obtained during laboratory analysis which was conducted in the Field R study area. Using the results of lithological, mineralogical, geochemical and petrophysical studies is an important part of lithotype interpretation on the basis of core, and they are further used as a training sample for machine learning algorithms. In addition, core is important, as it is "hard data" characterizing a formation section. The lithotypes were identified on the basis of core using two sets of data obtained during mineralogical and geochemical laboratory analyses. The mineralogical core analysis is focused on the percentage of accessory and authigenic minerals within the Bazhenov Formation core interval. These minerals include quartz, calcite, dolomite, chlorite, illite, kaolinite, plagioclase, muscovite, alternating-layer mineral (ALM), feldspar, siderite and pyrite. Variations of mineral values that best reflect different lithologic characteristics are observed only for 4 components, such as quartz, dolomite, calcite and argillite, which are a mixture of clay minerals (illite, chlorite, kaolinite) and mica (muscovite and ALM). Core can be divided into 4 lithotypes that correspond to the well log data curves. The lithotypes interpreted on the basis of core (silicites, argillite, carbonates, transition zone) were also interpreted with the help of a geochemical analysis and mineralogical analysis. The parameters obtained from the pyrolysis of core samples, such as S1, S2, TOC, HI, OI, were used as a foundation for lithotype interpretation. It is interesting to note that parameter S2 repeats the TOC curve, which indicates the high oil-sourcing potential of the Bazhenov Formation in the area under study.

One of the well logging data interpretation problems is finding a relationship between the parameters of physical fields measured in a well during well logging and reservoir properties determined during a laboratory analysis of core. Also, it would be convenient to forecast geochemical properties, one of which being TOC, in order to determine possible future oil- and gas-bearing areas within the Bazhenov Formation

It is impossible to obtain a reliable relationship between porosity and standard types of logs measuring a physical field being responsible for the same porosity (neutron and sonic logs) on the whole interval of the Bazhenov Formation. It is noteworthy that porosity log measurements reflect the porosity of rock less effectively due to the specificity of the Bazhenov Formation. Deep resistivity data (LL) were used to find relationships between core porosity and a geophysical field. The absence of deep resistivity in other wells of the field R area section can be compensated for by the recalculation of induction log values to deep resistivity log values. The most reliable dependence is derived on the basis of rock with porosity values obtained via the thermogravimetric analysis. Due to the lack of essential data, petrophysical and geochemical relationships can be obtained only for two lithotypes: 1 and 3. The correlation coefficient for porosity and lateral log resistivity is equal to 0.78. The permeability values measured via the thermogravimetric method also correspond to the thermogravimetric porosity correlation coefficient being equal to 0.75. TOC has a good correlation only with the total gamma activity due to the uranium nature of organic content with the correlation coefficient being equal to 0.8.

Applying the theory of machine learning to the task of interpreting lithotypes and recognizing them in other wells is the second global stage of the study. The whole learning process is based on core data, since core is non-changeable evidence that directly characterizes the Bazhenov Formation section in the area under study. A training sample can be created on the basis of the lithotypes identified from core and it can be used to determine lithotypes in other wells via a reference well. It represents a transfer of the research scale from core analysis to the reference well and to the entire field. In general, the problem of machine learning is divided into many types of problems, but there are two specific objectives the algorithms of which were used in this study. These objectives comprise supervised learning and unsupervised learning. Automatic classification deals with raw data, a so-called training set. The training set is a finite set of objects or events selected from the general population of precedents to solve machine learning problems. The basis of the training sample is formed by three well logs selected as the key ones in lithotype interpretation. A class can be assigned to each lithotype. Thus, the training set is a set of classes corresponding to the lithotypes that can be identified throughout the Bazhenov Formation section.

Clustering with the k-means algorithm is the most common method used to solve unsupervised learning problems. The method is based on the fact that all raw data are divided into several disjoint groups, whereas objects from the same group have similar characteristics. The characteristics of objects from different groups differ from each other. Groups where objects have similar characteristics and differ from other objects in other groups are called clusters. Data clustering was performed by dividing clusters into 4 groups, as four lithotypes were interpreted from core. To control clustering quality, average distances in each cluster should have approximately small equal values. It testifies to high-quality clustering and the absence of a residual point. If a high average value is obtained, it means that one of the groups is characterized by unsatisfactory clustering quality. Clustering results are compared with manual core interpretation data. If the ratio of truly and falsely interpreted clusters is 50/50, it means unsatisfactory clustering.

Due to the fact that clustering shows the low convergence of the results and low quality interpretation, supervised algorithms should be used. Before the application of supervised methods, we should describe the specificity connected with the lack of data. Core was sampled only from the half-interval of the Bazhenov Formation and this core allowed characterizing only 4 lithotypes. Therefore it was decided to extend the lithotypes that were not characterized by core to the lithotypes that were interpreted from core. As a result, the Bazhenov Formation is characterized by only 4 lithotypes. It is convenient to use machine learning algorithms. These lithotypes are described in Table 3. Potential reservoirs in this case are lithotype 2 and lithotype 3. Before applying the algorithms of machine learning, the training set was divided into test and training sets to select more stable algorithms and an optimal setting for them. Train data are a range of values (in case of well R4 – GR, LL, NGL) with an assigned class. Test data are a range of values used as raw data without any classes assigned, but actually a specific class is determined for a test set. It is used to compare learning quality with training data. Quality control was performed with the ROC analysis where the quality of classification can be numerically estimated with the AUC (area under curve) parameters. If AUC is equal to 1, then the algorithm is considered high-quality. If this parameter is equal to 0.5, then the classifier does not work properly. The most stable algorithms can be selected during the test/training procedure. They are kNN, RF and NB. The CT and SVM algorithms show unsatisfactory results, as they are very sensitive to different sets of data. For supervised learning, about 50-60% of the data from the entire training set can be sufficient for reliable learning. Also, more sensitive parameters having a high influence on the area under curve were selected for each algorithm. When the algorithms are set up, machine learning can be used for the reference well and then for other wells of the field.

The kNN algorithms show the best machine learning results in terms of percentage (74.1% of the truly classified parameters). Also, they ensure more reliable visual interpretation. Therefore, the kNN algorithms should be used to apply machine learning in other wells. Other wells should be adjusted to the reference well in order to apply the machine learning procedure. Further, the kNN algorithm should be used. For lithotypes 2 and 3, it is possible to create a thickness map and determine a potential zone for further drilling. Using the algorithms of machine learning, it is also possible to distribute petrophysical and geochemical properties. A porosity and permeability map was constructed for lithotypes 2 and 3. A TOC map can also be obtained. It is possible to perform quality control for machine learning of petrophysical and geochemical property prediction by comparing values obtained via machine learning with values obtained with the help of the linear regression prediction of properties. Linear regression relationships, as it has already been mentioned, were obtained only for the interval of lithotype 3. Figure 2 shows that the machine learning procedure makes a reliable prediction of properties, as the average values for linear regression prediction and machine learning are almost the same.

3. Conclusion

The following results were obtained during the study:

1. Main lithotypes were interpreted on the basis of well logging and core data. Eight lithotypes were obtained from well logging data, whereas core data allowed obtaining only 4 lithotypes. The remaining lithotype intervals that were not characterized by core were extended to the interpreted core lithotypes.
2. A machine learning procedure was applied. First, optimal settings for the algorithms were selected. Second, the authors determined the most stable algorithm showing good automatic interpretation results.
3. The automatic interpretation of the Bazhenov Formation was performed for other wells. Lithotypes with the highest development potential were identified. A potential reservoir map was created. Promising zones for drilling were selected.
4. A machine learning procedure was used to predict petrophysical and geochemical properties. Porosity, permeability maps for the potential reservoir and a TOC map for all fields were created. Promising zones were selected.
5. The analysis of petrophysical and geochemical data prediction was made. Machine learning techniques allowed obtaining reliable results.

References

[1] Alekseev A D 2014 The Bazhenov Formation: In Search of Big Shale Oil in Upper Salym *Rogtecmagazine*
[2] Pavlova M A, Kamkina A D and Suhorukova K V 2012 *Geology-Geophysical Model of the Bazhenov Formation* Report (Novosibirsk)
[3] Bukhanov N, Belozerov V B, Demyanov V V and Evdokimov A V 2015 Classification of Bazhenov Formation Using Seismic Traces, EAGE/SPE Joint Workshop *Exploration of Shale Oil Resources and Reserves* (Moscow)
[4] Rotatu A and Rukavishnikov V S 2015 Description and Comparison of Bazhenov Reservoir Models *EAGE Conf.*(Tyumen)

IOP Conf. Series: Earth and Environmental Science **33** (2016) 012035 doi:10.1088/1755-1315/33/1/012035

Selection of the optimal completion of horizontal wells with multi-stage hydraulic fracturing of the low-permeable formation, field C

A M Bozoev[1] and E A Demidova[2]

[1] Gazpromneft Muravlenko, Muravlenko, Russia, [2]Tomsk Polytechnic University, Tomsk, Russia

E-mail: [2]DemidovaEA@hw.tpu.ru

Abstract. At the moment, many fields of Western Siberia are in the later stages of development. In this regard, the multilayer fields are actually involved in the development of hard to recover reserves by conducting well interventions. However, most of these assets may not to be economical profitable without application of horizontal drilling and multi-stage hydraulic fracturing treatment. Moreover, location of frac ports in relative to each other, number of stages, volume of proppant per one stage are the main issues due to the fact that the interference effect could lead to the loss of oil production. The optimal arrangement of horizontal wells with multi-stage hydraulic fracture was defined in this paper. Several analytical approaches have been used to predict the started oil flow rate and chose the most appropriate for field C reservoir J1. However, none of the analytical equations could not take into account the interference effect and determine the optimum number of fractures. Therefore, the simulation modelling was used. Finally, the universal equation is derived for this field C, the reservoir J1. This tool could be used to predict the flow rate of the horizontal well with hydraulic fracturing treatment on the qualitative level without simulation model.

1. Introduction

Most oil companies are beginning to develop field with hard to recover reserves due to growing demand. However, development of these types of deposits cannot be achieved an uneconomical production rate by conventional method (i.e. vertical wells with hydraulic fracturing treatment). The new technology (horizontal wells with multi-stage hydraulic fracturing treatment) replaces the previous one due to high production rate.

Further, many scientists have analyzed this issue since 1984 years, for example, such scientist as Giger et al. suggested the first analytical solution for evaluation of horizontal wells intersecting fractures; however, this equation could not manage with flow in the fracture and in the reservoir properly. Moreover, Hujun Li (1996) and Economides and Martin (2010) presented the more practical application of mathematical models. In addition, the next mathematical model was introduced by Boyun Gou and Xiance Yu (2008). The proposed equation includes several flows (the reservoir radial and linear flow, fracture linear and radial flow respectively).T.M. Herge (1994) presented accurate methods of determining a productivity of multifractured horizontal wells as well as Hong Yuan (2010) showed the new approach for calculation of horizontal wells productivity with and without fractures [1-8].

IOP Conf. Series: Earth and Environmental Science **33** (2016) 012035 doi:10.1088/1755-1315/33/1/012035

In addition, these models are intended for horizontal wells with transverse fractures. The horizontal well with transverse fractures will be used for field C. Geological and petrophysical properties of the field C reservoir J1 are presented in the Table below.

Table. Petrophysical properties of the field C reservoir J_1.

Parameter	Unit	Value
Average depth (TVDSS)	m	3006
Average total thickness	m	3.3
Average net pay thickness	m	1.6
Porosity	fr.	0.15
Oil saturation factor	fr.	0.43
Permeability	mD	2.5
Net to Gross ratio	fr.	0.85

Unfortunately, none of these methods give good result due to the following reasons:

- Fracture radial flow is not taken into account
- Low permeable reservoir
- Interference effect

However, Economides method demonstrates best convergence in comparison with other methods. Moreover, the wells that have been exposed by the interference effect can be detected on the following chart (only 16 wells) (Figure 1). The coefficient of correlation showed a good result. It equals 71.4 % without wells with interference effect.

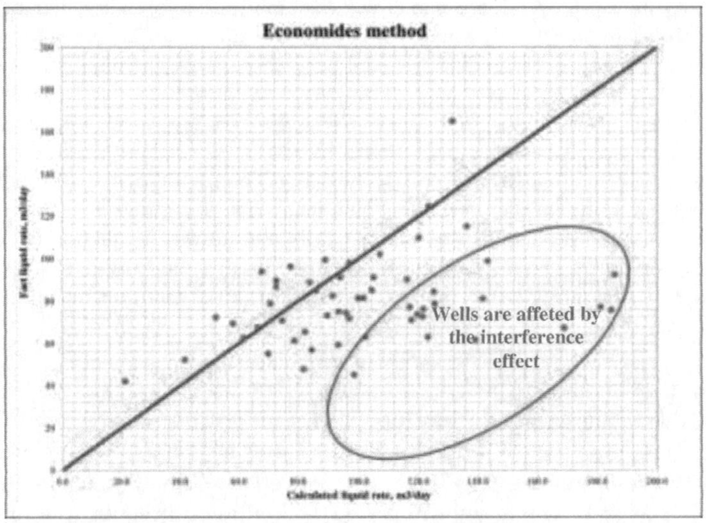

Figure 1. Economides method with interference wells.

Furthermore, none of these analytical solutions which were listed above, take interference effect between adjacent fractures into account. Therefore, the simulation model seems to be an appropriate approach to solving this problem (optimum completion of the wellbore).

2. Simulation approach

The size of this sector model is 1400 meters in length (X-direction) and 800 meters in width (Y-direction), moreover the number of the cells in the X-direction, Y-direction and Z-direction is 28, 16 and 32 respectively. Also, the dimensions of the sector model were selected with respect to the well pattern.

After that, the horizontal well (length of horizontal section is around 600 meters) was added in the center of the sector model with fractures along the horizontal wellbore. Moreover, the fractures were set by the function WFRA (well fracture) in the Tempest software. In addition, there is no pressure support, so horizontal well is on depletion drive during three years (normal lifetime of the fractures after multi-stage hydraulic fracturing treatment)

The simulation model was provided for the next cases:

- 1-7 hydraulic fractures along the horizontal wellbore (the maximum 7 fractures, because the minimum distance between fractures should not be less 80 meters due to technological constraints).
- The mass of proppant varies from 30 ton to 167 ton per one stage (there is a risk of breakthrough in the overlying water-saturated reservoirs above 167 ton of proppant per one stage).

Location of frac ports relative to each other should be uniform because uniform distribution of frac ports along horizontal part of the well gives the effective involvement of the reserves by a well and exclusion of interference effect.

In general, forty seven models have been done with different fractures and mass of proppant. The cumulative oil production schedule is presented with 100 ton of proppant per one stage for different number of fractures.

Based on this graph, it is possible to conclude that interference effect can be observed with increasing number of fractures (reduction of rate of cumulative oil production). Moreover, this effect also could be detected at the start of the well production. The following chart demonstrates that the production rate does not increase linearly, the production rate remains the same with 7 fractures in comparison with 6 fractures (Figure 2). In addition, the flow rate from the simulation model with 5 fractures (Qliquid =76.86 m3/day) is approximately the same in comparison with the real data, for example the well H5 (L=593 m, Qliquid =78 m3/day).

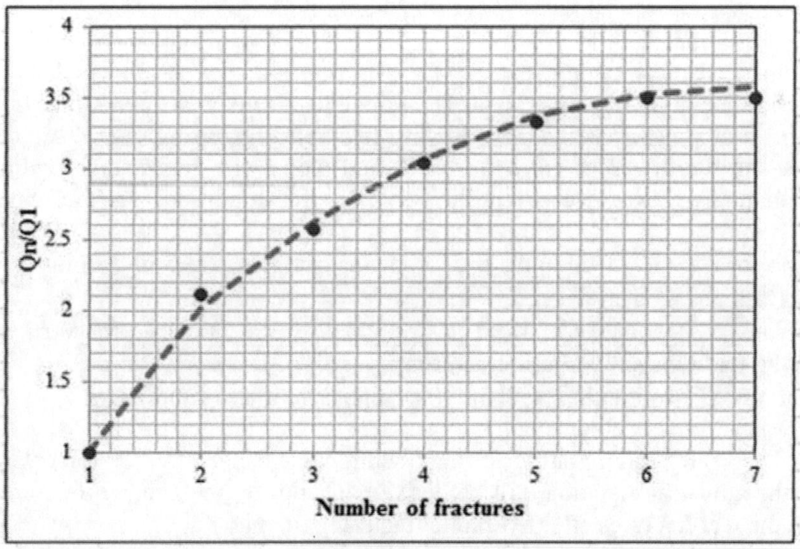

Figure 2. Influence of number fractures on the production rate of the horizontal well with 100 tons of proppant per one stage.

Selection of the optimal horizontal well arrangement is directly dependent on the economic results. As the rate of cumulative oil production reduces with fracture number increase, however, it is still the highest in comparison with others. However, the expenses also increase with the number of fractures [9-11]. The chart of NPV was constructed for different cases. The obtained results are demonstrated in Figure 3.

IOP Conf. Series: Earth and Environmental Science **33** (2016) 012035 doi:10.1088/1755-1315/33/1/012035

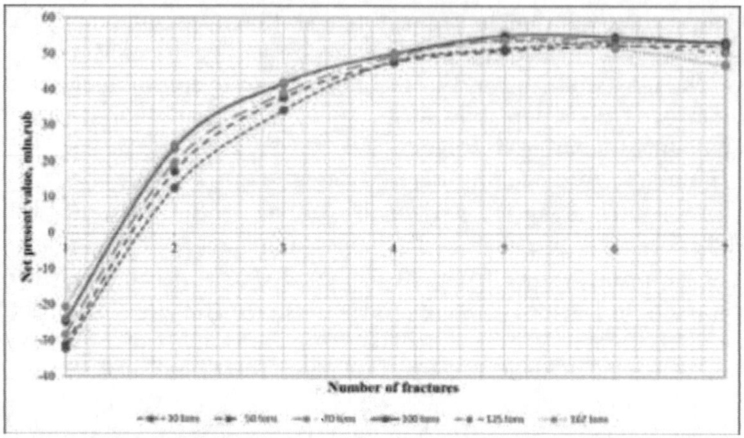

Figure 3. Net Present Value for different cases.

The optimum number of fractures of horizontal well is five with 100 ton of proppant per one stage based on the NPV results. Moreover, the fractures with higher mass of proppant per one than 100 ton show lower NPV. It could be explained by insufficient increase in fracture geometry (width and length) that can cover expenses.

Finally, the geometry of the fracture (100 ton of proppant) is listed below:

- Propped Half-Length 151 meters
- Total propped Height 56 meters
- Average Frac Width 3.65 mm

3. The universal equation

When the recommended distance between fractures is determined, the universal equation could be derived through the simulation modeling data. The principle of universal equation is the representation of fractures (horizontal well) as vertical wells with hydraulic fracturing [12]. This equation can be used to determine the starting flow rates for horizontal wells with multi-stage hydraulic fracturing treatment on a qualitative level. However, there are several conditions for the application of this equation:

- The presence of a vertical well in the proposed area of horizontal well drilling;
- Hydraulic fractured vertical wells;
- The properties and geometry of the fracture should be identical as for vertical well and for horizontal well with multi-stage hydraulic fracturing;
- Flow from fractures into the wellbore (or negligible flow from matrix into the horizontal section).

First of all, the geometry and properties of the fracture were chosen for 100 ton of proppant per one stage for which the universal equation will be derived. Therefore, the average fracture properties are almost similar to the real fractures after hydraulic fracturing treatment.

Secondly, the correspondence of the interference effect on the multiplicity of the starting flow rates (QL120 /QLn<120) were obtained from the simulation model (5 fractures), with a distance between the fractures being less than 120 meters (limiting distance).

Thirdly, the starting liquid flow rate depends on three variables:

, (1)

where – the starting liquid flow rate of vertical well with hydraulic fracturing, N- number of fractures of the horizontal well, – reservoir pressure in the region where the planned horizontal well will be drilled.

In addition, this function does not depend on the reservoir properties (KH), because the depositional environment is shallow marine so the reservoir properties do not tend to vary significantly. The following chart demonstrates the multiplicity of the starting liquid rate with different number of fractures in comparison with the starting flow rate with one fracture at different reservoir pressure (Figure 8). According to this chart, the equation was calculated based on the linear regression through the least-square deviation method that was derived in the special software Kompas.

— (2)

Finally, the universal equation of determining the starting liquid flow rate for horizontal well with multi-stage hydraulic fracturing was derived by combining these two equations:

——— (2)

If $L \geq 120$ than $L=120$

There are two main factors leading to a production decline at Field A, i.e. liquid loading and condensate banking.

To compare the obtained results (through the universal equation) with the starting liquid flow rate of horizontal well (with Hydraulic fracturing) 47 wells were chosen (Figure 4).

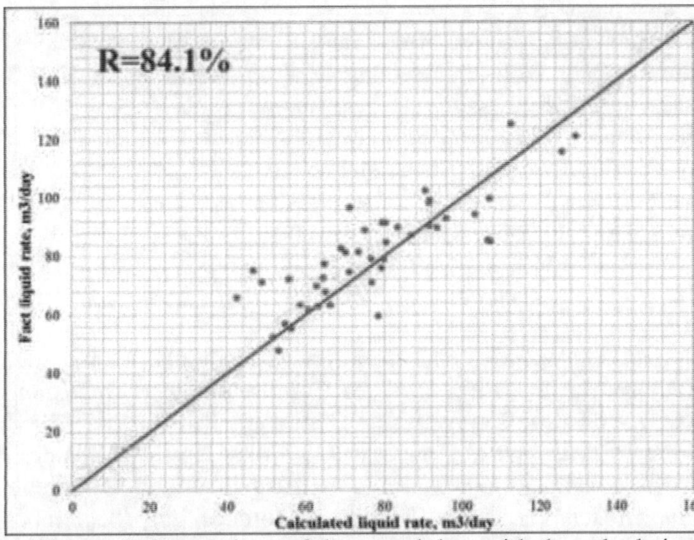

Figure 4. Comparison of the actual data with the calculation through the universal equation.

According to this chart, the linear relationship is observed and the correlation factor is approximately 84.1%. In conclusion, this tool could be used to predict the flow rate of the horizontal well with hydraulic fracturing treatment on the qualitative level without simulation model if the listed conditions have been already satisfied.

5. Conclutions

1. The horizontal wells with multi-stage hydraulic fracturing seem to be more preferable than vertical wells with hydraulic fracturing treatment in the Field C, reservoir J1.

2. The technology (Frack ports or Ball dropping process) of multi-stage hydraulic fracturing is applied in the Field C, reservoir J1

3. Analytical approaches do not take into account the interference effect. However, Economides method should be used with caution, despite the fact that the possible wells with interference effect are observed

4. The simulation model tends to be an appropriate approach to choosing the optimum completion of the wellbore

5. The optimum arrangement of the horizontal well is five fractures with 100 ton of proppant per one stage based on the NPV results. The geometry of the fracture (100 ton of proppant) is listed below:

- Propped Half-Length 151 meters
- Total propped Height 56 meters
- Average Frac Width 3.65 mm

6. The universal equation gives the appropriate correlation factor (R=84,1%) and could be used to predict the flow rate of the horizontal well with hydraulic fracturing treatment on the qualitative level without simulation model

References

[1] Herge T M and Leif Larse 1994 Productivity of Multifractured Horizontal Wlls *SPE 28845, European Petroleum Conference* (London, UK October 25-27)

[2] Hong Yuan, Desheng Zhou and Xian Shiyou 2010 A New Model for Predicting Inflow Performance of Fractured Horizontal Wells *SPE 133610, SPE Western Regional Meeting, Anaheim*, California, (USA, May 27-29)

[3] Guo B and Schechter D S 1997 A Simple and Rigorous Mathematical Model for Estimating Inflow Performance of Wells Intersecting Long Fractures *SPE 38104 SPE Asia Pacific Oil and Gas Conference and Exhibition* (Kuala Lumpur Malaysia April 14-17)

[4] Boyun Guo and Xiance Yu 2008 A Simple and Accurate Mathematical Model for Predicting Productivity of Multifractured Horizontal Wells *SPE 114452 CIPC/SPE Gas Technology Symposium 2008 Joint Conference* (Cakgary, Alberta, Canada, June 16-19)

[5] Economides M, Valko P, Oligney R Unified Fracture Design (London. IBT Global) p 262

[6] Wei Y and Economides M J Transverse Hydraulic Fractures From a Horizontal Well *SPE 94671 SPE Annual Technical Conference and Exhibition* (Dallas, Texas, USA, October 9-12)

[7] Daal J A and Economides: M J Optimization of Hydraulic Fractured Wells in Irregularly Shaped Drainage Areas *SPE 98047 SPE International Symposium and Exhibition of Formation amage Control* (Lafayette, LA, February, 15-17)

[8] Economides M J, Martin A N, Services B J 2010 How to Decide Between Horizontal Transverse, Horizontal Longitudinal and Vertical Fractured Completion *SPE 134424 SPE Annual Technical Conference and Exhibition* (Florence, Italy, September 19-22)

[9] Mirza Talha Baig, Sami Alnuam, Muzammil Hussain Rammay and King Fahd Productivity Increase Estimation for Multi Stage Fracturing in Horizontal Wells for Tight Oil Reservoirs *SPE-178030-MS SPE Saudi Arabia Section annual Technical Symposium and Exhibition* (Al-Khobar, Saudi Arabia, April 21-23)

[10] Crosby D G, Yang Z, Rahman S S 1998The Successful Use of Transverse Hydraulic Fractures from Horizontal Wellbores *SPE 50423 SPE International Conference on Horizontal Well Technology* (Calgary, Alberta, Canada, November 1-4)

[11] McDaniel B W and Keith Rispler 2009 Horizontal Wells with Multistage Fracs Prove to be Best Economic Completion for Many Low-Permeability Reservoirs *SPE 125903 SPE Eastern Regional Meetin* (Charleston, West Verginia, USA, September 23-25)

[12] Hemanta Mukherjee and Michael J Economides 1991 A Parametric Comparison of Horizontal and Vertical Well Performance *SPE-18303-PA*

Fluid Flow Prediction with Development System Interwell Connectivity Influence

M Bolshakov[1], T Deeva[2] and A Pustovskikh[3]

[1,3]Gazpromneft-ntc, Saint Petersburg 190000, Russia
[2]Tomsk Polytechnic University, Tomsk 634034, Russia

E-mail: [2]Deyevata@hw.tpu.ru

Abstract. In this paper interwell connectivity has been studied. First of all, literature review of existing methods was made which is divided into three groups: Statistically-Based Methods, Material (fluid) Propagation-Based Methods and Potential (pressure) Change Propagation-Based Method. The disadvantages of the first and second groups are as follows: methods do not involve fluid flow through porous media, ignore any changes of well conditions (BHP, skin factor, etc.). The last group considers changes of well conditions and fluid flow through porous media. In this work Capacitance method (CM) has been chosen for research. This method is based on material balance and uses weight coefficients lambdas to assess well influence. In the next step synthetic model was created for examining CM. This model consists of an injection well and a production well. CM gave good results, it means that flow rates which were calculated by analytical method (CM) show matching with flow rate in model. Further new synthetic model was created which includes six production and one injection wells. This model represents seven-spot pattern. To obtain lambdas weight coefficients, the delta function was entered using by minimization algorithm. Also synthetic model which has three injectors and thirteen producer wells was created. This model simulates seven-spot pattern production system. Finally Capacitance method (CM) has been adjusted on real data of oil Field Ω. In this case CM does not give enough satisfying results in terms of field data liquid rate. In conclusion, recommendations to simplify CM calculations were given. Field Ω is assumed to have one injection and one production wells. In this case, satisfying results for production rates and cumulative production were obtained.

1. Introduction

The aim of the project will be a selection of methods for production rates forecasting with the influence of interwell connectivity in the development scheme and theirs application on Field Ω.

The objectives of the study will include the following items:

- Investigate all methods which are used for production rates prediction.
- Select the most suitable method for the field Ω.
- Apply the chosen method on the different synthetic models.

All methods can be divided into three groups: statistically-based method, material propagation-based method and potential change propagation-based method. The last method is based on the material balance and superposition principle. Unlike simulation-based methods, capacitance method (CM) doesn't require geophysical and geological data to generate the proxy model. Moreover, CM is less time-consuming than simulation modelling.

Field Ω is a sandstone reservoir. The average permeability and porosity are 7.7 mD and 15%, respectively. Average initial oil saturation equals 53%. The field is developed by inverted 7-spot pattern.

2. Capacitance method

Capacitance method was checked on synthetic model, which has size 70x70x10 cells, the size of one cell is 10x10x2 m, porosity φ=15%, permeability κ=10 mD, anisotropy k_v/k_h=0.1.
The proposed formula for the calculation of average production is expressed as:

$$\bar{q}_j(t_n) = \frac{\tau_{pj} q_j(t_0)}{t_n - t_{n-1}} \left[e^{-\frac{t_{n-1}-t_0}{\tau_{pj}}} - e^{-\frac{t_n - t_0}{\tau_{pj}}} \right] + \sum_{i=1}^{I} \lambda_{ij} \left[\tau_{ij} \left(\frac{w'_{ij}(t_{n-1}) - w'_{ij}(t_n)}{t_n - t_{n-1}} \right) + w_i(t_n) \right], where \quad (1)$$

$$w'_{ij}(t) = \sum_{m=1}^{n} \left[e^{\frac{t_m - t}{\tau_{ij}}} - e^{\frac{t_{m-1}-t}{\tau_{ij}}} \right] w_j(t_m) \quad (2)$$

$$p'_{ij}(t) = \sum_{m=1}^{n} \left[e^{\frac{t_m - t}{\tau_{ij}}} - e^{\frac{t_{m-1}-t}{\tau_{ij}}} \right] p_{wf,kj}(t_m) \quad (3)$$

where $\bar{q}_j(t_n)$ - average production rate of well j, τ_{pj}- time constant, w'_{ij} – convolved injection rate equation, λ_{ij} – weight coefficient which depends on well location and boundary.

$$[\lambda]_{N_p \times N_i} = \bar{A}_{prod}^{-1} \left(\frac{\bar{1}_{N_p \times N_p} \bar{A}_{prod}^{-1} \bar{A}_{con}^{T} - \bar{1}_{N_p \times N_i}}{\bar{1}_{1 \times N_p} \bar{A}_{prod}^{-1} \bar{1}_{N_p \times 1}} - \bar{A}_{con}^{T} \right) \quad (4)$$

where A_{prod}. -influence matrix of producer wells

$$\bar{A}_{prod} = \begin{bmatrix} a_{11} & \cdots & a_{1N} \\ \vdots & \ddots & \vdots \\ a_{N1} & \cdots & a_{NN} \end{bmatrix} + \begin{bmatrix} S_1 & \cdots & 0 \\ \vdots & \ddots & \vdots \\ 0 & \cdots & S_N \end{bmatrix} \quad (5)$$

$a_{11} \ldots a_{NN}$ – influence functions which depends on well location and reservoir boundary, S_{NN} – skin factor of producer.
$[A_{con}]$- matrix accounting for influence of injection wells on producer.

$$\bar{A}_{con} = \begin{bmatrix} a_{11} & \cdots & a_{1N} \\ \vdots & \ddots & \vdots \\ a_{N1} & \cdots & a_{NN} \end{bmatrix} \quad (6)$$

In the case of one producer and one injector λ_{ij} equals 0.904. So equation (1) has the form:

$$\bar{q}_j(t_n) = \frac{\tau_{pj} q_j(t_0)}{t_n - t_{n-1}} \left[e^{-\frac{t_{n-1}-t_0}{\tau_{pj}}} - e^{-\frac{t_n - t_0}{\tau_{pj}}} \right] + \lambda_{ij} \tau_{ij} \left(\frac{w'_{ij}(t_{n-1}) - w'_{ij}(t_n)}{t_n - t_{n-1}} \right) + w_i(t_n) \quad (7)$$

Flow rates calculated from the synthetic model and CM application are presented in Figure 1 and 2. As we can see, capacitance method works well with relative error less than 1%.

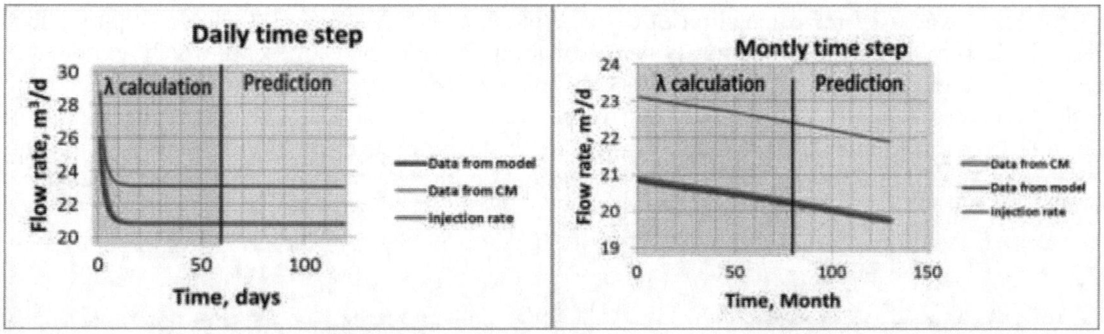

Figure 1. Sensitivity analysis of method at daily time step – left figure. Sensitivity analysis of method monthly time step – right figure; blue curve shows injection rate, green curve shows production rate calculated by CM, red curve shows production rate from synthetic model.

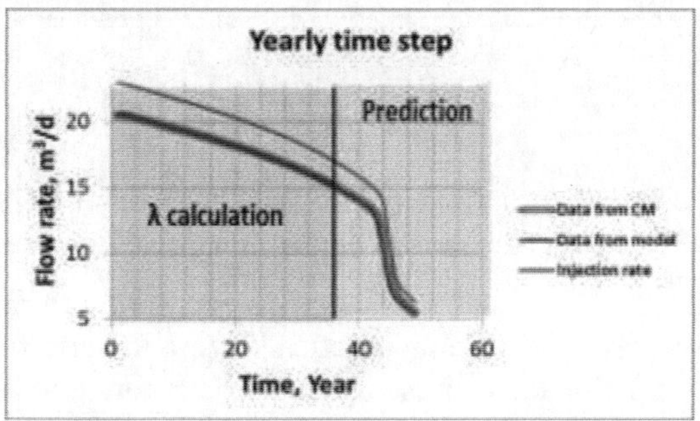

Figure 2. Sensitivity analysis of method at yearly time step; blue curve shows injection rate, green curve shows production rate calculated by CM, red curve shows production rate from synthetic model.

Difference of rates of synthetic model and CM has been revealed in initial period of time. It is evidence of transient flow regime which cannot be evaluated by capacitance method.

Sensitivity analysis of different time steps has been considered and results were represented in Figure 1. It was concluded that duration of time steps didn't influence on final results.

Another synthetic model with 7-spot pattern and semi permeable barrier was created. The location of wells and barrier are shown in Figure 3. Permeability of the barrier is fifty times less than cell permeability. The main idea of this model is resulted in the determination of barrier influence on weight coefficients λ_{ij} and time constants τ_{ij}.

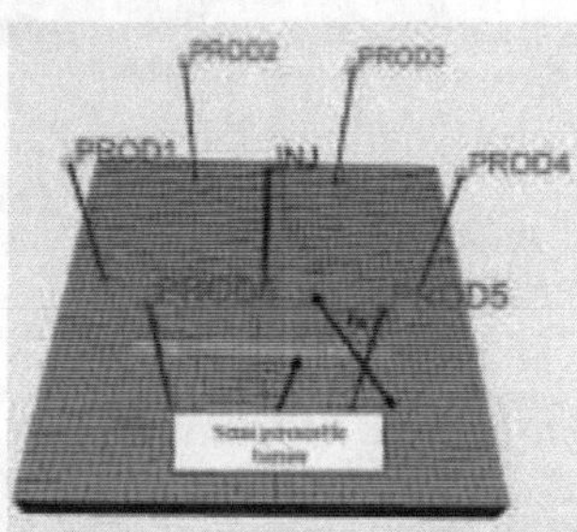

Figure 3. Location of wells and semi-permeable barrier.

Liquid rate's prediction of well PROD3 is quite similar for synthetic and capacitance models, but liquid rate of well PROD6 has the high difference between CM and synthetic model because of barrier's influence on rates and reservoir pressure.

South-western sector of field Ω has been chosen for CM's application. It consists of 10 production wells and 3 injection wells.

IOP Conf. Series: Earth and Environmental Science **33** (2016) 012036 doi:10.1088/1755-1315/33/1/012036

Equation 1 was used for calculation of average production rates. Figure 4 shows graph of liquid rates, which were obtained from monthly operational report (red curve) and data which was obtained from CM for typical production well 161. As can be seen, flow rate, calculated using the formula (1), has the trend that coincides with the field data of the well 161. It's obvious that cumulative production from field data and cumulative production from CM are the same in spite of difference of production flow rates. Data from CM converge with the field data with determination coefficient R^2 equals to 0.998.

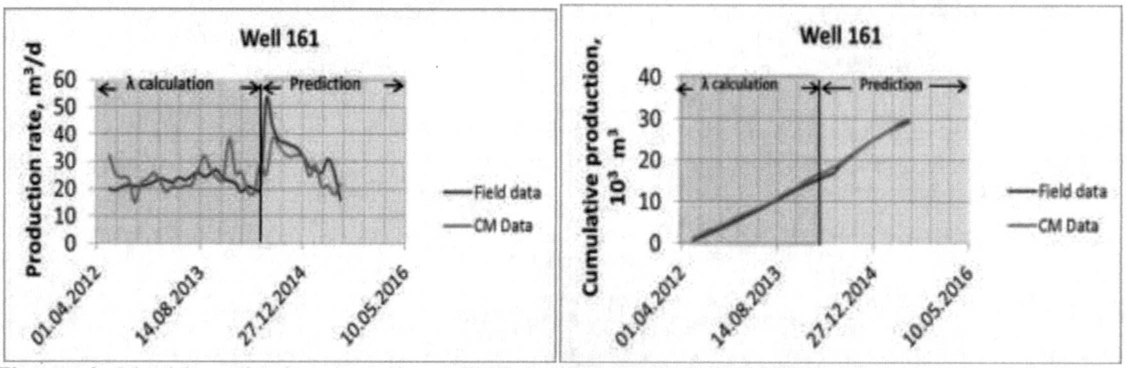

Figure 4. Liquid production rates from field (red curve) and from CM (blue curve) - left figure. Cumulative production from field data (red curve) and from CM (blue curve) – right figure.

It should be mentioned that wells which are located near the margins have lower determination coefficient R^2 for cumulative production and practically zero determination coefficient for liquid production rate. It can be explained by presence of production and injection wells outside the considered sector. Correlation for the wells located in the middle of the sector is much better.

Algorithm of fitting of λ_{ij} and τ_{ij} for large number of wells and time intervals requires a lot of time. All production wells can be assumed like one well and all injectors can be assumed like one well for the simplicity of task.

Average values of λ_{ij} and τ_{ij} were used in this case. Figure 5 shows data which was calculated by capacitance method. As can be seen on Figure 5, liquid production rate which was calculated by using analytical method has similar trend with field data.

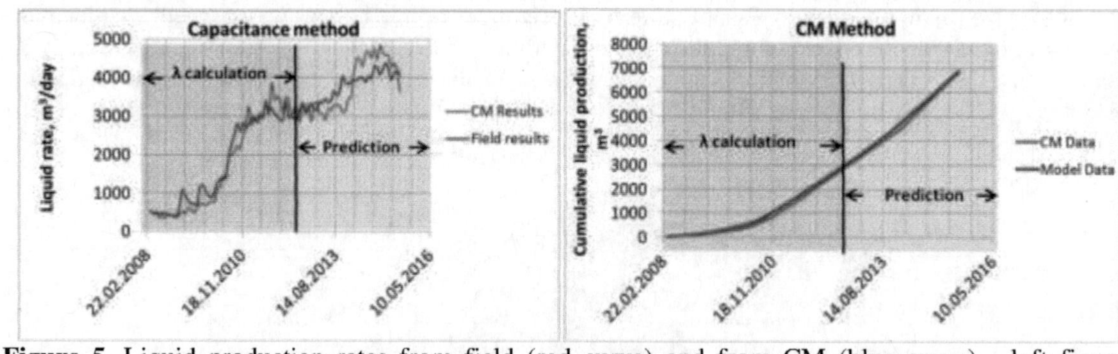

Figure 5. Liquid production rates from field (red curve) and from CM (blue curve) - left figure. Cumulative production from field data (red curve) and from CM (blue curve) – right figure.

Figure 6 shows correlation of liquid production rates from field data and CM .Coefficient R^2 has high value and it means that coefficients λ_{ij} and τ_{ij} were estimated right.

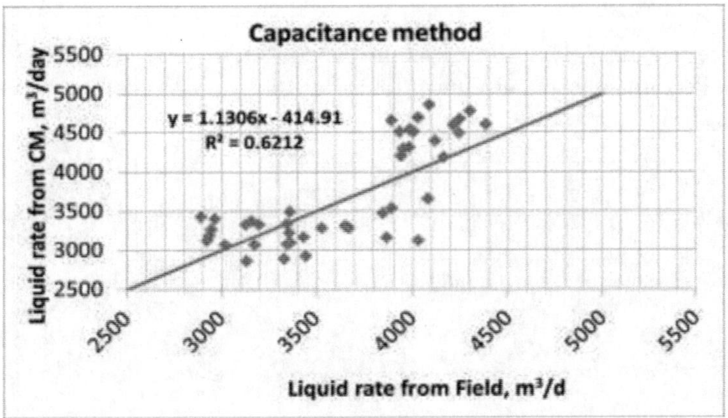

Figure 6. Correlation of liquid production rates from field data and CM.

Cumulative liquid production which was calculated by using CM and field data has similar trend. Both curves (red and blue) have good convergence which means that variance of data is small. Weight factors λ_{ij} for CM is 0.15 and time constant τ_{ij} equals 20 days.

So in case of one development object and small field, oilfield can be considered like a simple model (tank) which has one injector and one producer.

Fractional flow model (equation 8) is used to estimate oil production at capacitance method.

$$q_{oil} = q_{fluid} \times WC = \frac{q_{fluid}}{1+\alpha \cdot w(t)^{\beta}}, where \quad (8)$$

q_{oil}, q_{fluid} – oil rate and liquid rate, respectively, w(t) – injection rate, α and β– coefficients, which were determined from field data.

3. Conclusion

- Parameters λ and τ depend on many parameters: fluid and reservoir properties, square of field, number of samples and noises of input data.
- Methods CM works better for field which has high diffusivity constant $>10^6$ bar*mD/cp
- Sensitivity analysis has shown that optimized time step will be 1 month for calculation by using analytical methods.
- The equations which have been used in this paper do not take into account transient flow.
- Large error is occurred when methods CM uses real production field data.
- Capacitance method works like "tank" (material balance) model and operates with liquid flow rates and cumulative liquid production. Fractional flow model should be used for oil production rates estimation.

References

[1] Yousef A A et al. 2005 A Capacitance Model To Infer Interwell Connectivity From Production and Injection Rate Fluctuations *SPE Annual Technical Conference and Exhibition* (9 – 12 October, Dallas, Texas, 2005)

[2] Mohammad Soroush *Interwell Connectivity Evaluation Using Injection and Production Fluctuation Data* The faculty of graduate studies in partial fulfilment of the requirements for the degree of doctor of philosophy

[3] Panda M N and Chopra A K 1998 ARCO Exploration and Production Technology. An Integrated Approach to Estimate Well Interaction *SPE India Oil and Gas Conference and Exhibition* (New Delhi, 1998)

[4] Kaviani D et al. 2008 Estimation of Interwell Connectivity in the Case of Fluctuating Bottomhole Pressures *Abu Dhabi International Petroleum Conference and Exhibition* (UAE,

2008)
[5] Sayarpour M et al. 2007 The Use of Capacitance-Resistive Models for Rapid Estimation of WaterfloodPerformance and Optimization *SPE Annual Technical Conference and Exhibition* (Anaheim, 2007)

Methods of study in the characteristics of elastomers' samples of mud motors at modeling the conditions of their operation

A V Epikhin[1], V V Barztaikin[2], V V Melnikov[3], Ye D Isaev[4] and O S Ulyanova[5]

[1, 2, 3, 4, 5]National Research Tomsk Polytechnic University, 30 Lenina Ave., Tomsk, 634050, Russia

E-mail: [1]epikhinav@mail.ru

Abstract. The paper describes the development of the methods of experimental research on rubber elastomers parameters of mud motors at modeling the conditions of their operation. The specific features of mud motors operation as well as main aggressive agents affecting the state and parameters of elastomers, such as drilling mud type, ambient temperature, and mechanical action of the rotor on the plate of the stator are shown in the paper. Due to described agents, new methods of investigations and new units of experimental bed are developed.
The experiments confirmed the relevance of the proposed methodology, with its main advantages - ease of implementation and speed of research.

1. Introduction

One of the conditions of drilling with modern bits is the use of high-torque mud motors capable of ensuring the rotation of the tool in the frequency range of 100 400 rpm to achieve high rate of penetration. This condition is satisfied by the construction and power parameters of mud motor. Mud motor takes up one of leading positions among drilling bit gears in the construction of oil and gas wells, as well as their workover. Approximately 50-70% of drilling is provided with mud motors in various regions of Russia [1, 3].

Despite a wide range of advantages, mud motors still have some disadvantages, which limit their application. One of them is small operating time of the motor under field conditions, particularly in horizontal wells. In this case, the operating time is equal to 130 – 250 hours relative to the estimates of 300 – 400 hours. The reason for mud motor failure is a wear of rubber cover of stator that leads to power decrease of mud motor or its complete failure. As a consequence, the numbers of additional round-trip operations increase as well as the cost price of 1 meter of a well, while run speed decreases. A key factor influencing the wear rate of elastomer is an active interaction between its material and drilling mud under high temperatures. According to field research, elastomer of stator in aggressive environments changes its geometrical parameters (bloats or shrinks), and strength characteristics significantly decrease (crumb formation or cracking). As practically proved, in mud motors elastomer wears more often in hydrocarbon drilling mud [2-3].

Talking about the influence of high temperature on mud motors one should bear in mind not only the change of rheological characteristics of drilling mud as a working agent that rotates mud motor rotor. According to field research, the higher temperature and the more aggressive and more abrasive drilling mud, the higher wear rate and elastomer wear that depends on solids and contains [5]. It is connected with low thermoconductivity of rubber, and, if the motor has not special cooling channels, rubber components wear and fail [4, 6].

Improving the efficiency and durability of mud motor from their design to operation is considered by a number of researchers in the following areas: optimization of working agent shapes; application of new materials and covers for stator and rotor, modification of stator; development of advanced manufacturing technologies [1-2, 7-9].

Firstly, some ways to decrease negative influence of high temperatures on mud motors are determined. Searching for and testing new rubber compounds that are wear-resistant under high temperature and are not prone to plastic deformation [4, 10].

Secondly, new operating rate of drilling equipment under high temperature conditions is developed. Then, it is important to make a model of temperature profile along wellbore at all stages of drilling [5] and to design effective drilling equipment. In addition, technological solutions for artificial decreasing the temperature of the drilling equipment and mud (cooling) are developed.

Another relevant area of the research is the development of alternative drilling mud formulation, which preserves its key parameters under minimum negative influence on mud motors, required to ensure the effective drilling, such as maintenance of suspension state and cutting transport; drill through tools cooling, preservation of bottom-hole zone permeability; effective transmission of hydraulic energy to bottomhole and feedback to surface; as well as prevention of complications during drilling. Therefore, the purpose of the paper has been chosen to design the research method of technical of elastomers' samples of mud motors at modeling the conditions of their operation.

2. Designing the research method of technical characteristics of the elastomers' samples

An experimental bed has been designed to conduct research on the mechanical loads on elastomer in various drilling muds. It is a loading unit where a vertical drill press is used to supply a load. The loading unit consists of a metal cartridge with a removable bottom endcap allowing to place and firmly fix rubber samples (Figure 1) and a loading element which is made in two variations (Figure 2). Loading elements are used to simulate the load on the elastomer that is similar to interaction between rotor and stator of the motor in kinematic relationship. The tool with a flat profile of loading (Figure 2a) which has circular perimeter in cross-section with clearance of 1 mm relative to cartridge's sides is used to study the effect of long-term cylindrical loads on the sample, as well as to evaluate the influence of solids mud on sample parameters. The tool with a pointed profile (Figure 2b) is used to produce increased load, and simulate abrasive effect on sample. The design of the cartridge and the loading element, if they are fixed on a vertical drill press, provides a cyclical rotational mechanical action with a sample of elastomer. With the right terms, it allows us to simulate the interaction of "rotor-stator" pair in mud motor.

a b

Figure 1. A metal cartridge for elastomer's samples testing: a – in the disassembled condition; b– in the assembled condition.

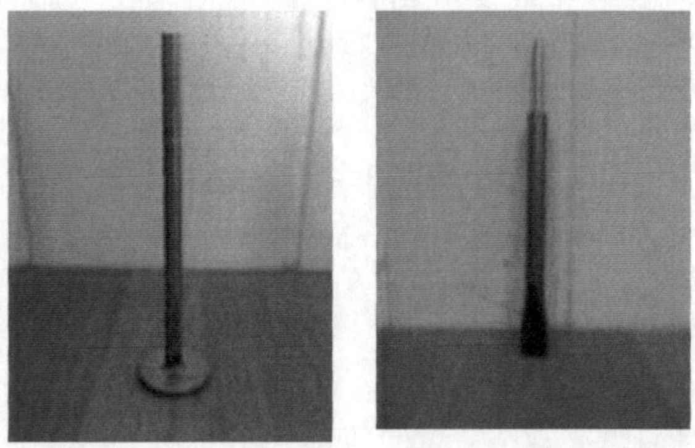

a b

Figure 2. A loading element: a – a tool with a flat profile; b – a tool with a pointed profile.

Experimental samples are made of rubber IRP-1226 with geometrical sizes suitable for fixing them in a cartridge-clip. The technology of their production includes the process of cutting the cylindrical samples with 10-12 mm in thickness by special milling cutter of drill press. The diameter of the sample is determined by the experimental conditions. After cutting the sample, a notch, by which change in the sample diameter is controlled during the experiment, is made on it. Also, a point of thickness is marked on the sample.

The load on the sample of elastomer ranges from 2 to 14 kilos at a pitch of 2 kilos and is provided by adding load to a system of the axial load of the drill machine. The second changing parameter is the time of the experiment: 1, 5, 10 and 15 minutes. In the course of experiments tool sticking (its sticking to the loading tool with a flat profile) is observed with the time of more than 5 minutes and with the load of more than 6 kilos. In this regard, the time of the experiment for the load with more than 6 kilos is corrected and equaled to 1, 2, 3 and 5 minutes. At the first stage of the experiment the spindle's rate of rotation remains unchanged.

The following parameters are recorded in the experiment: the sample mass before and after the experiment, the sample deformation at the end of the experiment and 24 hours later. In addition, visual verification of samples is provided. According to the experimental results, the changes in mass of the sample during the experiments are insignificant and data on this indicator can be ignored.

The second research method is a static experiment in which samples in the form of plates sized 30x100 mm with rectangular cross-section (Figure 3) are kept in geometrical containers filled with drilling mud for different periods of time.

Two opposing holes are perforated in each sample to allow the loading unit. After completion of the specified time interval, samples are removed and installed in the system of loading where their upper end is fixed with static hook, while the lower end of the sample is inserted into the hook with loads hung on it. Such parameters as level of deformation under load (in different time intervals) as well as permanent deformation in 24 hours are measured. Special cylindrical valves are used to decrease heighten load on perforated holes. The use of temporary deformation loads qualitatively illustrates the working "rotor – stator" pair in mud motors because they can be remotely compared with cyclic loads of long duration.

3. Designing the method of elastomer testing under the influence of temperatures

To study the effect of temperature on the characteristics of the elastomers in various mud types, special experiment has been developed, in which the dimensions and durability of the rubber samples IRP-1226 have been observed in different mud types in the temperature range from 25 to 100 degrees. In addition, it is planned to research the influence of subfreezing temperature on elastomers that characterizes the operating conditions of mud motors in Western Siberia in the winter. The main problems in this case are the freezing elastomer or its cover with ice that, in any case, lead to a decrease in mud motors performance.

a b

Figure 3. Elastomer's sample after the experiment: a – deformation under cyclic deformation load; b – abrasive wear.

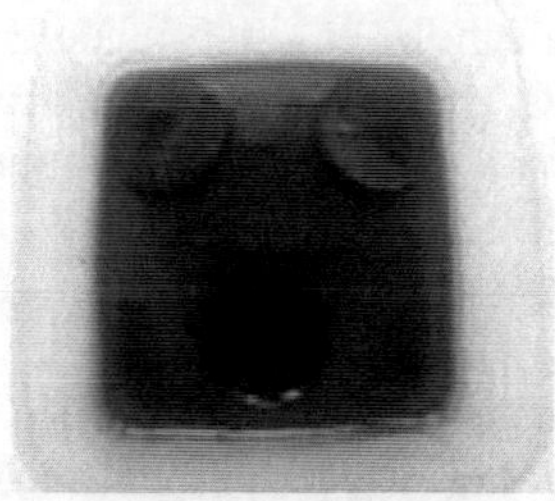

Figure 4. Darkening of diesel fuel under temperature influences (100° C) on elastomer's samples

When the static experiment has been finished, research on the effect of temperature on engineering specifications of rubber IRP-1226 is conducted in several ways: the direct mechanical action at high temperatures, (for this purpose it is planned to develop a separate pilot unit), while postponed mechanical action (after experiments on static exposure medium samples of drilling mud). This defines the mentioned above method of samples preparation for experimental research.

5. Conclusion

The developed method of experimentation has been tested in a series with application of service water as the drilling mud. The research method to study the effect of temperature is used for the base of drilling mud environments (diesel fuel, saturated salt solution, alkaline solution and biodegradable drilling mud). Each of described techniques allows us to observe visually the impact of the drilling mud environment on the parameters and characteristics of the samples. This was expressed in the sample deformation under cyclic deformation load (Figure 3), darkening of the mud (Figure 4), and splitting of samples in experiments evaluating the effect of temperature.

To sum up, in the course of the experiment:

– the problems of mud motors and technological factors limiting their application have been revealed;

– experimental units have been designed, and a method of experimental study of cyclic loading, abrasive and deformational effects on samples of elastomer has been developed;

– the technique of the experiment on the effect of temperature on the characteristics of the rubber of elastomer has been developed.

References

[1] Baldenko D, Baldenko F and Dvoynikov M Design solution to improve the working of downhole PDM *Journal "Burenie i neft"* URL: http://burneft.ru/archive/issues/2013-02/10

[2] Simonyants S L 2007 *Know-how to drill wells by hydraulic down-hole motors* (Moscow: the I.M. Gubkin oil & gas Russian State University) 160

[3] Baldenko D and Korotaev Yu The current status and Russian PDM perspectives of development *Burenie i neft* URL: http://burneft.ru/archive/issues/2012-03/1
Adamson K, Birch G, Hand S, Macdonald C, Mack D and Quadri A 1999 High-pressure, high-temperature well construction *Oilfield Review* URL: http://www.slb.ru/userfiles/file/Oilfield%20Review/1999/autumn/3%20pressure.pdf

[4] Esman B I, Dedusenko G Ya and Yaishnikova E A 1962 *Vliyanie temperatura na process bureniya glubokih skvazhin* 152 [in Russian]

[5] Gidravlicheskie poteri v zabojnyh dvigatelyah vliyanie temperatury na ih rabotu *Techology of wells drilling* URL: http://teplozond.ru/termogidravlika-pri-burenii-skvazhin/gidravlicheskie-poteri-v-zabojnyx-dvigatelyax-i-vliyanie-temperatury-na-ix-rabotu.html [in Russian]

[6] Fufachev O, Goldobin D, Plotnikov V and Traprznikov S New design of startors of screwy wownhole engines made by VNIBT-drill instrument CO. LTD *Journal "Burenie i neft"* URL: http://burneft.ru/archive/issues/2010-06/16 [in Russian]

[7] Anyanwu O N, Klotz C, Labrecque D and Ulrich C 2012 Optimized downhole mud motor delivers outstanding performance improvement in Alaska coiled tubing drilling *Society of Petroleum Engineers – Coiled Tubing and Well Intervention Conference and Exhibition* 88–100

[8] Ross M, Anyanwu O N, Klotz C and Ulrich C 2012 Rib-steered motor technology: The revolutionary approach extends the coiled tubing drilling application scope *Society of Petroleum Engineers – Coiled Tubing and Well Intervention Conference and Exhibition* 101–15

[9] Thompson J and Rossing M 2011 High-temperature downhole motor facilitates cleanout of obstructed thermal well, saving production downtime *SPE Automotive and Composites Divisions – Coiled Tubing and Well Intervention Conference and Exhibition* 455–61

The integrated method to select drilling muds for well construction in difficult geological conditions

V S Horev[1], A Y Dmitriev[2], I A Boyko[3], T R Rakhimov[4], Pandey Sushil Kumar[5]

[1, 2, 3, 4, 5] National Research Tomsk Polytechnic University, Tomsk, Russia

E-mail: [1]Horevvs@tpu.ru, [2]dmitrievau@tpu.ru, [3]Boykoia@niitek.ru, [4]rtr@tpu.ru, [5]sushil_pandey@live.com

Abstract. The article is concerned with the integrated approach to choose drilling mud composition for pay horizons penetration. An optimal choice of drilling mud composition would not only mean ensuring basic fluid properties, but also should minimize the impact on the pay zones when penetrating the pay horizons for the first time. To carry out better assessment of drilling mud impact on pay horizons, it is reasonable to study both technological parameters and filtration analyses, which would allow us to estimate the level of drilling mud impact on the pay horizon.

1. Description of the drilling mud and the experimental data

As it was mentioned above, several types of drilling mud were proposed as initial formulations. The parameters of the obtained drilling muds were studied in terms of thermal stability. The colmatation ability of drilling muds was determined; the return permeability tests under formation conditions were carried out [1, 2].

The following types of drilling muds were considered as the initial formulations:

Polymer mud. Being used at the beginning of the 90-s, this formulation is considered to be obsolete nowadays. Though, it is still widely used due to its relatively low cost. There is a considerable amount of bentonite, which adversely affects the reservoir properties. Thus, while penetrating the pay zone the drilling mud pollutes the bottom-hole zone significantly. It should be noted that large amount of bentonite and solids content increase flow resistance in the circulation [1, 3] It adversely affects not only the colmatation parameter, but also well cleaning and brining the well on to stable production during its development.

Oil-based mud (OBM). Hydrocarbon-based drilling muds have positive properties due to the fact that the liquid phase is diesel fuel, mixed with service water in various proportions to form an emulsion. This emulsion is an inert low density solution.

Along with the inertia, this drilling mud is highly tolerant of contamination (during operating time), so it can be re-used many times after negligible solution treatment. This type of mud being clay-free, it influences positively the reservoir properties while penetrating the pay zone as well as it is less exposed to colmatation by acid insoluble components. [4, 5].

*Biopolymer-carbonate mineralized mud (BCMM).*Being clay-free, this type of mud has a big advantage at the stage of pay zone penetration.

The solid phase of the drilling mud consists of acid soluble colmatation particles, which reduces the required degree of (technological) colmatation of reservoir horizon.

The developed drilling mud formulation has a high ability to inhibit clay deposits. It ensures better factors/conditions to increase penetration rate in the complicated intervals. Bio-polymeric reagents used in the drilling muds degrade if being in situ for a long period after well completion. While well-bore perforation it improves reservoir properties because bio-polymeric reagents lose their original colmatation properties.

To illustrate and prove feasibility of using the proposed drilling mud formulations, thermal stability of drilling muds have been tested [6].

To perform the experiment, 3 time spans, 24, 48 and 72 hours, were used. The timing was based on statistical observations as the common time periods, during which the drilling mud is in the borehole without further processing. The data on thermal stability of the solutions are presented in Tables 1-3.

Table 1. Process parameters of polymer mud after the aging chamber.

Parameters	units	Parameters' values after 24 hours	Parameters' values after 48 hours	Parameters' values after 72 hours
Mud density	g/cm^3	1.08	1.08	1.08
Relative viscosity	St	54	46	45
Shearing strength 1/10	$pound/100\ pounds^2$	4/18	4/7	4/6
Water loss by API	$cm^3/30$ min	6	7	7
Filter cake	mm	0.5-0.7	0.5-0.7	0.5-0.7
Plastic viscosity (PV)	cps	20	16	15
Yield point (YP)	$pound/100\ pounds^2$	16	16.5	18
pH	-	9	9.5	9.5

Table 2. Process parameters of oil-based mud after the aging chamber.

Parameters	units	Parameters' values after 24 hours	Parameters' values after 48 hours	Parameters' values after 72 hours
Mud density	g/cm^3	1,08	1,08	1,08
Relative viscosity	St	45	45	45
Shearing strength 1/10	$pound/100\ pounds^2$	4/6	4/6	4/6
Water loss by API	$cm^3/30$ min	2.1-3	2.1-3	2.1-3
Filter cake	mm	0.5-0.6	0.6-0.7	0.6-0.7
Plastic viscosity (PV)	cps	21	22	21
Yield point (YP)	$pound/100\ pounds^2$	10	10	10

Table 3. Process parameters of biopolymer-carbonate mineralized mud after the aging chamber at temperature of 65 °C.

Parameters	units	Parameters' values after 24 hours	Parameters' values after 48 hours	Parameters' values after 72 hours
Mud density	g/cm^3	1.08	1.08	1.08
Relative viscosity	St	60	52	52
Shearing strength 1/10	pound/100 $pounds^2$	4/5	4/5	4/5
Water loss by API	$cm^3/30$ min	5.0	5.0	5.5
Filter cake	mm	0.5	0.5	0.4-0.5
Plastic viscosity (PV)	cps	17	16	16
Yield point (YP)	pound/100 $pounds^2$	20	18	18
pH	-	9	9	9

The data obtained show that the original formulations have high thermal stability, however, the process parameters of the polymer mud have changed after long staying in the aging chamber. It can be explained by a change in behavior of polymeric additives under high temperature.

2. Filtration study

The next stage of the laboratory research was to determine the colmatation ability of the drilling muds. The filtration studies aimed at assessing the dynamics of core samples permeability exposed to drilling muds were carried out under formation conditions using a laboratory setting PIC-PS.

Schematic diagram of the experiment is shown in Figure 1. With irreducible water saturation, core samples were placed in a core holder of the setting that simulated the conditions similar to downhole environment and confining lithostatic pressure. Kerosene was filtered through the core samples in the "forward" direction (from the formation into the wellbore). When permeability to oil was measured, the tested drilling mud was allowed to flow through the core sample in the “reverse” direction (from the wellbore to the formation) in the mode of dynamic filtering imitating the penetration stage.

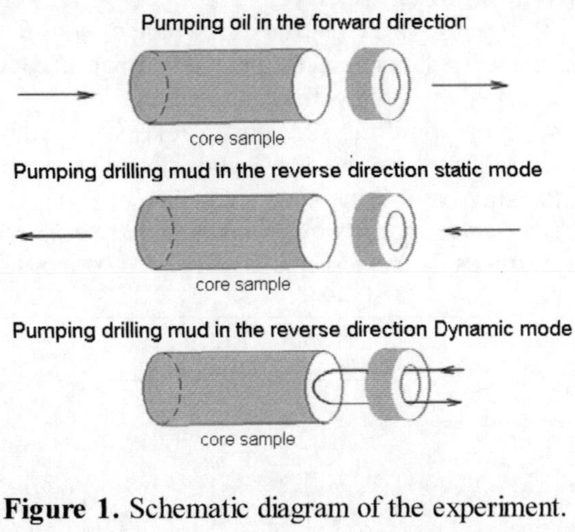

Figure 1. Schematic diagram of the experiment.

IOP Conf. Series: Earth and Environmental Science **33** (2016) 012038 doi:10.1088/1755-1315/33/1/012038

The filtration experiment was finished by pumping kerosene again in the "forward" direction to measure core sample permeability to oil before and after contact with the drilling mud. The experiment resulted in calculating the coefficient of return permeability to oil in % (formula 1).

$$K = \frac{K_{oil_1}}{K_{oil_2}} * 100\% \qquad (1)$$

where K – coefficient of return permeability to oil of core samples, %; K_{oil_1} – coefficient of return permeability to oil of core samples before contact with drilling mud, mD; K_{oil_2} – coefficient of return permeability to oil of core samples after contact with drilling mud mD.

The composition and basic parameters of the core material used in the laboratory tests are listed in Table 4. The results of the filtration studies are presented in Table 5.

Table 4- Core samples for laboratory analysis.

№	Lithological composition	Permeability, 10^{-3} mcm^2
1	Sandstone, fine-grained, silt, clayey	from 5.237- to 5.851
2	Siltstone light gray with a cream shade, sandy, clayey, micaceous, non-calcareous cemented	
3	Grey sandstone, fine-grained, quartz-feldspar, calcareous cemented, non-layered, homogeneous. Cement clayey.	
4	Sandstone, fine-grained, silt, clayey	
5	Sandstone light gray, fine-grained, silt, clay, cemented, non-calcareous, slightly micaceous. Cement clay	
6	Siltstone light gray with a cream shade, sandy, clayey, micaceous, non-calcareous cemented	
7	Sandstone light gray with a cream shade, fine-grained, slit, clayey, non-calcareous, cemented	
8	Grey sandstone, fine-grained, clayey, quartz-feldspar, non-calcareous, cemented, with inclusions of fossils of flora	from 2.807 – to 2.552
9	Grey sandstone, fine-grained, quartz-feldspar, non-calcareous, cemented, non-layered, homogeneous. Cement clayey.	
10	Grey sandstone with greenish-brown shade, fine-grained, with a mixture of grains of medium-grained fraction, clayey, quartz-feldspar, micaceous, non-calcareous, cemented, patchy weakly-cemented. Cement clayey.	
11	Sandstone light grey, grey, consertal, silty, quartz-feldspar, weakly micaceous, non-calcareous, cemented, non-layered, homogeneous. There is dense packing of the grains and cementless fixation	
12	Grey sandstone, fine-grained, weakly clayey, quartz-feldspar, weakly micaceous, non-calcareous, cemented, non-layered, homogenous. Cement clayey.	

Table 5 - Results of filtration studies.

Experiment	Interval	Drilling mud type	*K* of permeability, 10^{-3} mcm^2	*K* of recovery, %
1	2815.00-2826.20	Polymer mud	75.7337	75.734
2	2816.37-2815.07	BCMM[1]	15.309	61.249
3	2504.0-2509.6	OBM[2]	45.494	41.583
4	2957.70-2970.70	Polymer mud	3.279	66.915
5	2815.0-2834.0	BCMM	8.025	69.238
6	2834.0-2853.0	OBM	182.9	73.356
7	2834.0-2853.0	Polymer mud	160.9	17.1
8	2834.0-2853.0	BCMM	215.461	32.98
9	2834.0-2853.0	OBM	505.449	86.857
10	2853.0-2867.5	Polymer mud	28.678	60.60
11	2853.0-2867.5	BCMM	36.858	93.49
12	2853.0-2867.5	OBM	39.916	108.92

The impact of drilling muds on colmatation of the bottom-hole zone is more obvious shown in Figure 2 (an example of core samples № 1 – 5).

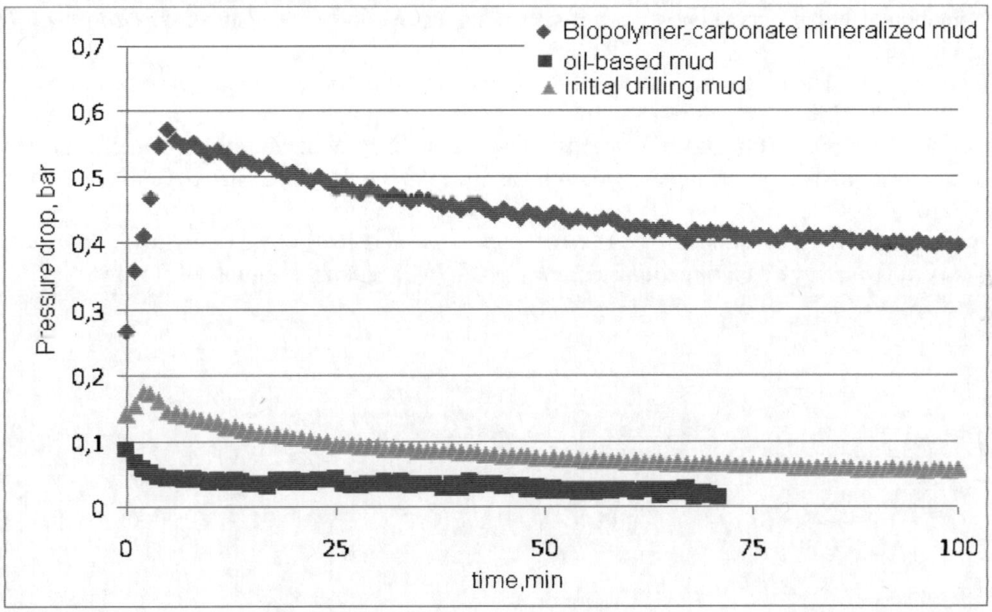

Figure 2. Dynamics of pressure drop while disrupting the crust of drilling muds for samples № 1 – 5.

Filtration study of drilling muds shows advantage of OBM over other drilling mud formulations in terms of coefficient of return permeability. Dependence of the coefficient of different mud formulations return permeability on sample permeability to gas shows strong tends to lower permeability to kerosene (low coefficient of return permiability). It was proved by the study of all drilling muds being filtered through the most permeable core samples. The analysis of oil/water-based drilling muds reveals that application of BCMM in the real condition demonstrates relatively high coefficients of return permeability for all studied suites. The oil-based formulation regularly shows the

[1] Biopolymer-carbonate mineralized mud
[2] Oil-based mud

best results. This fact can be explained by the minimal impact of filtrate penetrating the sample pore space on phase permeability. Basic polymer mud and biopolymer carbonate mud showed almost equal efficiency.

3. Conclusions

The proposed algorithm of drilling mud formulation selection showed its feasibility and objectivity. It is aimed at minimizing the decrease in permeability. The research was carried out to develop integrated approach to drilling mud formulation selection with regard to particular geological conditions. It proved the method to be reasonable and promising. The method allows us not only to select the required process parameters, but also to evaluate its thermal stability, the degree of colmatation effect and quality of reservoir cleaning after acid treatment.

References

[1] Myslyuk M A and Malygin YU M 2003 O vybore optimal'noy retseptury obrabotki burovogo rastvora *Stroitel'stvo neftyanykh i gazovykh skvazhin na sushe i more* **4** 35 – 38

[2] Lind YU B, Kletter V YU, Akhmatdinov F N and Mulyukov R A 2009 Optimizatsiya sostava burovykh rastvorov i operativnoye upravleniye ikh svoystvami *Neftyanoye khozyaystvo* (M.: VNIIOENG) **5** 90 – 93

[3] Medentsev S, Kretsul V and Kuskov V 2003 «Sildril» rastvor na vodnoy osnove s ingibiruyushchimi sposobnostyami rastvora na uglevodorodnoy osnove *Tekhnologii TEK* **6**. 43 – 44

[4] Gaydarov M M and Rogov Ye A 2010 *Burovoy rastvor na uglevodorodnoy osnove* Patent Rossii № 2388784 Byul № 13

[5] Parshin G YU, Vasil'chenko S V and Ryabtsev P L 2014 Modifikatory reologicheskikh svoystv burovykh rastvorov na uglevodorodnoy osnove *Sovremennyye naukoyemkiye tekhnologii* **2** 103 – 104

[6] Ovchinnikov V P, Aksenova N A, Grosheva T V and Rozhkova O V 2013 *Sovremennyye sostavy burovykh promyvochnykh zhidkostey* (Tyumen': TyumGNGU) p 156

The integrated method to select drilling muds for abnormally high pressure formations

V S Khorev[1], A Yu Dmitriev[2], I A Boyko[3], N S Kayumova[4] and T R Rakhimov[5]

[1,2,3,5] National Research Tomsk Polytechnic University, Tomsk, Russia

E-mail: [1]Horevvs@tpu.ru, [2]Dmitrievau@tpu.ru, [3]Boykoia@niitek.ru, [4]Kaumovans@nipineft.tomsk.ru, [5]rtr@tpu.ru

Abstract. The article describes the method for choosing a drilling mud for drilling abnormally high pressure formations. A carefully selected drilling mud formulation would not only enhance an array of interrelated fluid properties, but also minimize the impact on the pay zones when the drill bit first penetrates the pay. To ensure a better assessment of drilling mud impact on the pay zone, it is reasonable to carry out the study focused on the analysis of technological parameters, involving filtration, acid and drilling mud tests, as well as formation damage analysis. This would enable evaluating the degree of mudding off, reservoirs acid fracturing effect and the risks of pipe sticking at significant depth. The article presents the results of the above-described study with regard to the currently used drilling mud and new experimental formulations developed at National Research Tomsk Polytechnic University (Drilling Mud and Cement Slurry Laboratory).

1.Description of drilling mud formulations and experimental data

As it was mentioned previously, several types of drilling mud have been proposed as initial formulations. The critical parameters and characteristics of these drilling muds were analyzed in terms of their thermal stability. The colmatation ability of drilling muds was determined; the return permeability tests were carried out. In addition, drilling mud inhibition properties were evaluated. As initial formulations, the following types of drilling muds were considered:

2. Materials and Methods

2.1Drilling muds with barium salts without additional acid soluble weighting additives

Despite wide application stipulated by low price, the type of drilling mud formulation that dates back to at least the beginning of the 1980s and even earlier is currently outdated. It involves a great amount of barite that adversely affects the reservoir properties. Thus, this drilling mud can cause significant pollution of the bottom-hole formation zone when the drill bit first penetrates the pay.

2.2 Drilling muds with heavy-soluble barium salts

Drilling muds containing siderite and heavy barium salts in low concentrations are considered next-generation formulations as the solid phase added into the drilling mud and negatively affecting reservoir properties is not significant. The use of highly mineralized brines should be mentioned due to their inhibitory ability and almost full absence of bentonite clay, which, in turn, has a positive effect on penetrating strongly cemented shales and other rocks of hard clay. [3]

2.3 Heavy biopolymer drilling muds with sodium and potassium formates

Drilling muds with sodium and potassium formates demonstrate no fundamental differences from the drilling fluids made of heavy salts except for one factor. Unlike heavy salts, formates increase the mud density without using additional weighting agents [4, 5].

The experiment was carried out during the 24-hour period, 48-hour period and 72-hour period. The time periods were defined on the basis of statistical observations, i.e. most commonly encountered time periods when the drilling mud is in the borehole without additional processing. The results of the mud thermal stability are presented in Tables 1-3.

Table 1. Process parameters of drilling mud with barite weighting agent.

Parameters	units	Parameters after 24 hours	Parameters after 48 hours	Parameters after 72 hours
Mud density	g/sm^3	1.68	1.68	1.68
Relative viscosity	St	40-45	56	60
Shearing strength 1/10	pound/100 pounds^2	15/19	13/15	12/13
Water loss by API	sm^3/30 min	5.0	5.3	5.4
Filter cake	mm	0.5-0.6	0.6-0.7	0.6-0.7
Plastic viscosity (PV)	cps	15-30	17	15
Yield point (YP)	pound/100 pounds^2	45-80	42	42
pH	-	9	9	9

Table 2. Process parameters of drilling mud with heavy-soluble barium salts after aging chamber.

Parameters	units	Parameters after 24 hours	Parameters after 48 hours	Parameters after 72 hours
Mud density	g/sm^3	1.68	1.68	1.68
Relative viscosity	St	60	56	60
Shearing strength 1/10	pound/100 pounds^2	15/19	13/15	12/13
Water loss by API	sm^3/30 min	4.0	4.3	4.4
Filter cake	mm	0.5-0.6	0.6-0.7	0.6-0.7
Plastic viscosity (PV)	cps	25	17	15
Yield point (YP)	pound/100 pounds^2	41	42	42
pH	-	9	9	9

Table 3. Process parameters of drilling mud with potassium formate after aging chamber.

Parameters	units	Parameters after 24 hours	Parameters after 48 hours	Parameters after 72 hours
Mud density	g/sm^3	1.68	1.68	1.68
Relative viscosity	St	57	68	69
Shearing strength 1/10	pound/100 pounds^2	2/3	2/3	3/4
Water loss by API	sm^3/30 min	6	6	6.3
Filter cake	mm	0.6	0.3	0.6-0.7
Plastic viscosity (PV)	cps	36	43	45
Yield point (YP)	pound/100 pounds^2	4	4	42
pH	-	10.5	10.5	10

The obtained formate-based drilling mud contains polysaccharide reagents for regulation of filtration and rheology properties. In addition, it includes marble chips for temporary colmatation of the bottom-hole formation zone. Compared to heavy inorganic salts, formates have a number of advantages, precisely: environmental safety, excellent inhibition capacity towards shale, increased thermostability, low corrosivity, compatibility with formation fluids. All these factors positively affecting the penetration conditions lead to the conclusion that the drilling muds formulated using potassium formate are the most suitable for abnormally high pressure formations (AHPF).

2.4. Filtration study

The next stage of laboratory research was to determine the colmatation abilities of the drilling muds. The filtration studies designated to assess the dynamics of core samples permeability exposed to drilling muds were carried out under formation conditions using a laboratory setting PIC-PS.

The schematic diagram of the experiment is shown in Figure 1. Having irreducible water saturation, core samples were placed in a core holder of the setting where the conditions similar to downhole environment were created by applying confining lithostatic pressure. Kerosene was filtered through the core samples in the "forward" direction (from the formation into the wellbore). When permeability to oil was measured, the tested drilling mud was allowed to flow through the core sample in the "reverse" direction (from the wellbore to the formation) in the mode of dynamic filtering imitating the penetration stage.

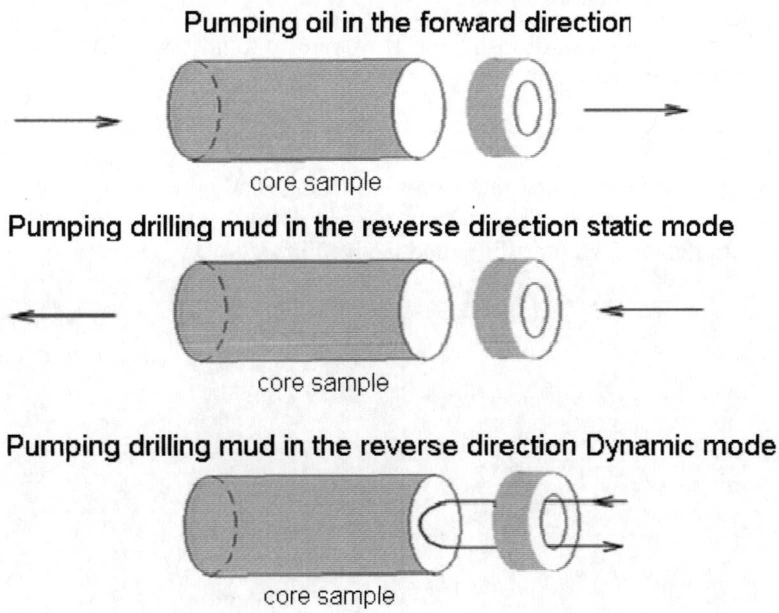

Figure 1. Schematic diagram of the experiment.

At the end of the filtration experiment, kerosene was pumped again in the "forward" direction in order to measure core sample permeability to oil before and after contact with the drilling mud. As a result, the coefficient of return permeability to oil was calculated in % (formula 1).

$$K = \frac{K_{oil_1}}{K_{oil_2}} * 100\% \quad (1)$$

where K – coefficient of return permeability to oil of core samples, %; K_{oil_1} – coefficient of return permeability to oil of core samples before contact with drilling mud, mD; K_{oil_2} – coefficient of return permeability to oil of core samples after contact with drilling mud mD.

The composition and basic parameters of the core material used in the laboratory tests are listed in Table 4.

The efficiency of core sample return permeability was investigated by filtrating formation water through the core samples before and after contact with drilling mud. The first experiment involved the drilling mud containing potassium formate and colmatant. The dependence of pressure drop on the flow rate of fluid filtered through the core sample before and after contact with drilling mud is given in Figure 2.

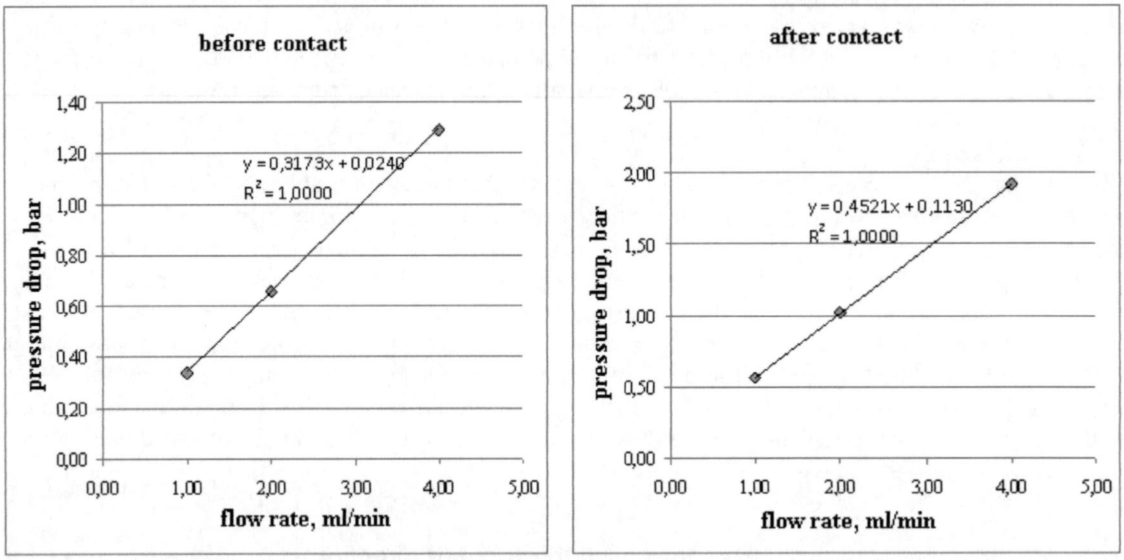

Figure 2. Dependence of pressure drop on the flow rate of fluid filtered through the core sample before and after contact with drilling mud.

The mud cake was disrupted when the reverse pressure drop was 0.213 MPa. Return permeability coefficient – 87.88%.

The second experiment was carried out with the drilling mud containing sodium formate without bridging particles. The dependence of pressure drop on the flow rate of fluid filtered through the core sample before and after contact with drilling mud is given in figure 3.

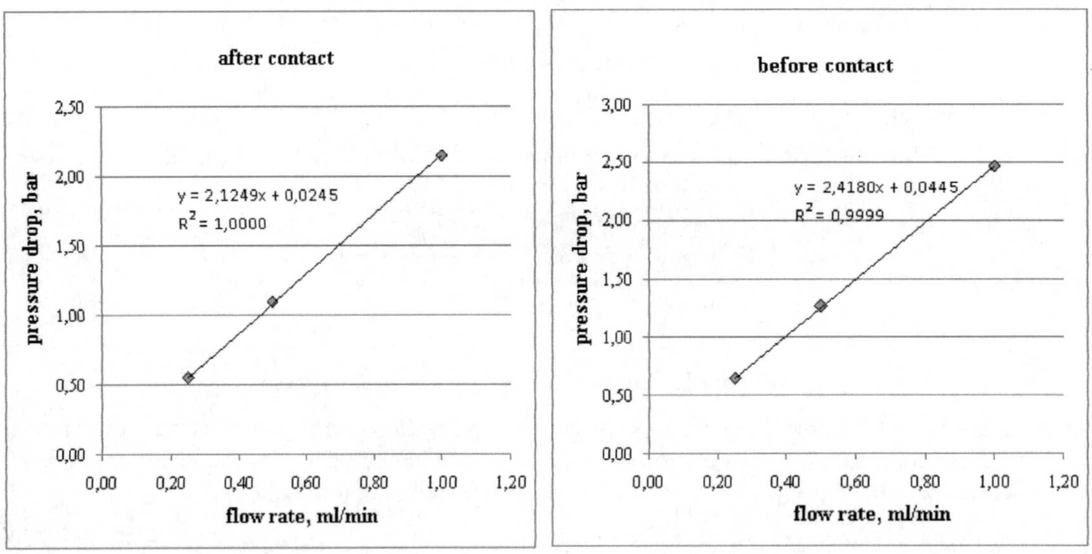

Figure 3. Dependence of pressure drop on the flow rate of fluid filtered through the core sample before and after contact with drilling mud.

3.Results and discussion

The comparative analysis of two drilling muds, the first containing potassium formate and the second with sodium formate colmatant without bridging particles, has showed a significant positive effect in blocking the pore channels by bridging particles. To evaluate the drilling mud inhibition capacity, swell capacity test instrument was used. The analysis results are presented in Table 4.

Table 4. Results of the experiments.

Type of drilling mud	Swell capacity, mm	Inhibitor additive	Temperature, °C	Composition of the sludge tablets
Mud with barium salts without inhibitor	2.8		63	aleurolite 50% + mudstone 50 %
Mud with barium salts with inhibitor	2.4	Shale inhibitor	64	aleurolite 50% + mudstone 50 %
Formate-based mud without inhibitor	2.1		64	aleurolite 50% + mudstone 50 %
Formate-based mud with inhibitor	2.0	Shale inhibitor	62	aleurolite 50% + mudstone 50 %

The obtained data on clay swelling have shown that the formate-based drilling muds demonstrate the lowest values. The difference in swell capacity can be explained by the absence of bentonite in the drilling muds with salts and formats and by a high content of chlorides that increase inhibition properties of shale. For illustration purposes, the data on clay swelling are presented in Figures 4-5.

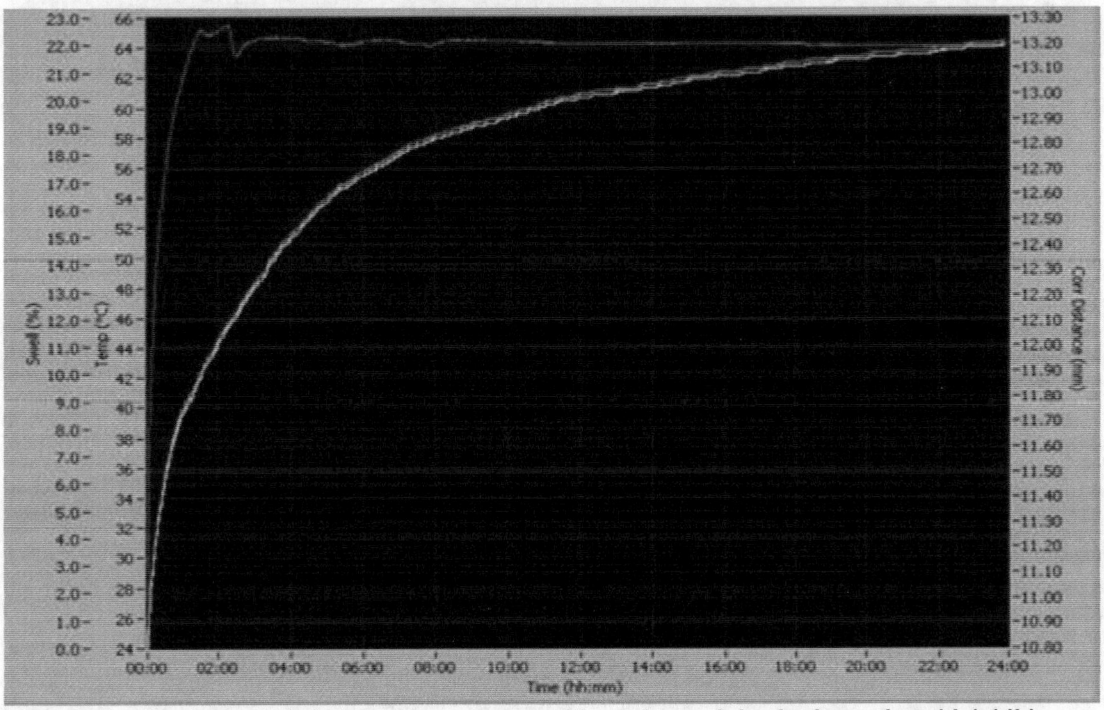

Figure 4. Swelling curve of sludge tablets contact with mud containing barium salts with inhibitor.

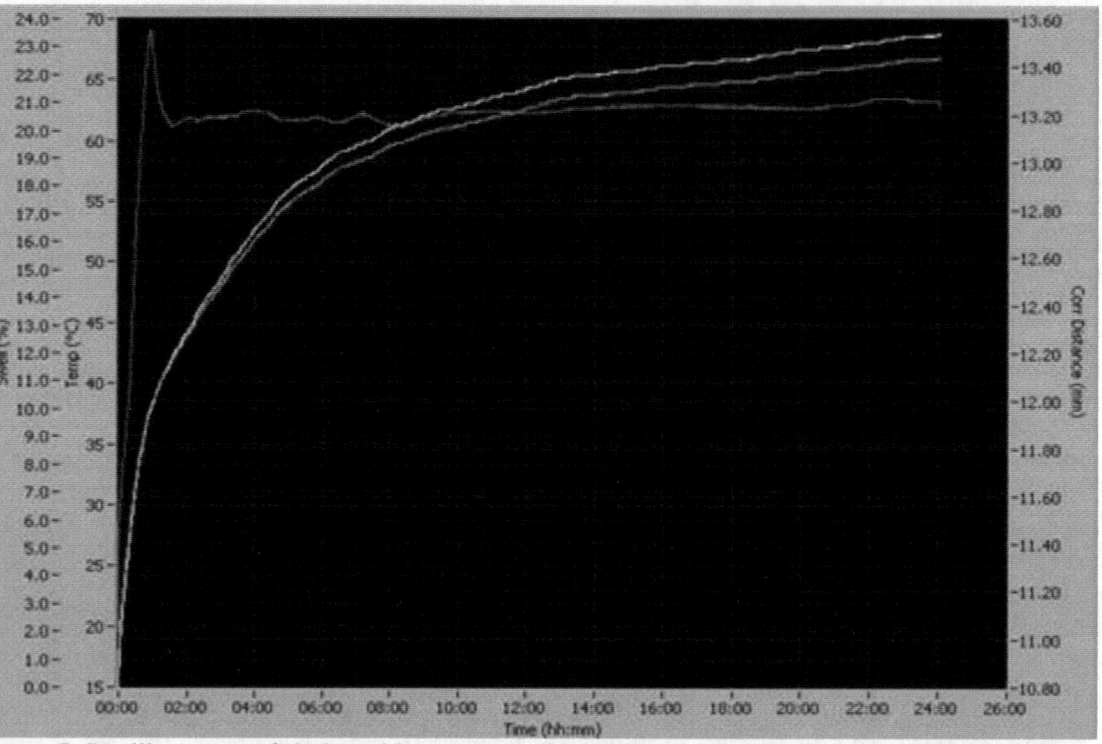

Figure 5. Swelling curve of sludge tablets contact with mud containing barium salts without inhibitor.

4.Conclusion

Compared to the conventional polymer drilling muds, the drilling mud containing formates and barium salts is also proved to be thermostable and efficient for abnormally high pressure formations. The results obtained by means of the swell capacity test instrument have shown that the proposed formulations demonstrate shale swelling reduction of 250% - 300% in comparison with the conventional muds.

Taking into account all the above-mentioned factors that positively influence the conditions of penetration, it can be concluded that the drilling muds containing potassium formate are the most appropriate to penetrate abnormally high pressure formations. Thus, it can be stated that the use of the described method allows engineers to choose the type of a drilling mud that would satisfy specific geological conditions.

References

[1] Avdeyev V D and Yepikhin A V 2012 Vozmozhnosti snizheniya agregatnoy tverdosti gornykh porod *Inzhener-neftyanik* **3** 24 – 29

[2] Rastegayev B A, Shishkov S N and Koshelev V N 2007 Water-based and non-water-based muds for reliable drilling with abnormal high pressure *Bureniye i neft'* **3** 18 – 21

[3] Kashbiyev P P and Urakov M Kh 2013 New type of inhibiting additives to drilling muds *Bureniye i neft'* **3** 40 – 41

[4] Yodgorov N and Magrupov A 2008 Burovyye rastvory na osnove modifitsirovannogo polimernogo reagenta MPR-1 *Neft', gaz i biznes* **2** 60 – 62

[5] Drilling Muds Division of SSK CJSC 2011 Drilling muds to the world standards *Bureniye i neft'* **9** 18 – 19

Application of microwave method for moisture determination of organic and organic-mineral soils

V V Kramarenko [1], A N Nikitenkov [2], V Yu Molokov [3], A V Shramok[4] and G P Pozdeeva[5]

[1, 2, 3, 4, 5] Tomsk Polytechnic University, 30 Lenin Avenue, Tomsk, 634050, Russia

E-mail: [1]kramarenko_v_v@mail.ru

Abstract. The problem of rapid drying arises when determining moisture, ash and organic matter content, as well as during many other soil tests. For highly-organic and organo-mineral peat soils the problem of advanced measurement of moisture content is of special importance, since after reweighing the dry sample increase in mass may be observed. The article examines the methods in determining the moisture content in peat and organic soils via microwave radiation, which will greatly speed up the process, simplify the complexity and cost of laboratory tests. The paper presents a detailed review of the methods determining moisture content in soils and characteristics, as well as application scope. The work contains the research results on moisture organic soils: drying in a microwave oven and the current domestic standards.

1. Introduction

The problem of rapid soil drying arises when determining moisture, ash and organic content as well as many other types of soil tests. For organic and organo-mineral peat soils, advanced moisture determination is of particular importance, since reweighing demonstrates the sample mass increase. Due to this fact, it is required to determine the smallest mass which is relevant to GOST 5180 [4].

The aim of this work is to study drying process in microwave ovens to determine the moisture content of organic and organo-mineral soils. The research objectives include: an overview of foreign and domestic scientific works on the testing techniques of moisture content in soils; sampling and laboratory tests to classify the indicators of composition and physical properties of peats and peat soils; measurement and comparison of their moisture by microwave and convection drying; measurement of organic matter loss after drying; development of brief instructions in using of microwave ovens for rapid drying of organic soils.

Literature and regulatory document reviews revealed variable methods for measuring soil moisture [1-14], and, accordingly, to current standards, a drying method in a convection oven is applied. In accordance with GOST 5180 [4] the soil sample weight of 15-50 g is dried to constant weight at a temperature of 105±2 °C for 5 hours, after which the drying continued to obtain the mass difference between two successive weighings (not more than 0.02 g), with a time interval of 2 hours of mass measurement . According to ASTM D2216 [10] drying dispersed and rock soils is conducted at a temperature of 110±5°C. To reduce gypsum dehydration in saline soils or decomposition of organic matter in peats, it is recommended to dry samples at 60° C or at room temperature in a desiccator. As defined in GOST 19723 [11] the peat portion of 15-20 g was dried within 4.5 hours at 105°C. The dried mass sample per hour is considered permanent if the difference in mass in two successive weighings does not exceed 0.02 g. Based on GOST 11305 [12] the peat portion of 5 - 10 g is placed in a heated oven to a 105-110°C and dried within 2.5 - 4.0 hours, after weighing peat portions are dried for 30 minutes. If the mass loss does not exceed 0.01 g, the test is completed. In the accelerated method [12] peat portions weighing 5-6 g are placed in a drying oven heated to 165-170 °C and then dried at 145-150 °C within 30 min. If the peat portions contain moisture (moisture is the ratio of water mass to the initial mass of wet peat) more than 55 % – 45 min are required. Current measurements are performed in the case of determining a single peat portion. In accordance with ASTM D 2974-14 [13] the moisture peat content is determined by two methods: A) by drying at a temperature of 110±5° C and b) by dewatering in two stages: after evaporation at room temperature and after drying in an oven at 110±5° C (if peat is used as fuel).

According to ASTM D 4643 [14], the drying soil process in a microwave oven is also used to determine the moisture content. A peat portion of 200 g (for soils containing up to 10 % of the fraction greater than 2 mm) is placed in a microwave for 3 minute defrosting. Then the peat portion is weighed at 1 minute interval until re-weighing demonstrates mass loss of less than 0.1 %. Optimal power of the oven may be higher than in defrosting. The time of initial drying of clay soils can be increased to 12 minutes. Moisture is calculated as the ratio of the water mass to dry soil mass within the accuracy of 0.1 %.

2. Materials and methods

The authors studied the soil samples from the territory of Tomsk oblast. During the tests, following samples were analyzed: peat samples of varying decomposition, low-ash, wood, grass, moss, and grass-moss groups and peat clay and loam. In accordance with the procedures [2, 4, 5, 7]: moisture content (m), degree of peat decomposition (Ddp), organic content (Ir) and moisture at the boundary of liquid and fluidity (wL and wp) of organo-mineral soils were determined. During the laboratory investigations the following equipment was used: drying oven, moisture-testing oven SSP–0.25–100 and a dry-air oven SSVL–80–Kasimov, microwave ovens (LG of 900 Watts and Welton WMO–1700GW with variable regulation power, muffle oven MIN–10 UE, scale with accuracy of 0.01 and 0.001 g. To compare convection and microwave drying results each soil sample was divided into halves. Based on the methods [4, 11] from each half, 4 to 8 parallel 15-50 g peat portions were selected. Peatmoss samples were taken with a weight of 5-10 g [12], 15-20 g [11], as well as heavier samples of 100-200 g [14], in order to reduce the variation between parallel determinations of water-saturated samples. Drying was conducted before obtaining the mass difference of soils within containers in two weighings (not more than 0.02 g) [4]. The drying time of the samples in a convection oven complies with the standard requirements [4, 11, 12]. The testing time in the microwave was determined by the moisture content and sample weight and ranged from 10 to 30 minutes for a 50 g peat portion, and for peat portion of 100-200 g up to 1.5 hour (at radiation power ~200 W). The intervals between weighings were selected empirically in respect to the sample weight, moisture content, evaporation area, as well as the inertia of the microwave heating which resulted in significant drying slowdown (the intervals between weighings - less than 1-2 min). The first weighing was carried out at 1-2 minute interval for a sample weight up to 50 g and 3-7 minute interval for the samples of more than 50 g.

It is necessary to note the specific features of drying organic and organo-mineral soils (Table 1, Figure 1) such as, high moisture content in peat, process duration and possible organic matter loss at high temperatures. Maximum moisture content (1500-3000 % or more) was observed in slightly decomposed grass-peat and moss samples; average value of 400-600 % was revealed in highly - and medium -decomposed peat of wood and grass groups, minimum value of about 25-70 % - for soils with low peat content, which is typical for these varieties. Accordingly, increasing moisture content values are increasing too. If the sample weight will be increased from 5 to 20 g, then the dispersion of measurements results decreases slightly (Figure 1, b, d; PL. 1). Considerable variation of parallel definitions are observed, if a peat moss portion of 100-200 g (Figure 1, a). Increasing sample number to 8-10 peat portions did not improve the results (Figure 1).

Table 1. Results of the moisture content determination of organic and organo-mineral soils.

	The name of the soil	Sample weight, g	The depth of sampling, m	Water content, % by drying in the oven: microwave, Minimum	microwave, Maximum	microwave, Average	convection, Minimum	convection, Maximum	convection, Average
Organic soils	Least decomposed peat moss group	100–200	0,2	3125	3255	3188	2412	3327	2994
	Least decomposed peat moss group	15–20	1,0	2253	2794	2527	2390	3003	2761
	Medium decomposed peat herbal group	5–10	1,0	2200	2754	2513	2250	3063	2816
	Highly decomposed peat wood group	15–50	1,0	603	617	609	574	598	587
	Least decomposed peat grass-moss group	15–50	1,0	353	464	397	369	485	416
	Least decomposed peat grass-moss group	15–50	1,0	1925	2333	2097	2340	2506	2447
	Highly decomposed peat woody-herbal group	15–50	3,0	1443	1506	1464	1465	1966	1762
	Least decomposed peat moss group	15–50	1,0	394	402	398	416	438	428
Organic-mineral soils	Plastic peat clay	15–50	2,0	56	60	58	58	61	60
	Peat loam	15–50	4,5	29	31	30	27	30	29

The analysis of moisture vs the drying time graphs showed that all varieties can be divided into three main stages: the first stage (I) – initial, with a relatively upward slope, during which the heated moistured sample with a slight acceleration of evaporation is observed; second stage (II) – stage of evaporation, when main moisture volume evaporates from the sample; third stage (III) – final drying, characterized by a relatively slow evaporation of the remaining free water.

At the same time, investigated types of soils have particular characteristics of drying. Most water saturated least decomposed moss and grass-moss peats (Figure 1, b, C, d) are characterized by prolonged stage of moisture heat-up and rapid evaporation. Medium-decomposed peats of wood, grass (Figure 1, g), wood-grass and grass-moss groups are characterized by a longer second stage, due to the fact that the number of samples, has a fairly large area of evaporation and ensures good penetration of microwaves into the samples. Therefore, they are characterized by a short time of final stage of drying, during which the remaining moisture is easily removed. Separately, a group of soils with low peat content (Figure 1, g) and relatively low moisture content should be identified These groups are unstable or have no initial stage (Figure 1, e) due to the rapid heating of the entire volume of the sample and intensive evaporation in the first minutes of the experiment.

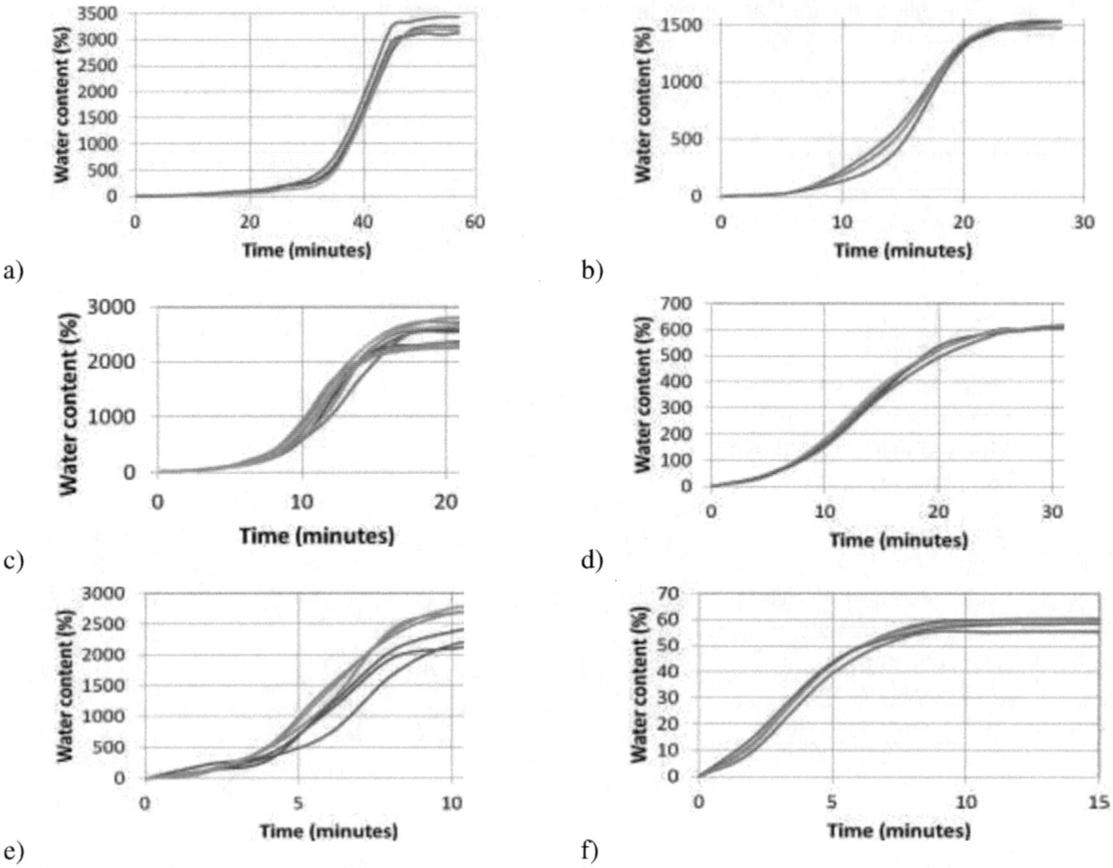

Figure 1. Dependence of soil moisture on drying time: a) least decomposed peat moss group (weight 200 g), b) least decomposed peat grass-moss group (15-50 g), c) least decomposed peat moss group (weight 15-20 g); d) medium decomposed peat grass group(15-50 g), e) least decomposed peat moss group (weight 5-10 g); f) clay with low content of peat (15-50 g).

This contributes to a relatively small sample surface area of evaporation and formation of a solid crust. As a consequence, high heating imposes restrictions on the material used for containers. In general, first two stages of drying take half of the testing time, during which up to 90% of the initial volume of moisture evaporates. The final stage of drying takes from 20 to 50 % of the experiment time.

To identify possible mass loss of soil due to the organic matter burning in convection and microwave ovens [10], methods of calcination were used in accordance with the applicable standards [13, 15-17]. Gravimetric method is based on determining organic matter in peat soils according to GOST 26213 and GOST 27784 [15, 16] after calcination at a temperature of 525±25°C. According to GOST 11306 [15] for the peat fuel production the ash content was determined by calcination at 800±25°C, and for agricultural and environmental at 525±25°C. In accordance with ASTM D 2974 [12] the ash content is determined by burning in a muffle oven: by "C" method at a temperature of 440±40°C and "D" method (for peat, used as fuel) at a temperature of 750±38 °C. The organic matter decomposition ranges between 229 and 579°C, with a maximum of 300-400°C [18]. Finally, for the sample calcination the following temperatures were selected: 800, 525 and 350°C. The conducted tests (Table 2, for clarity, values are rounded to integers) showed a good correlation of the results of organic content determination after drying and other methods. Thus, it can be stated that organic matter loss is the same both for the application of microwave and convection ovens (2-3 %). It should be noted that lower the temperature is, higher the ash content is.

Table 2. Organic matter content in organic and organic-mineral soils.

Name of the soil	Organic matter Content I_r, % after drying					
	Convection oven			Microwave oven		
	800°C	525°C	350°C	800°C	525°C	350°C
Least decomposed peat moss group	96	94	91	95	91	90
Highly decomposed peat wood group	90	92	85	90	92	87
Loam with low contents of peat	14	13	12	16	14	13

3. Results and discussion

The research results demonstrate that the variations in the moisture content of peats obtained at microwave drying are not higher than in convection drying in most cases (Table 1, Figure 1). In both cases variations fail to meet the standard requirements (allowable difference in moisture content over 100% is 5% [4], moisture – 1 % [11]). Mineral soils have little variation moisture content [14], as it has been proved. Variations of the obtained values are accounted for uniformity coefficient of the soil composition, irregularity in the distribution of organic remains and mineral inclusions, humus content of different fibering, and variable absorbing and water-holding capacity of plant remains. It should be noted that it is quite difficult to conduct testing in undrained peat areas, as well as to transport large naturally preserved samples. The moisture loss in these processes significantly exceed the difference of parallel measurements, therefore, it is rational to determine the physical properties of a single peat portion [12],by increasing the number of core samples (to 15-20 pieces) from each of the engineering-geological soil elements.

It was possible to make brief recommendations on the methodology for moisture content determination via microwave ovens for organic soils.

1. Appliance of a microwave oven excludes the use of metal containers, impenetrable to microwaves and resulting in a short circuit. Samples should also have a low content of electrically conductive minerals.

2.. The optimal heating power of the microwave oven for drying organic soils is 200-500 watts (for small samples less heating power is recommended to strike a balance between optimal drying speed and heating). For drying highly-decomposed organic, organic with high ash content and organo-mineral cohesive soils, porcelain or ceramic containers withstanding high temperature drop are recommended. Plastic containers are not advised for highly decomposed organic samples as they melt in the final stages of drying. However, for least decomposed water-saturated peats, household plastic containers can be applied. Setting microwave power over 200W is recommended only for porcelain containers, as in the third (III) drying phase extensive heat occurs that can destroy most materials.

3. The optimal sample mass is determined by the initial water saturation of the soil and required weight range of 15-50 g for medium- and highly-decomposed peats and organic soils; peat portions up to 100-200g are recommended for least decomposed peat moss, peats with large plant remains, fibers or different fibering peats .

4. To reduce the moisture content variations, it is desirable to balance between the soil mass and the number of samples in the oven. To analyze medium- and highly-decomposed peats and peat soils, samples of 15-50 g. with the number from 3 to 6 pieces are recommended. For least decomposed water-saturated peats, peat portions of over 100 g are recommended for the number of samples not more than 2-3 pieces. For multiple moisture determination it is efficient to use an accelerated method according to [11].

5. The drying time for peat portions of more than 100 g is 30-40 minutes, for peat portions of 30-50 g (and for peat soils) is 10-15 minutes, for peat portions of 5-10 g is 3-5 minutes. The time intervals between reweighting are: 3-5 minutes for sample weighting more than 50 g; 1-2 minutes for peat portions of more than 5-10 grams and for soils with low peat content. Expanding the area of tested

sample evaporation intensifies drying process as well as reweighing via sample blending (e.g., for cohesive soils by rolling them into a thin layer and surfacing channels with a glass spreading rod).

4. Conclusions

Thus, the study confirmed that household microwave ovens are an efficient means for rapid moisture content determination of organic and organo-mineral soils, as the testing is carried out in short time, and the obtained data are as accurate as in applying the drying ovens. After being tested on moisture content, the soil calcination at different temperatures indicated that the organic matter loss is comparable as well. The positive aspect of this paper is that, reweighing a sample weight does not increase after microwave drying (Figure 1), which, in its turn, increases the accuracy of moisture content determination. The authors expect that the proposed recommendations could be used for low-budget equipment in engineering-geological surveys more extensively, as well as in the field and laboratory environment.

References

[1] GOST 23061–2012 Soils. Methods of radioisotope measurements of density and moisture
[2] GOST 30416–2012 Soils. Laboratory tests. General provisions
[3] GOST 30672–2012 Soils. The field tests. General provisions
[4] GOST 5180–84 Soils laboratory determination of physical characteristics
[5] ASTM D 1558–10. Standard Test Method For Moisture Content Penetration Resistance Relationships Of Fine-Grained Soils
[6] ASTM D 4944–11. Standard Test Method For Field Determination Of Water (Moisture) Content Of Soil By The Calcium Carbide Gas Pressure Tester
[7] ASTM D425–88(2008) Standard Test Method for Centrifuge Moisture Equivalent of Soils
[8] ASTM D6938–10 Standard Test Method for In-Place Density and Water Content of Soil and Soil-Aggregate by Nuclear Methods (Shallow Depth)
[9] Device Comparison for Determining Field Soil Moisture Content Ernest S. Berney IV, James D. Kyzar, and Lawrence O. Oyelami Geotechnical and Structures Laboratory U.S. Army Engineer Research and Development Center 3909 Halls Ferry Road Vicksburg, MS 39180–6199
[10] ASTM D 2216–10. Standard Test Methods For Laboratory Determination Of Water (Moisture) Content Of Soil And Rock By Mass
[11] GOST 19723–74 Peat. Method for determination of moisture content in the deposits
[12] GOST 11305–2013 Peat. Methods for determination of moisture
[13] ASTM D 2974–14 Standard Test Methods for Moisture, Ash, and Organic Matter of Peat and Other Organic Soils
[14] ASTM D 4643–08. Standard Test Method for Determination of Water (Moisture) Content of Soil by Microwave Oven Heating
[15] GOST 11306–2013 Peat and products of its processing. Methods for determination of ash content
[16] GOST 26213-91 Soil. Methods for determination of organic matter
[17] GOST 27784-88. Soil. Method for determination of ash content of peat and peat soil horizons
[18] Marcos E, R Tarrega and Luis E 2007 Changes in a humic cambisol heated (100–500°C) under laboratory conditions: The significance of heatingtime *Geoderma* **138** 237 – 243
[19] Kramarenko V V, Nikitenkov A N and Molokov V Y 2005 On the applicability of microwave method for determining the moisture content of sandy soils *Modern problems of science and education* **1**1 – 12 URL: http://www.science-education.ru/121-18451

Clay swelling of Quaternary and Paleogene deposits in the south-eastern flanks of West Siberian iron ore basin

V V Kramarenko[1], A El- Shinawi [2], I A Matveenko[3] and A A Shramok[4]

[1, 3, 4] Tomsk Polytechnic University, 30 Lenin Avenue, Tomsk, 634050, Russia
[2] Geology Department, Faculty of Science, Zagazig University, Zagazig, Egypt

E-mail: [1]kramarenko_v_v@mail.ru

Abstract. Revealing soil swelling and estimation of swelling rate are of great importance at the initial survey stages as well as the bases for further determination of more accurate factors used for selection of recovery methods and facility design. The paper states briefly the most conventional prediction express-methods and determination of swelling indicators, the results of laboratory research in clay composition and properties, free swell index of Quaternary and Paleogene clays in the south-east of West Siberian iron-ore, gives the estimates of the swelling rate. The results have been statistically analyzed, revealing the relationship of properties, on the basis of which the correlation dependencies are suggested to predict the free swell index as well as to apply it for frost heave prediction.

1. Introduction

When changing the temperature-humidity conditions swelling clay poses a serious problem for construction as, when steeping, one observes a transition from solid and semi-solid phases to plastic one, density reduction, several time decrease in strength properties, and increase in maximum swelling pressure up to 2 MPa that can result in partial or complete facility destruction. In accordance with the normative techniques used in geological engineering survey [1], swell index is determined for soils of natural humidity, which depends on the test time and may be significantly different depending on weather conditions at different time of the year. Even slight humidity changes are enough for the same soil to be classified differently in terms of relative swelling ε_{sw} according to [2]. Therefore, revealing of swelling clays is one of the most important tasks of a geological engineer for further detail study.

The purpose of the paper is to reveal the distribution of soils prone to swelling in the profile of south-eastern part of West Siberian iron ore basin.

The objectives include the review of previously conducted research in geological engineering conditions of the area to establish the presence of smectite and hydromica minerals, study the prediction methods of expansive soil behavior and its classifications, selection of express-methods to determine swelling characteristics, sample preparation and laboratory soil tests; statistical data processing and establishment of relationship between the values of their composition and properties.

2. Materials and methods

The review of previous local research [4-6] as well as some authors' studies [7, 8] has proved the presence of swelling minerals in the deposits. Swelling is known to develop in smectites and mixed-layer illites as well as silt-loam soil (non-swelling at water saturation) while steeping with chemical flows or process solutions. The maximum swelling is typical for montmorillonite-hydromicaceous minerals, the least one – for kaolinite-quartz-hydromicaceous minerals, but for the former swelling process is 3-6 times longer than for the latter [9].

The upper part of the profile includes genetic stratigraphic sequences of quaternary, Paleogene, and Cretaceous deposits. Cretaceous deposits of marine and coastal-marine genesis of Coniacian-Maastricht formation in the upper part includes Kuznetsov, Ipatov, and Gankin suits, in composition of which clay minerals of smectite, kaolinite, hydromica groups as well as mixed lattice minerals are found. Paleogene deposit includes Lyulinvorsk marine, Urkov coastal marine, montmorillonite, and hydromica. In stratigraphic-genetic sequence of lacustrine-alluvial deposits of Lagerno-Tomsk suit the value of relative clay swelling changes from 0.01 до 0.1 at the humidity 0.27-0.36 unit fraction [4]. Engineering geology of the USSR [4] presents the data on content of hydromica, montmorillonite (with the exception of higher level), kaolinite, and illite in all Quaternary deposits. In Upper Neopleistocene alluvial terrace deposits fine-dispersed fraction consists of hydromica as well as kaolinite, and, probably, montmorillonite [4], the recent alluvial deposits of flood-plain terraces contain hydromica with admixed montmorillonit, sometimes halloysite [5], in subaerial blanket formation clays include hydromica with admixed kaolinite [4].

Presence of clays is revealed by X-ray diffraction, chemical, spectral, differential thermographic analysis, electron microscopy, determination of cation exchange capacity, soil examination in thin sections and immersion, and other methods. In geological survey the mineral clay composition is not a mandatory value [9] due to the high cost of mentioned analyses, labour intensity, and low equipment availability. In terms of the methods used [1], tests of expansive soil are performed by soil free swell devices and odometers *at natural humidity* to determine the indicators: free swell ε_{sw}, swelling under pressure ε_{swp} and swelling pressure p_{sw}. At the initial survey stages the prediction is made using natural humidity and dry soil density [11]. Hence, these soil research methods do not take into account probability of humidity change, therefore, the methods of swelling soil research including dry condition (in powder) are of particular interest [11-13].

The following swelling indicators are used abroad: Free Swell Index [14], Potential Volume Change (PVC) и Swell Index, lb/ft^2 [15]; Percent swell CBR (%) [16]; COLE and LE (%) [17], Expansion Index [18]. There are also indirect prediction methods of soil volume change in terms of Atterberg's limits – Liquid Limit (LL), Plastic Limit (PL), Shrinkage Limit (SL), plasticity index, colloidal content, soil behavior, dry soil humidity and density, sum water intake value and many other factors [12, 22-26]. One should note the difference in methods for determining the consistency index in domestic and foreign standards, for comparison of the results of which it is recommended to apply dependencies given in the work [26]. To estimate the expansion rate for one or several indicators mentioned above the classification have been developed. These prediction methods do not account for soil mineral composition, therefore, cannot predict their swelling adequately. Free Swell Index *FSI*, unit fraction or % allows revealing expansive soils more accurately yielding additional information on the behavior of clay minerals [10]. The method of *FSI* determination was suggested by Holtz and Gibbs (1956) [14]: dry soil of 10 cm^3 volume (V_0) is sieved through the mesh № 425 μm, filled in the cylinder of 100 cm^3 volume, poured with water, then in 24 hours soil volume (V) is measured and swelling percentage is determined:

$$FSI = \frac{V - V_0}{V_0}\, 100.$$

The authors suggested that soils with free swell index lower than 100 % can cause serious structural damage producing maximum pressure on foundation. If the index is lower than 50%, such soil can hardly cause a volume change even at low pressure [14].

According to IS: 2720 [26] *FSI* is defined according to the following formula:

$$FSI = \frac{V_d - V_k}{V_k} * 100.$$

where V_d – the volume of dry soil deposit of 10 g weight sieved through the mesh 425 μm and placed in 100 ml measuring cylinder with distilled water, V_k – deposit volume of 10 g soil placed in the cylinder with kerosene (carbon tetrachloride). If soils are heavily swelled, 5 g of soil sample or cylinder of 250 ml volume are taken.

The methods yield negative values of *FSI* for soils rich in kaolinite [10], the other disadvantage consists in rough measurement of initial volume. To eliminate the problem the researchers offer modified free swell index (*MFSI*) equal to the relationship of deposit volume (V_d) of the dry soil of mass 10 g in 100 ml cylinder of water to the dry soil mass

$$MFSI = \frac{V_d}{10}.$$

The research was based on experiment works at soil composition and properties obtained in accordance with methods of exsiting specification documents [8, 27]. The following parameters were defined: grain-size distribution, water content (w), clay soil water content at plastic and liquid limits (w_L and w_p), density (ρ), solid particle density (ρ_s), dry soil density (ρ_d), porosity factor (e), porosity and maximum water capacity were calculated. To determine *FSI* the method was chosen [14] with insignificant difference in procedure: soil dried in the furnace is grained in the mill, then filled in the flask of 3 ml (V), poured with distilled water up to 10 ml mark, thoroughly mixed and stirred up. The volume of soil deposit (ΔV) is measured in a day or more, depending on the suspension transparency. Finally, the swell index is determined: $FSI = \Delta V / V_0$, unit fraction or %. In authors' opinion, negative values of the kaolinite clay index can be used to predict soil frost heaving, which is rather relevant for the West Siberian area.

3. Results and discussion

To solve the set problems more than 200 samples are selected from clay soil from the depth of 160-200 m of Quaternary and Paleogene deposits. *In clays of Quaternary age* the fraction content is less than 0.002 mm, on average amounts 48.5%, changing from 25 to 52% (10 samples); in loam clay from 18,9 to 27%, on average 22% (6 samples). Water content of loam clay amounts from 20.5 to 30%, the minimal value of clay is 26.5%, maximum – 40%. Water content at the loam clay liquid limit is from 37 to 41%; clay from 42 to 51%. Water content at loam clay plastic limit is from 20.2 to 26.4%, clay from 19.4 to 31%. The liquidity index of loam clay is from 0.02 to 0.33%, that of clay is from 0.25 to 0.53%, respectively, loam clay is in semi-solid and low-plastic state, clay consistency is from low- to high-plastic. Loam clay density changes from 1.89 to 1.97 g/cm^3; clay – from 1.82 to 1.96 g/cm^3. The loam clay particle density amounts from 2.71 to 2.77 g/cm^3; clay – from 2.63 to 2.75 g/cm^3. Dry soil density of clay loam is from 1.47 to1.57 g/cm^3, clay – from 1.31 to 1.53 g/cm^3.

In Paleogene deposits the fraction conent in clay is less than 0.002 mm, amounts on average 34.6%, changing from 28.2 to 51.8% (60 samples), in loam clay on average 19,0%, ranging from 10.0 to 29.9% (60 samples), in dust sand and sand clay it amounts 1.6 to 14.7% of fraction content, the average content being 3.9% (40 samples). The clay water content has the minimal value – 11.9%, maximal – 44.5%, that of loam clay is from 19.1 to 61.7%; that of sand clay – from 19 to 42.3%; that of dust sand – from 14.7 to 40.4. Water content at the clay liquid limit is from 42 to 77%; loam clay – from 28.5 to 59%; sand clay – from 33 to 48%. Water content at clay plastic limit changes from 18.7 to 47%; loam clay – from 18.7 to 45%; sand clay – from 26.7 to 42.6%. The clay density changes from 1.70 to 2.04 g/cm^3, loam clay – from 1.73 to 2.15 g/cm^3; sand clay – from 1.77 to 2.06 g/cm^3; sand – from 1.73 to 2.35 g/cm^3. The clay particle density changes from 2.50 to 2.82 g/cm^3; loam clay – from 2.53 to 3.03 g/cm^3; sand clay – from 1.58 to 2.96 g/cm^3; sand – from 2.50 to 3.09 g/cm^3. Dry soil density of clay loam accounts for from 1.16 to 1.77 g/cm^3; clay – from 1.18 to 1.65 g/cm^3; sand clay – from 1.27 to 1.65 g/cm^3, sand – from 1.25 to 1.86 g/cm^3. Hence, Paleogene dispersive soil is characterized by higher density in contrast to soil of quaternary deposits, clay soil consistency is various and includes all types from solid to liquid. According to [9], swelling is found at clay particle content in the amount of more than 40–60%, dry soil density – more than 1.5–1.7 g/cm^3, water content – less than 0.20–0.30 unit fraction. Hence, the data obtained allow determining soil composition and properties as low- and medium swelling.

As a result of experiments the following *FSI* values were obtained:

- in clay of quaternary deposits they vary from 0.23 to 0.87 %, the average is 0.59, in loam clay it is from 0.10 to 0.73, the average – 0.42% (4 samples of every type),

- in clay of Paleogene deposits they vary from minus 0.03 to 0.23 %, the average is minus 0.10% (20 samples), in loam clay – from minus 0.13 to 0.47, the average – 0.23% (20 samples), in sand clay they are from minus 0.17 to minus 0.03, the average being minus 0.10, in dust sand they are from minus 0.10 to 0.33, the average being 0.02%.

In general, *FSI* distribution in the section is shown in Figure 1 where expansive soils in green claystone-like deposits of Lyulinvorsk suit are clearly seen, in the interlayer of bluestone and loam clay of Novomikhailovsk and Lagerno-Tomsk (swell index is up to 0.50 unit fractions), three maximum values are observed in Quaternary deposits. Negative values of *FSI* are established in soils of Novomikhailovsk, Lagerno-Tomsk, and Urkov suits, therefore, these soils shouls be studied in terms of relative heaving deformation. According to the classifications [8, 26] the studied soils are referred to low and medium swelling indexes. According to the classification [12] the results obtained are of serious concern, as at *FSI* < 10 soils are non-expansive, low-expansive within the range 10-15%, medium-expansive at 15-20%, high-expansive soils are within the range 20-40%, and more 40% - very high.

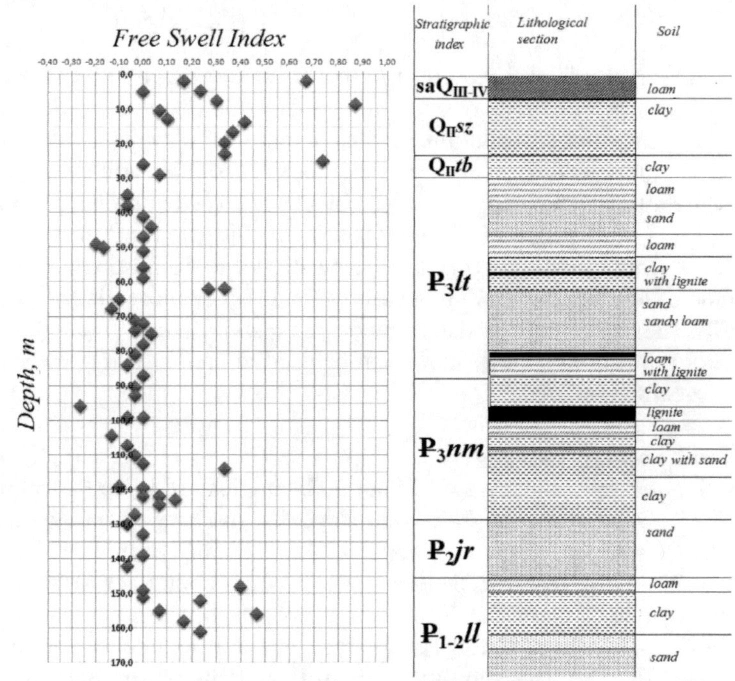

Figure 1. Changes in free swell index FSI with depth.

The statistical analysis has revealed the relationship between the soil properties and *FSI*, the closest dependencies are registered between *FSI* and plasticity index *IP* (correlation coefficient $r = 0.66$), particle number of size less than 0.002 mm $C_{<0.002}$ ($r = 0.63$), more than 0.05 $C_{>0.05}$ ($r = -0.34$), within the range 0.05–0.002 mm $C_{0.05-0.002}$ ($r = 0.24$), water content at plastic limit *PL* ($r = -0.43$), water content at liquid limit *LL* ($r = 0.21$), liquid indicator *IL* ($r = -0.23$), internal friction angle φ ($r = -0.26$). Based on the established relationships of *FSI* with composition and properties the regression equations are obtained:

$$FSI = -0.262+0.021\cdot IP,$$
$$FSI = -0.223 + 0.014\cdot C_{>0.002},$$
$$FSI = -0.149+0.006\cdot LL,$$
$$FSI = 0.492-0.014\cdot PL.$$
$$FSI = 0.263-0.009\cdot\varphi.$$

4. Conclusion

Hence, the research results confirmed the common occurrence of expansive soil in the sections of Quaternary and Paleogene deposits at the South-Eastern flanks of West Siberian iron ore basin. Their properties are not always taken into consideration when choosing a method of mineral deposits development which might result in undesirable effects because the soil swell index increases 2-3 times at soil structure disturbance [9]. The work performed is the first step in research of clay properties including problem statement and determination of relationships to predict swelling. The next step should be a more detailed study in swelling soil behavior allowing their qualitative estimation based on comparison of the obtained results with more detailed soil tests. Simultaneously, the database would be enlarged that would become a base for local classification in terms of free swell index. Similar procedures would be performed to study soil heaving properties with negative values of *FSI*. Finally, it is important to note that *free swell index should be a mandatory value* at any geologic survey.

References

[1] GOST 12248-2010. *Soils. Laboratory methods for determining the strength and strain characteristics*

[2] GOST 25100–2011. Soils. Classification

[3] Kryventsov A V et al. 1986 *Geological and hydrogeological structure of Parabel-Parbig Ob Region. Report of group complex geological, hydrogeological and engineering-geological survey of 1:200 000 scale by Parbig party of 19801–1986* (Tomsk)

[4] Egorov B A, Skogorev A I and et al. 2001 *Compilation and analysis of geological, hydro-geological, hydrogeological and geotechnical materials on the territory of Tomsk in order to support mapping of 1: 25000 scale* The report of JSC "Tomskgeomonitoring."(Tomsk) p 274

[5] Silverstov V N 1997 *The stratigraphy of Quaternary sediments of the Tomsk region, scale 1: 500,000* (Tomsk) p 12

[6] *Engineering Geology of the USSR* 1976 (Moscow: MGU) V. II p 494

[7] El- Shinawi A, Zabin V Y and Kramarenko V V 2011 Features of the physical properties of the soil the lower reaches of the territory *Bulletin of the Tomsk Polytechnic University Earth sciences* **319** 1 178–182

[8] Code of practice 47.13330.2012 *Engineering survey for construction. Basic principles*

[9] Code of practice 11-105-97 part 3. *Engineering geological site investigations for construction*

[10] Sridharan A and Prakash K 2000 Classification Procedures for Expansive Soils, Geotechnical Engineering *Proc. ICE* (UK) **143** 235–240

[11] Snethen D R, Townsend F C, Johnson L D, Patrick D M, Philip J'Vedros 1975 *A review of engineering experiences with expansive soils in highway subgrades* Offices of Research and Development Federal Highway Administration U. S. Department of Transportation Washington, D. C. 20590 p 177

[12] Engineering properties of expansive clays *Transactions of ASCE* **121**

[13] Rauh F, Thuro K and Spaun G 2006, The powder swelling test 1– advantages and limitations, in Engineering Geology for Tommorrow's cities *Geological Society London, Engineering Geology Special Publication 22 IAEG2006* **472**

[14] Holtz W G and Gibbs H J 1956 Engineering properties of expansive clays. *Transactions of ASCE* **121** 641–663

[15] Lambe T W and Whitman R V 2000 *Soil mechanics, SI version* (New York:John Wiley & Sons)

[16] Yoder E J and Witczak M W 1975 *Principles of pavement Design* (New York: John Wiley and sons, Inc.)

[17] Brasher B R, Franzmeier D P, Vallassis V T and Davidson S E 1966. Use of Saran resin to coat natural soil clods for bulk density and water-retention measurements *Soil. Sci* **101** 108

[18] United States Bureau of Reclamation. Earth Manual 1973 USBR Federal center (Denver)

[19] Seed H B, Woodward R J and Lundgren R 1962 Prediction of swelling potential for compacted clays. *J.Soil Mechanics and Foundations Division ASCE* **88** 53–87

[20] Carter M and Bentley S P 1991 *Correlation of soil properties* (London: Pentech Press)

[21] Skempton A W 1953 The colloidal "activity" of clays *Proc. 3rd Int. Conf. Soil Mechanics and Foundation Engineering (*Zurich) **1** 57–61

[22] Van Der Merwe D H 1964 The prediction of heave from the plasticity index and percentage clay fraction of soil *Civil Eng. S. Afr.* **6** 103–106

[23] Snethen D R 1979 *An evaluation of methodology for prediction and minimization of volume change of expansive soils in highway subgrades* Research Report No. FHWA-RD-79-49. U. S. Army Eng. Waterway Exp. Sta., Vicksburg, MS
[24] Chen F H 1975 *Foundations on Expansive Soils* (Elsevier Amsterdam)
[25] Bellendir E N, Kagan A A and Vekshina T Yu 2005 On the classification of clayey soils in russian and foreign standards *Soil Mechanics and Foundation Engineering* **42** 1 22–27
[26] IS: 2720. *Indian Standard Methods of Test for Soils: Part 40—Determination of Free Swell Index of Soils* 1977 (BSI, New Delhi)
[27] GOST 5180-84. *Soils. Laboratory methods for determination of physical characteristics*

Landslide hazard assessment in city under construction Innopolis (Russia)

A Latypov[1], N Zharkova[2] and I Nuriyev[3]

[1, 2, 3] Vernadsky Institute of Geochemistry and Analytical Chemistry of the Russian Academy of Sciences, Moscow, Russia Kremlevskaya str. 4/5, Kazan (Volga region) Federal University, Kazan, Russia

E-mail: [2] kazannad@rambler.ru

Abstract. The paper reflects the main results of exploring the landslide hazard threating Innopolis city, precisely field and laboratory studies, assessment of the slope stability. A regional short-term landslide hazard forecast was performed using a digital model of geological medium and the method of engineering-geological analogy: an engineering-geological zoning was conducted; data from standard areas, the critical values of surface slopes were calculated for the identified types of soils and different types of humidity; the obtained data were extrapolated, verified and corrected. On the basis of the forecast, development of a landslide monitoring network is proved.

1. Introduction

The estuary area of the Sviyaga river (located in Verkhneuslonsky District of Republic of Tatarstan) is currently under active development: since 2005 there has been created and constantly reconstructed a unique ski sports complex (SSC) "Kazan", and in summer of 2012 there has began the construction of new city - IT-center with extensive business infrastructure (industrial parks, research centers, etc.) called Innopolis.

Despite good location of Innopolis and SSC (proximity to Kazan city and M-7 highway), the investigated area is characterized by highly complex engineering and geological conditions. The presence of a thick layer of clay soil (of natural and technogenic origin), complex hydrogeological conditions, as well as the large ruggedness of terrain (the Sviyaga river slopes and numerous gullies and ravines systems) stipulate the development of landslide processes.

In 2012, under a contract with the Ministry of Ecology and Natural Resources of the Republic of Tatarstan, Kazan (Volga) Federal University performed the integrated assessment of landslide hazard [1], which was of particular urgency in terms of ensuring safe operation conditions on the territory of city "Innopolis" under construction and "Kazan" ski sports complex.

2. Results and discussion

Main stages of the work are as follows:

Collection, consolidation and analysis of published and fond materials. As part of the work, there has been studied different sources of information of previous years, such as the reports of the regional search works, hydrogeological reports on groundwater reserves assessment, geological, hydrogeological and engineering-geological surveys, geological maps of Quaternary sediments, and etc. Taking into account the fact that the territory of Innopolis city is undeveloped, there are no any reports of detailed engineering-geological investigations. The area of SSC "Kazan" is the only exception as there are 46 reports on engineering surveys (performed by SUE RT "Tatinvestgrazhdanproekt" in the period from 2002 to 2008).

Interpretation of aerial and satellite materials. This kind of work was carried out by the employees of the State Unitary Enterprise "NPO Geocenter RT" using Erdas Imagine, ArcGIS, MapInfo softwares. Despite the presence of a large number of satellite imageries of the study area, the images of required quality were taken only in 2005, 2011 and 2013, thus engineering and geological interpretation was carried out only during this period. The total square of the study area is 21 km^2. Thus, the comparative analysis of landslide dynamics within the study area was conducted for the period of 2005-2013.

Visual inspection of the area to assess the conditions of landslide activation. The visual inspection was carried out along the elements of erosion and hydrographic network (the slopes of the Sviyaga river and Morkvashka creek, slopes of numerous gullies and ravines), because gravitational processes are tend to develop in such areas. 10 observation points were planned at 1 km of the route within the slopes of the terraces, river banks and erosion systems, and 4 points of observation were chosen within the watersheds, gentle slopes, as well as at the adjacent territory.

A passport was developed for each landslide area. It includes geological and geomorphological description of the slope (height, steepness, shape, exposure, genesis), the nature of landslide bodies' boundaries, the type and subtype of landslides, their age, the phase of sliding process, type of sliding surface. In addition, geometric parameters of landslide bodies (width, length, rupture surface, amplitude of displacement) were measured, and the main reasons for landslide were identified.

In addition to description of the slopes and slope processes, the natural and artificial rock outcrops were documented in the course of route surveys. Particular attention was given to Quaternary sediments (usually dominant in the composition of landslide bodies) and technogenic soils: using a number of direct and indirect methods, the age and genesis of sediments were determined, the thickness (if possible) was estimated, structural and textural features were determined (grain size, the size and the number of rudaceous fraction (clay soils), the type and amount of filler, petrographic composition of the debris, the degree of roundness (for coarse clastic soils). Also, in the field conditions the degree of pore filling with water and consistency were approximately estimated in coarse-grained soils and clay, respectively. In technogenic soils, density (qualitatively) and age of the mound (when possible) were estimated [2]. Typification of ground strata was performed according to the data (Table 1).

Table 1. Groundwater strata typification scheme.

Top	**A.** Loam and clay			**B.** Loam and clay with gravel and scree <10 %			**C.** Loam and clay with gravel and scree >10 %			**D.** Technogenic clay soils with gravel and scree		
	Thickness, m											
Bottom	<1,5	1,5–4,0	> 4,0	<1,5	1,5 – 4,0	> 4,0	<1,5	1,5 – 4,0	> 4,0	<1,5	1,5 – 4,0	> 4,0
	1	2	3	1	2	3	1	2	3	1	2	3
P_2kz ***I***	-	-		-	-	-			-	-	-	-
P_2ur ***II***	-	-				-		-	-	-		
P_3sd ***III***	-	-				-	-	-	-	-		
dp Q_{3-4} ***IV***	-	-	-		-	-	-	-	-	-		

All buildings and structures located in the study area were examined on the presence of deformations, indicating the development of landslides (cracks, distortions of door and window openings, deviation from the vertical, etc.).

During the period of field surveys 14 visual inspections were conducted (364 observation points, the total length - 53 km).

Comprehensive study of the geological structure of bedrock and landslide bodies, identification of their contours, composition, structure and physic-mechanical properties of soils in the most dynamic and dangerous areas of landslides. Geodetic survey, including aerial photography, was carried out on landslide-prone areas. These areas were identified during the route monitoring. Then, to obtain the reference section for the most characteristic areas (landslide depressions, axial strips of large landslides, inter-landslide ridges, other large and most typical for the region forms of relief), the mine workings were carried out (wells and bore pits) to collect monoliths of soil (326 pcs.) and water samples (20 pcs.).

Further, to clarify the boundaries of landslide bodies and the engineering-geological elements, as well as position of the slip boundary and groundwater level within the study areas, the complex of geophysical surveys, including VES (using a 4-electrode setting AMNB) and geo-radar survey (GPR «ProEx, MALA GeoScienceinc) was conducted [3, 4].

Identification of physical and mechanical characteristics of soil properties and chemical analysis of water samples were performed at Institute of Geology and Petroleum Technology KFU in the Laboratory of Soil and Soil Mechanics. Deformation and strength properties of soils were determined by ASIS NPP "Geotech" in compression and shear instruments. Filtration coefficients were determined on undisturbed samples of a cylindrical shape with a diameter of 100mm, height of 65mm on Wille Geotechnik D 3325S filtration plant.

Assessment of landslide-prone slopes stability using the latest domestic and foreign techniques.

As a result of the joint analysis of visual observations, geodetic measurements and topography, 19 measurement points were constructed on the territory of Innopolis city, 10 measurement points were constructed on the territory of SSP "Kazan" on the so-called reference stations which were selected during route surveys. On one hand, these areas are characterized as potentially landslide-prone slopes from geometric and geological point of view, and on the other hand, they are located within the boundaries or in the vicinity of projected and existing buildings.

Stability calculation was performed using GEO 5 FINE software package. As a result, the most probable sliding surface and coefficients of stability for the natural state, as well as for the predicted water saturation, were obtained for each measurement station [5].

Implementation of short-term forecasting of landslide processes under the influence of natural and technogenic factors. The slope stability evaluation on the standard areas and the extrapolation of the results to the entire territory (with preliminary determination of surface slopes critical value for two modes of soils moisture, Table 2) based on maps of the slopes and soil types were followed by engineering-geological zoning aimed at predicting landslide hazard using digital model of geological environment [6].

Table 2. The critical values of surface slopes for different types of soils at the natural moisture (W_n) and full water saturation (W_{sat}).

Type of soil strata	Stability factor									
	< 1.0		1.0÷1.1		1.1÷1.2		1.2÷1.3		>1.3	
	W_n	W_{sat}	W_n	W_{sat}	W_n	W_{sat}	W_n	W_{sat}	W_n	W_{sat}
	21°	19°	20°	18°	19°	17°	18°	16°	16°	14°
	39°	37°	38°	37°	37°	36°	35°	34°	32°	31°

	36°	33°	34°	31°	33°	30°	32°	29°	30°	27°
	36°	34°	35°	33°	32°	30°	32°	30°	30°	27°
	34°	32°	33°	30°	32°	30°	30°	28°	28°	26°
	43°	40°	41°	39°	40°	38°	37°	36°	35°	33°
	63°	59°	62°	58°	60°	56°	57°	54°	52°	50°
	47°	44°	46°	43°	45°	42°	43°	40°	41°	38°
	35°	34°	34°	33°	33°	32°	31°	30°	30°	29°
	32°	31°	32°	30°	30°	29°	28°	26°	25°	23°
	28°	27°	27°	26°	27°	25°	26°	24°	25°	22°
	32°	31°	32°	30°	30°	29°	28°	26°	25°	23°

Then, subsequent verification and correction of the results obtained using data from satellite images interpretation, seasonal instrumental geodetic measurements, as well as detailed engineering-geological survey (using information on the stage of the landslide process) were carried out. The main result is a detailed engineering-geological map with elements of forecast (Figure 1).

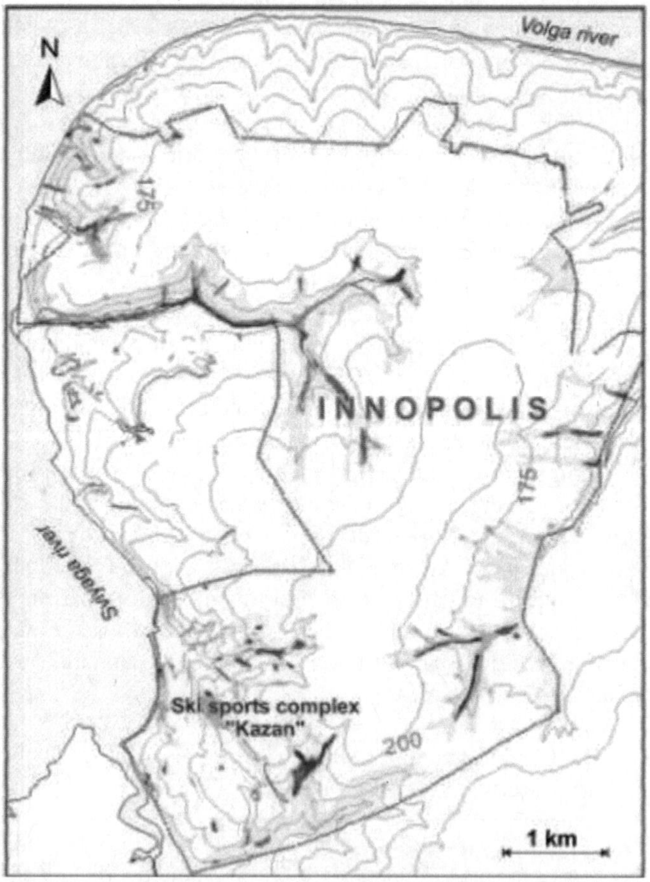

Legend to the landslide hazard forecasting map in table 3.

Figure 1. Landslide hazard forecasting map.

Development of high-priority and long-term anti-landslide measures to reduce the negative impact of landslide processes. According to the described forecast of landslide processes, high-priority (mechanical fastening of slopes with the help of massive retaining walls, sheet piling, anchoring, injection curtains, ramparts, terracing, slope flattening, removal of landslide mass) and long-term (drainage measures, planting of vegetation, reinforcement of slope surface by geosynthetic grids and frames) anti-landslide measures, aimed to reduce geological risk, were developed.

Table 3. Legend to the landslide hazard forecasting map (Figure 1) and the criteria of landslide displacements probability.

Area km^2 (%)	Probability of displacement	Safety Factor (SF)	
		Natural moisture of soil	Fully water-saturated soil
0,10 (0,44)	Inevitable throughout the year	$SF \leq 1{,}0$	
0,06 (0,26)	Inevitable in spring and autumn, possible in summer	$1{,}0 \leq SF < 1{,}1$	$1{,}0 \leq SF < 1{,}1$
0,20 (0,88)	Inevitable in spring, possible in autumn, unlikely in summer	$1{,}1 \leq SF < 1{,}2$	$1{,}1 \leq SF < 1{,}2$
0,48 (2,09)	Possible in spring, unlikely in autumn, impossible in summer	$1{,}2 \leq SF < 1{,}3$	$1{,}2 \leq SF < 1{,}3$
1,62 (0,70)	Unlikely in spring, impossible in autumn and summer	$SF \geq 1{,}3$	$SF \geq 1{,}3$
2,15 (0,94)	Impossible under current engineering-geological conditions	$SF \geq 1{,}3$	

Substantiation and establishment of observational network for landslide processes on the territory of Innopolis and "Kazan" ski complex. The necessity of a monitoring network is substantiated for the territory of Innopolis city. In addtion, 9 observation stations were identified (including the territory of the ski complex), 10 ground frames were constructed, instrumental geodetic observations and remote methods were recommended. It was proposed to equip the observation areas with hydrogeological and directional wells, wall marks and ground reference points. The further use of satellite imagery in the infrared spectrum and SAR-monitoring was recommended.

3. Summary

Detailed maps (1: 10 000) were compiled according to the results of the study: map of factual material, engineering-geological map with elements of forecasting the development of landslide processes; large-scale maps (1: 25 000): geological and hydro-geological map, map of Quaternary sediments, map of soil types, map of landslide processes development, monitoring scheme of slope processes; the digital database of landslide and erosion processes distribution and formation conditions, containing facts of their impact on economically valuable objects.

Acknowledgements

The work was financially supported by Kazan (Volga-region) Federal University in order to improve its competitiveness among the world's leading research and education centers.

References

[1] Zharkova N I and Latypov A I 2013 *A comprehensive assessment of landslide hazard on the territory of under construction satellite city of Kazan "Innopolis" and ski sports complex "Kazan"* (FSAEI HPE "Kazan Federal University", Inv. № 2168), Territorial fund of geological information on Volga Federal District of the Russian Federation) p 298

[2] Latypov AI and Zharkova N I 2013 Landslide hazard assessment on the territory of under construction city Innopolis (Tatarstan) for the organization of geodynamic monitoring system *J. Engineering surveys* ***10*** 56-59

[3] Zharkova N and Latypov A 2014 Regularities of slope processes forming on the "Kazan" ski complex area (Russia) *Int. Multidiscip. Sci. GeoConf. SGEM* (Sofia: STEF92 Technology

Ltd) 755-759

[4] Zharkova N I and Latypov A I 2014 Regularities of slope processes forming on the territory of "Kazan" ski complex, 2014 "Sergeev Readings. Evolution of academician Sergeev's scientific ideas at the present stage" *Annual session of the Scientific Council RAS on the problems of Environmental Geoscience, Engineering Geology and Hydrogeology* (Moscow: PFUR) 269-273

[5] Latypov A I, Zharkova N I, Nuriyev I S and Cherniychuk G A 2014 Modeling of landslide processes for prediction of landslide hazard on the territory of "Kazan" ski complex and Innopolis city *Scientists notes of Kazan University, series of "Natural Sciences"* **156** (Kazan: K(VR)FU press) 148-162

[6] Zharkova N, Latypov A, Nuriev I and Cherniychuk G 2015 Short-term forecast of landslide processes development using the method of engineering-geological analogy on the territory of Innopolis city (Russia) *15th Int. Multidiscip. Sci. GeoConf. SGEM* (Sofia: STEF92 Technology Ltd) 649-656

Geotechnical conditions contributing to negative geological process development in urban areas (the case of Kemerovo-city)

A V Leonova[1], R R Khabibullin[2] and A V Baranova[3]

[1, 2, 3] National Research Tomsk Polytechnic University, 634050, Lenin Ave., 30, Tomsk, Russia

E-mail: [1]avleonova@tpu.ru, [2]ravil.khabibullin.94@mail.ru, [3]baranskikh@yandex.ru

Abstract. The paper addresses the issue of intensive urban development in the area of Kemerovo-city. Underestimation of geotechnical conditions of the area at the project and construction stages results in negative geological processes such as erosion, waterlogging, soil subsidence, and underflooding. These processes can lead to deformation and failure of buildings and constructions.

1. Introduction

Modern urban areas are creating conditions conductive to the development of hazardous geological processes such as landslides, gulleys, frost penetration etc. Underflooding, which has been recently growing in scale, belongs to this array of negative processes. Underflooding results in flooded basement and negative changes in soil physical properties. Flooded landfills and catch pits are a source of chemical contamination of groundwater. What is more, underflooding can cause the development of negative geological processes related to groundwater behavior, such as subsidence and swelling. As a rule, urban underflooding is technogenically conditioned. It results from interference with natural dynamic water balance of the area. Human activities, such as urban residential development, lead to such interference which is developed by two stages: while construction activities and building maintenance. Technogenic underflooding has been only recently faced with and studied, so the preventive measures have not been fully developed and put into practice yet [1, 7, 8, 9, 10, 11, 12].

2. Materials and methods

Kemerovo city is the administrative center of the Kuzbass region. There are a lot of high-rise apartment blocks and underground constructions are being built in the city nowadays. The increase in building density and anthropogenic load makes it necessary to ensure safety, social and environmental comfort for the residents, as well as to improve economic efficiency of urban development. The constant growth of geotechnical investigations resulted in a large amount of geotechnical data which require to be analyzed and integrated for further use while carrying out current geotechnical and engineering activities, forecast mapping, and monitoring geological processes for the city to be developed. We have considered several cases of underflooding in Kemerovo.

The first is the construction site of Kuzbass technopark. The construction site of the designed Technology Transfer Center with an exhibition hall is situated in Rudnichny district, Kemerovo city, bounded by three streets: Institutskaya, Tereshkovoy, and Sosnovaya [2]. In terms of geomorphology the site is located within the bounds of the right side valley of the river Tom'. General topographic south-east inclination is observed, the altitude is 223-228 m.a.s.l [2]. The results of previous geotechnical investigation made up to the depth of 10 meters show the following profile (from top to bottom): sedentary clays and clay loams with sedimentary debris and gravel, medium fine gravel, and sandstone on argillaceous cement (block zone of weathering crust) (Table1).

Table 1. Values of soil physical parameters.

Soil type	Density, ϱ, g/cm^3	Porosity factor, units	Soil modulus E, mPas	Angle of internal friction, φ, deg.
Sedentary clays and clay loams (e P_2 - Q)	2.0-2.17	0.4-0.73	25	
Medium fine gravel and cobble sedentary soils (eP_2)	2.15		30	
Sandstone on argillaceous cement (P_2)	2.35		50	45

During the investigation made by "Geotekhika" Ltd. (December 2009, July 2010, March 2011) up to the depth of 10 meters the groundwater was not found [2].

The investigations made in November and May 2012 revealed temporary perched groundwater in the northern part of the area. Its depth varies from 2.3 m to 4.2 m. The horizon is a spreading groundwater dome in the area intensively fed with technogenic water. The temporary perched groundwater dome is formed in dense layers of the lower part of eluvium (eQ_{II}), with low filtration properties – coefficient of permeability k = 0,001 m/day. The groundwater conditions are unstable, which is characteristic of technogenic water-fed horizons. The annual cycle fluctuation range of ground water is predicted within 1.0-1.5 m. The highest water table is in May and June. In terms of chemical composition the ground water is sulfate-chloride-hydrocarbonate, calcium-sodium. The chemical composition is characterized by high content of nitrates, chlorides, sulfates, and ammonium, which is typical of technogenic water sources [3].

The investigation made in February 2013 recorded groundwater level at the depth of 5.2-6.2 m in some parts of the site, which corresponds to 219.45–221.00 m.a.s.l. The groundwater occurs as a sporadic spreading dome in the zone fed by technogenic water most intensively and having the loosest soils of the weathering crust. Technogenic origin of the groundwater is indirectly proved by the chemical analysis result (values of hydrocarbons, nitrates, and nitrites). The groundwater regime is unstable and conditioned by the volume of technogenic water losses and infiltration of precipitation. The results of special geotechnical survey carried out within the Tekhnopark area show that the temporary perched ground water is technogenically contaminated as well[4].

The second area is a construction site of a business-incubator building. Administratively, the construction site belongs to Rudnichny district, Kemerovo, and is situated on the crossing of Tereshkovoy and Institutskaya streets. In terms of geomorphology the studied area is confined to the right side valley of the river. The altitudes are 227–229 m.a.s.l. The geo-lithologic profile of the area was studied up to the depth of 10 meters and consists of the following layers (from top to bottom): fill ground (layer 1), loess loam (layer 3), and sedentary clays weathered to argillaceous and debris rocks (layer 4, 5a); sandstones of different degree of weathering were discovered at the depth of 5.8 – 6.5 m (Table 2).

Table 2. Values of soil physical parameters.

Soil type	Density, ϱ, g/cm^3	Porosity factor, units	Moisture content, S_r, units	Soil modulus E, mPas	Angle of internal friction, φ, deg.
Sedentary clays with debris inclusions		0.38-0.58	0.73-1.0	28	
Fine sandstone	2.35			50	45

The groundwater was not identified while the survey (from July to December 2009) being conducted up to the depth of 10 meters.

The third area to consider is a site designed for dwelling house construction. Administratively, the projected house № 25 belongs to the 12th micro-district, Rudnichny district, Kemerovo city.

In terms of geomorphology, the site is located on the slope of Tom' watershed. The land surface is undulated and slightly inclined south-eastwards to the Tom' valley. The altitudes are 247.27-248.49 m.a.s.l. A foundation pit had been excavated and some piles been driven by the time the geotechnical investigation took place on the site (the piles were driven approximately in 2008 year) [5]. Later, the building construction was abandoned, the pile field has not been conserved and the foundation pit has been filled with water by now.

According to the survey made in September 2007 up to the depth of 25 meters, the geological profile of the site consists of the following layers (from top to bottom): deluvial-proluvial loams from stiff to very soft ones, deluvial stiff loams, eluvial-deluvial loams and clays from hard to very soft ones (Table 3). The groundwater level was found at the depth of 1.8-2.45 m, which corresponds to the altitude of 245.22-245.29 m.a.s.l.

Table 3. Values of soil physical parameters.

Soil type	Density, ϱ, g/cm^3	Moisture content, S_r units	Soil modulus E, mPas
plastic clay loams (d-pr Q_{III})	1.90-1.95	0.95-1.0	4.9
stiff loam (d Q_{III})	2.0-2.09	0.95-1.0	11.9
plastic and stiff loams (ed Q_{II-III})	1.92-2.11	0.99-1.0	10.5
stiff loam (ed Q_{II-III})	1.93-2.07	0.95-1.0	19.6
firm clay (ed Q_{II-III})	2.03-2.08	0.89-1.0	

The survey made in June 2013 showed the groundwater level at the depth of 2.2-2.6 m, the altitudes are 245.07-245.74 m.a.s.l. The groundwater is recharged from infiltration of precipitation, surface water during river floods, as well as from confined water of underlying Upper Permian formation. The groundwater is drained through the local hydrographic network. The groundwater regime is characterized as unstable [5].

The next sampling site is administratively situated in the 12th micro-district, Rudnichny district, Kemerovo city. It is a construction site of projected apartment block №24.

In terms of geomorphology, the site is located on the slope of Tom' watershed. The land surface is slightly inclined, undulated, and lowering to the south-east towards the Tom' valley. The altitudes are 246.67-248.93 m [5].

Piles had been driven and the zero cycle construction had been completed by the time the geotechnical investigation took place on the site (the piles were driven approximately in 2008 year). Later, the building construction was abandoned, the pile field has not been conserved and the foundation pit has been filled with water by now.

In 2007 "Geotekhika" Ltd. made a survey for the construction project development. According to the survey the geotechnical profile of the site up to the depth of 25 m comprises the following soil layers (from the top to bottom: deluvial-proluvial loams from stiff to very soft ones; deluvial loams from semi-heard to stiff ones; eluvial-deluvial loams and clays from hard to soft ones (Table 4).

Table 4. Values of soil physical parameters.

Soil type	Density, ϱ, g/cm^3	Moisture content, S_r, units	Soil modulus E, mPas
plastic clay loams (d-pr Q_{III})	1.89-1.98	0.98	5.6
stiff loam (d Q_{III})	2.0-2.04	0.95-0.99	13.3
plastic and stiff loams (ed Q_{II-III})	1.94-2.07	0.99-1.0	11.2
stiff loam (ed Q_{II-III})	1.96-2.11	0.91-1.0	

In September 2007, the level of groundwater was found at the depth of 2.0-2.6 m from the land surface, which corresponds to the altitude of 244.85-246.10 m.a.s.l. In May 2013, the groundwater level rose up to the depth of 1.5-2.3 m, which corresponds to the altitude of 245.17-245.66 m.a.s.l. [6].

3. Conclusions

The surveys prove that the hydrogeological conditions of the construction sites have dramatically changed. The survey conducted in 2009 showed no groundwater to the depth of 10 meters, however, the surveys of 2012-2013 identified groundwater, with some evidences of its technogenic origin. Groundwater increases humidity in basements and ground floors of buildings and leads to underflooding of foundations and basements. Moisture increase results in changes in soil physical properties and reduces soil stability, which can have negative impact on building sustainability and reliability. Therefore, it is essentially important to assess the impact of exogenous geological processes (EGP) on urban area, which, in its turn, should be taken into account while choosing a site for construction in terms of its feasibility study. We also consider it reasonable to apply GIS-technologies to assess hazardous geological processes. It necessitates not only developing a clear system of input, filing and keeping the data obtained by geotechnical investigations, but also deep understanding of particular geological processes typical for a studied area.

References

[1] *Underflooding forecast and calculation of drainage system on developed urban areas* URL: http://www.gosthelp.ru/text/PosobiekSNiP2061585Progno.html Manual to SNiP 2.06.15-85 (construction rules and regulations) (in Russian)
[2] *Report on geotechnical investigation. Site: «Kuzbasskiy tekhnopark, Kemerovo city (site №1). Production-laboratory department «Ecology and Environmental Management»* 2011 Code 29-11(in Russian)
[3] *Report on hydrogeological investigation. Site:Project correction of land planning to identify the first stage of construction of Kusbasskiy tekhnopark, Kemerovo (site №1)* Code 280-2012 (in Russian)
[4] *Report on geotechnical investigation. Site: Kuzbasskiy tekhnopark, Kemerovo city (site №1). Technology Transfer Center with Exhibition hall* Code 02-2013 (in Russian)
[5] *Report on geotechnical investigation. Site: Apartment Block № 25 in micro-district № 12, Rudnichny district, Kemerovo city* Code149-13 (in Russian)
[6] *Report on geotechnical investigation 2013. Site: apartment block № 24 25 in micro-district № 12, Rudnichny district, Kemerovo city* Code 102-13 (in Russian)
[7] Strokova L A 2010 Chislennye raschety osadok muzykal'nogo teatra g Kemerovo *Inzhenernye izyskaniya* **9** 50 – 57 (in Russian)
[8] Strokova L A *Nauchno-metodicheskie osnovy chislennogo prognoza deformirovaniya gruntovykh osnovaniy* Dissertation Theses (in Russian)
[9] Strokova L A 2015 Modeling of tunneling-induced ground surface movement *IOP Conf. Ser.: Earth Environ. Sci.* **24** (1)
[10] Pokrovskiy V, Pokrovskiy D, Dutova E and Nikitenkov A 2014 The research underflooding processes of architecture monuments on the territory of Tomsk with using GIS-technology *Int. Multidiscip. Scien. GeoConf. Surveying Geology and Mining Ecology Management, SGEM.* **2** 935 – 941
[11] Belousova A P 2011 Risk assessment of underflooding of areas by groundwater during floods *Water Resources* **38** (1) 39 – 46
[12] Zhukov M and Tishaieva A 2010 Compositional model of data distribution in problems of underflooding detection by means of remote sensing *Geoinformatics 2010 - 9th Int. Conf. Geoinformatics: Theoretical and Applied Aspects*

Analysis of rock strength properties in Kuznetsk Basin coal-bearing formations

V E Olkhovanenko

Tomsk State University of Architecture and Building (TSUAB)
2 Solyanaya Square, building 2, Tomsk , Russia

E-mail: igg@tsuab

Abstract. The formation features and rock strength properties in Kuzbas coal-bearing basin have been described. The influence of granulometric, mineralogical and chemical composition on rock strength has been determined. The change of rock strength property has been identified at different stages of lithogenetic alterations.

1. Introduction

Determining the rock strength properties in Kuznetsk Basin (Kuzbass Basin) is of profound importance in evaluating the geotechnical conditions for coal strip mining. Pit wall stability, in most cases, depends on the specific physico-mechanical properties developed during lithogenesis and their further forecasting [1-3, 6-8].

Kuzbass Basin embraces well-developed coal formations from Balakhonian ($C_{2\text{-}3\ bl}$), Kolchugino (P_2) and Tarbagansky ($I_{1\text{-}2}$) stages, which, in its turn, include sandstones, aleurolites and agrillites. Identifying the rock strength properties in Kuzbass coal-bearing formations is of great importance because of the necessity to evaluate the geotechnical conditions for coal strip mining and determine the pit wall stability of these massive coal formation intervals. This, in itself, urges the analysis of the rock composition and its physico-mechanical properties within Balakhonsky and Kolchugino series during different diagenetic transformation stages [4, 5].

2. Influence of granulometric rock compositions on their strength in Balakhonian and Kolchugino coal-bearing formations

Granulometric composition indicates the sedimentation conditions and, in many cases within Kuzbass coal-bearing formations, these conditions vary depending on different geologic-genetic rock complexes. According to the granulometric composition analysis, the rock granulometric composition of one and the same lithologically formed rock type, involved in identical transformation stages and formed in similar geotectonic zones, practically does not affect the rock strength properties. Only early-catagenetic Kolchugino aleurolites and adjacent coal beds (open burning and parabituminous coals) under conditions of increasing aleurolite concentration revealed a decreasing tendency of tensile strength in uniaxial compression and shearing resistance angle, whereas adhesion insignificantly changed. And, conversely, the increase in arenaceous concentration results in the increasing shearing resistance angle and limited adhesion decrease. As diagenetic transformation continues the granulometric composition influence on rock strength sharply decreases. The research results indicated that there is no correlation between tensile strength in uniaxial compression and the concentration percentage of varied-size clastic particles. This could be explained by the fact that during diagenetic transformation new rock structural bonds were formed as a result of the replacement of feldspar clastics and volcanic rocks by carbonates and / or clay matter by carbonate cement. Simultaneously, the dissolution of quartz fragments could have been observed followed by the re-sedimentation of the latter, resulting in regeneration nodes and rims. At the same time, conformed-generation granoblastic and micro-stylolitic structures were formed. As the transformation stages progress, the more intensive the involved processes are. In consequence of newly-formed structures and structure bonds, the tensile strength in uniaxial compression differences of rocks are balanced due to the difference in the granulometric composition.

3. Influence of clastic composition on rock strength

The research determined that during the early transformation stages of Kolchugino aleurolites, volcanic rock fragments and sedimentary clastics significantly influence the strength properties; whereas their increase sensibly decreases the tensile strength in uniaxial compression and adhesion. It is conversely in the mid diagenetic transformation stage which embraces bituminous coal in Balakhonian sandstones being located within the intensive linear Prisalaire folded zone. These rocks show an intensive replacement of feldspar clastics by carbonates, corrosive quartz grains, formed generation nodes on quartz fragments and replacement of clay cement by carbonate matter. The more intensive the processes are, the greater is the extent that diagenetic carbonates form as clastics replacement. Simultaneously, new highly strong structure bonds between rock clastic particles and cement are formed. As a result, the concentration of pseudomorphous carbonates in the rocks could be indicators of intensive diagenetic processes and the further formation of highly strong structure bonds. The high concentration of well-preserved quartz and feldspar clastics in the rocks indicates the fact that the diagenetic processes are decelerated resulting in a strength decrease. An excellent example could be the composition of Balakhonian sandstones hosting coking and bituminous coals. According to the research data, it should be noted that sandstones with a high quartz and feldspar concentration have the lowest tensile strength in uniaxial compression and adhesion values. If the latter decreases and carbonate clastics content increases, the strength properties intensify.

4. Influence of chemical composition on rock strength

The influence of rock chemical composition on structure bond strength involved a detailed examination of the chemical composition and %-content determination of SiO_2, Al_2O_3, Fe_2O_3, MgO_3, CaO, K_2O, Na_2O in lithologically different rock types from Kuzbass Basin. These rocks formed during various catagenesis stages and were pervasive throughout different geotectonic zones. The strength indicators (σ_c, σ_p) were defined and the interrelation between these indicators and the chemical composition characteristics were also identified. The analysis data showed that various oxide concentrations in rocks differently affect the rock strength. It should be noted that more than 200 correlation relationship pairs (bonds) were investigated, and, only in 4 cases, there existed a strong correlation relationship (bond), where the correlation coefficient exceeded 0.75. Kolchugino aleurolites, associated to gas coals, revealed a positive $\boldsymbol{\sigma}_c$ increase and increasing Fe_2O_3 content in the rocks during the transformation stage; however there appeared to be a reversed pattern σ_c decreased to an increase of Na_2O. This could be explained by the fact that during early diagenesis and catagenesis siderite was formed within Kuzbass rocks. Na_2O content could indicate insignificant changes of plagioclase itself during these early stages, whereas, it could have been replaced by carbonate matter in the later formation stages.

5. Influence of cement content on the chemical-mechanical rock properties

Kolchugino rocks include well-formed argillaceous, argillaceous-carbonate, and carbonate cement. It was established that the composition and percentage content of above-mentioned cement types significantly depends on the conditions (i.e. stages) and intensity of diagenetic transformation itself. Argillaceous cement in early transformation rocks (open burning coals) was predominately kaolinite clay, but sometimes, montmorillonite or hydromica. During the late transformation stages secondary hydromica (hydromuscovite) was formed resulting in the subsequent vanishing of kaolinite clay and montmorillonite, which, in its turn, was not involved in the formation stages of semi-anthracite and anthracite coals. Such secondary minerals as sericite and chlorite could be found in the cement matter.

Clay strength cementing of clastic particles was the lowest under identical transformation conditions in regard to parabituminous coals. Although rock strength increases proportionally to the transformation stages, it does not exceed the strength factor of rocks with carbonate cement.

Carbonate cement is predominately dolomite, rarely calcite. Well-formed rocks also could include siderite cement. Originally, carbonate cement is catagenetic, especially during the late rock transformation stages. The strength factor of rocks with carbonate cement under identical conditions is the highest. Moreover, the strength factor within this or that transformation stage and within different geotectonic zones is governed by one and the same behavior patterns as those of rocks with argillaceous cement. However, the latter strength factor is lower than the former one. The rocks with argillaceous-carbonate and carbonate clay cement reveal intermediate strength factors resulting from clay cementing matter being replaced by carbonate cement.

As a result of catagenesis processes the content of cemented clastic Balakhonian rocks significantly changes. However, as in the case of Kolchugino rocks, there are three main cement types: argillaceous, argillaceous-carbonate and carbonate. Comparable to Kolchugino rocks, the cement content in Balakhonian rocks is predominately hydromica (hydromuscovite) and sercite, as well as chlorite, and no kaolinite or montmorillonite. The latter could be explained by the significant secondary alterations of Balakhonian rocks during catagenesis. Carbonate cement includes catagenetic minerals, in most cases, dolomite, rarely calcite and/or siderite. The most coherent clastic cemented rocks are those with carbonate cement, while the less coherent are those rocks with argillaceous cement.

Based on above-mentioned factors, it could be stated that the cement content itself significantly influences the structure bond strength of rocks from Kuznetsk Basin. In this regard, to determine the rigidity of rocks on the open-pit walls, the rocks (i.e. rocks from Kuznetsk Basin) classification should include the rock cement content factor.

6. Changes of rock strength properties affected by catagenesis and regressive lithogenesis

Catagenetically, Kuzbass rocks alter as a result of high pressure, temperature and/or the intensive involvement of pore solutions and ground waters. Ground waters play a significant role in the diagenetic transformation of rocks. It is the composition of groundwaters and pore solutions that defines the plagioclase replacement intensity and formation of secondary clay minerals and carbonates. As a result, not only the content and behavior of rock structure bonds change, but also it initiates the formation of hydrocarbonate- sodium groundwaters in Kuzbass coal deposits. Carbonate clastics could be an important source of carbonate matter, whereas, interacting with $C0_2$, this carbonate dissolves and passes into the groundwaters. Besides early catagenetic transformation of rocks and free pore-filling carbonate crystallization, the intensive corrosion of clastic matter occurs. Recrystallization of carbonates under significant pressure occurs in Kuzbass rocks during late catagenesis, resulting in clay matter replaced by carbonate cement and further formation of monocrystal carbonates. Simultaneously, plagioclase, volcanic rock clastics and shale are replaced by carbonate matter. In response to high pressure new structure bond types as micro-stilolites are formed, resulting in significant changes of rock strength (Table 1).

It was established that cemented clastic Balakhonian rocks, especially sandstones with carbonate cement, have the highest strength factor in Kuzbass coal-bearing formations. Significant tectonic activity within the intensive linear Prisalaire folded zone (Kuzbass) during the post-inversion period notably decreased the structure bond strength, but did not have a strong effect on the rock properties. Thus, it can be stated that Balakhonian rocks, being abundant in this zone and formed during late catagenesis, have a rather low strength factor comparable to the rocks hosting coking coals. However, these Balakhonian rocks are found in Prigornoshor zone, Kuzbass monocline.

Table 1. Changes of physico-mechanical rock properties during different lithogenesis stages.

Lithogensis stages		Rock types	Density, gr/cm^3	Porosity, %	Resistance of uniaxial compression, MPa	Shearing resistance angle, °	Cohasion, MPa
Late diagenesis		Sandstones	2.01	36.0	0.16	21	0.05
		Aleurolites	2.10	33.6	0.13	29	0.04
Catagenesis	Early	Sandstones	2.36	14.0	3.13	41	
		Aleurolites	2.38	13.2	3.16	41	
	Mid	Sandstones	2.47	8.2	4.91	41	
		Aleurolites	2.47	9.5	4.05	39	
	Late	Sandstones	2.53	6.2	7.57	38	
		Aleurolites	2.49	5.3	6.54	38	
Metagenesis		Sandstones	2.56	4.8	5.26	35	1.34
		Aleurolites	2.50	5.0	5.08	36	1.08

7. Conclusion

Research data results have been used in determining the rock stability on the open-pit walls within Kuznetsk Basin. According to the calculations the highest rock stability level was established in Prigornoshor monocline zone, while the lowest- in intensive linear Prisalaire folded zone (Kuzbass).

References

[1] Deev P V, Sammal' A S and Baryshnikov V D 2012 Estimating remaining life of underground tunnel concrete lining by convergence measurements. *J. Mining Science* 49-54

[2] Savichev O G and Reshetko M V 2015 Evaluation of plain river channel deformation in the absence of observation data *IOP Conf. Series.: Earth Environ. Sci.* **24** : Scientific and Technical Challenges in the Well Drilling Progress, 24–27 November 2014, Tomsk, Russia

[3] Fotieva N, Sammal A, Deev P and Bulychev N 2005 Design of shallow tunnel linings under seismic effects of earthquakes *Proc. 16th Int. Conf. Soil Mechanics and Geotechnical Engineering: Geotechnology in Harmony with the Global Environment, September 12-16 2005* Osaka, Japan vol **3** 1607-1610

[4] Olkhovatenko V E 2007 *Inzhenernaja geologija ugol'nyh mestorozhdenij Sibiri i Dal'nego Vostoka* Tomsk

[5] Olkhovatenko V E and Trofimova G I 2013 Zakonomernosti formirovanija fiziko-mehanicheskih svojstv gornyh porod uglenosnyh otlozhenij Erunakovskogo rajona Kuzbassa *KSTU Vestnik* **5** 136–142

[6] Strokova L A 2015 Modeling of tunneling-induced ground surface movement *Scientific and Technical Challenges in the Well Drilling Progress IOP Conf. Seri.: Earth Environ. Sci.* **24** *(24–27 November 2014)* Tomsk, Russia

[7] Strokova L A 2009 Numerical model of surface subsidence during subway tunneling *Soil Mechanics and Foundation Engineering* vol **46** (3) 117–119

[8] Strokova L A 2010 Methods of estimating surface settlement during driving of urban tunnels *Soil Mechanics and Foundation Engineering* vol **47** (3) 92–95

Prevention and remediation of rockslide at left portal of north tunnel of Da Nang – Qu ang Ngai expressway in Quang Nam, Vietnam

Phi Hong Thinh[1], Pham Quoc Tuan[2] and Pham Ngoc Minh[3]

[1]University of Transport and Communications, Hanoi, Vietnam Vietnam
[2] Society of Rock Mechanics, Hanoi, Vietnam
[3] Transport Engineering Design Incorporation, Hanoi, Vietnam

E-mail: [1]phthinh.tomsk@gmail.com

Abstract. Da Nang - Quang Ngai Expressway is one part of the North - South Expressway of Vietnam being 130km in length, 24.5m in width. The Tunnel is a section of 4 construction packages with station of 22+500km – 23+000km. The tunnel is designed to go through a small hill, altitude of about 120m above sea level and 500m long and has two bounds – called North and South Tunnel. The North Tunnel has two portals – left and right. Bedrock is sedimentary rock (alternation of sandstone and conglomerate). Authors have investigated the field, detected inaccuracy in geotechnical investigation report, applied discontinuity model and failure criterion for anisotropic rock, analyzed stability of rock slope and proposed a solution to prevent rockslide at the Left Portal of the North Tunnel of the Expressway.

1. Introduction

Da Nang - Quang Ngai Expressway is one part of the North – South Expressway of Vietnam, being 130km in length, 24.5m in width. The tunnel is a section of 4 construction package with station of 22+500km – 23+000km. The tunnel is located at Duy Son and Duy Chung communes, Duy Xuyen district, Quang Nam province, Vietnam. The tunnel has two bounds - North and South. The North tunnel has two portals - Left and Right.

2. Geotechnical investigation and seismic survey

2.1 Geotechnical investigation results

The tunnel is designed to go through a small hill with altitude of about 120m above sea level and 500m long (Figure 1). Bedrock is a sedimentary rock (alternation of sandstone and conglomerate). Line of strike of the terrain is Northeast - Southwest with angle of about 60^0 from the North, bedding of bedrock dipping 30° to the Northwest.

- At Right Portal: many blocks of unstable rock located on the surface with different sizes (from 0.5 m to 5.0 m in diameter); plants grow poorly.

- At Left Portal: bedrock exposed on the surface at some positions, rock is highly weathered, plants grow well. However, surface failure (mass movement) is found near the tunnel: the new is about 15m in width x 30m in length on the South bound tunnel, the older is about 30m in width x 50m in length on the North bound tunnel.

- Geological condition of Left Portal: A layer of extremely weathered rock is found over the bedrock with different thickness of from 4 m to 0.65m. It is residual soil (sand with gravel, reddish brown, medium dense). Conglomerate and sandstone alternate below, the proportion of conglomerate and sandstone is 4:1. Particles of conglomerate is rounded and size of 2 to 20mm, fresh to lightly weathered condition. Permeability of the fracture zone is quite small (nearly 0.5mm/min with the pressure 1 kG/cm^2), compression strength of lightly weathered rock is high, approximately 50MPa in average [2].

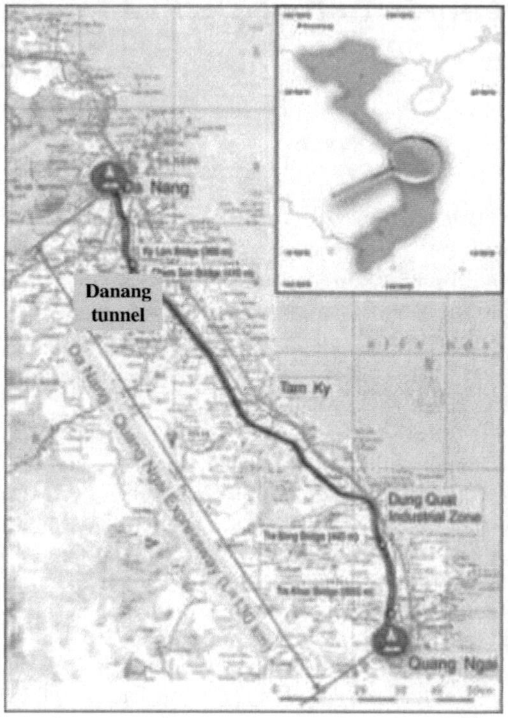

Figure 1. Location of Da Nang - Quang Ngai Expressway [7].

2.2 Results of seismic survey

Results of data seismic survey processing determined the velocity of the geological layers. At the same time, the positions of low longitudinal wave velocity zone were identified (in fracture and fault zones). The elastic parameters were determined according to each profile. The results of seismic survey in the North bound and South bound profiles showed many similar properties. The stratigraphic classification of the survey area is as follows [2]:

Layer 1: covering layer, rather thin, from 0 to 4.5m. Longitudinal wave velocities are between 400 to 1000m/sec. Modulus of elasticity is from 4 to 25 (1000 kG/cm^2); modulus of deformation is from 0.25 to 2.5 (1000 kG/cm^2).

Layer 2: average thickness is about 4m; at some positions in the profile of the North bound Tunnel there is a thickness of more than 6.5m. Longitudinal wave velocities are from 900 to 1800m/sec, almost above the groundwater level. Modulus of elasticity is from 20 to 81 (1000 kG/cm^2). Modulus of deformation is from 1.8 to 10 (1000 kG/cm^2).

Layer 3: average thickness is about 4m. Longitudinal wave velocities are from 2100 to 2300m/sec. Modulus of elasticity is from 122 to 133 (1000 kG/cm^2). Modulus of deformation is from 10 to 15 (1000 kG/cm^2).

Layer 4: thickness is from 7 to 9m in average, a few locations are thinner, about 5m. Longitudinal wave velocities are from 2700 to 3500m/sec. Modulus of elasticity is from 184 to 312 (1000 kG/cm^2). Modulus of elasticity is from 25 to 60 (1000 kG/cm^2).

Layer 5: longitudinal wave velocities in this layer are from 3700 to 6000m/sec, average value is 4200m/sec. Modulus of elasticity is from 349 to 932 (1000 kG/cm^2). Modulus of deformation is from 75 to 400 (1000 kG/cm^2).

Damage zones, faults and geological anomalies have been observed in six locations of each survey line. Longitudinal wave velocities are 2000 to 3200m/sec. Modulus of elasticity is from 100 to 260 (1000 kG/cm^2). Modulus of deformation is from 10 to 40 (1000 kG/cm^2).

3. On-site investigation results

According to on-site investigation results, the authors detected inaccuracy both in the geotechnical investigation report itself and studied and analyzed intact rock in this area. In situ intact rock has 2 fault systems with grade V. Intact sandstone location in Left Portal of the North Tunnel has 3 faults system and bedding system. The bedding system has direction angle of 15^0- 20^0; fault system perpendicular to bedding system has direction angle of 75 ° -80°; fault system that overlaps rock slope face has direction angle perpendicular to tunnel axis.

In addition, according to the results of topographical survey, intact rock in study area is tectonic system of grade II and is divided into river delta and hilly regions (Figure 2).

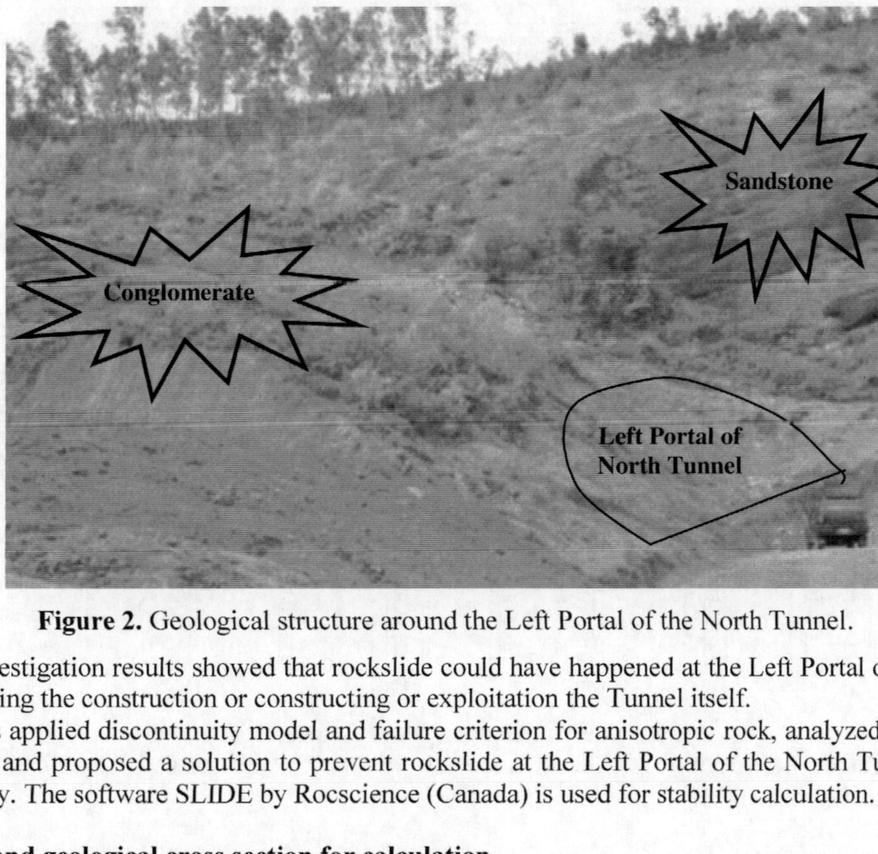

Figure 2. Geological structure around the Left Portal of the North Tunnel.

The investigation results showed that rockslide could have happened at the Left Portal of the North Tunnel during the construction or constructing or exploitation the Tunnel itself.

Authors applied discontinuity model and failure criterion for anisotropic rock, analyzed stability of rock slope and proposed a solution to prevent rockslide at the Left Portal of the North Tunnel of the Expressway. The software SLIDE by Rocscience (Canada) is used for stability calculation.

4. Profile and geological cross section for calculation

Base on documents of project and investigation results on the site, authors established profiles and geological cross sections for study area, as follows:

Table 1. Geological cross section.

No.	Name of section	Location	Remark
1	Profile of axis North Tunnel with study area Left Portal		
2	Cross section MC01	Km23+031	
3	Cross section MC02	Km23+041	Left Portal

Geological cross sections with dimension, arrangement and structure of the tunnel updated from drawings of construction contractor are shown in Figures3 to 4.

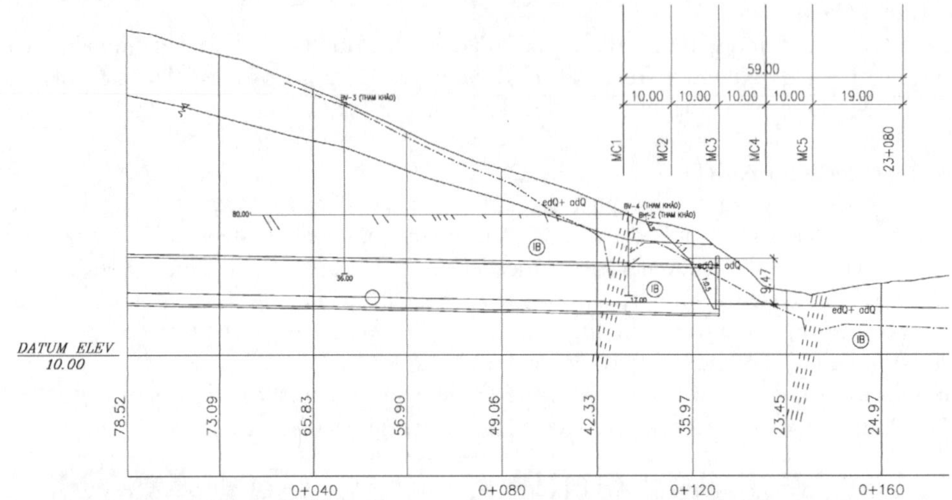

Figure 3. Profile of axis North Tunnel with study area Left Portal.

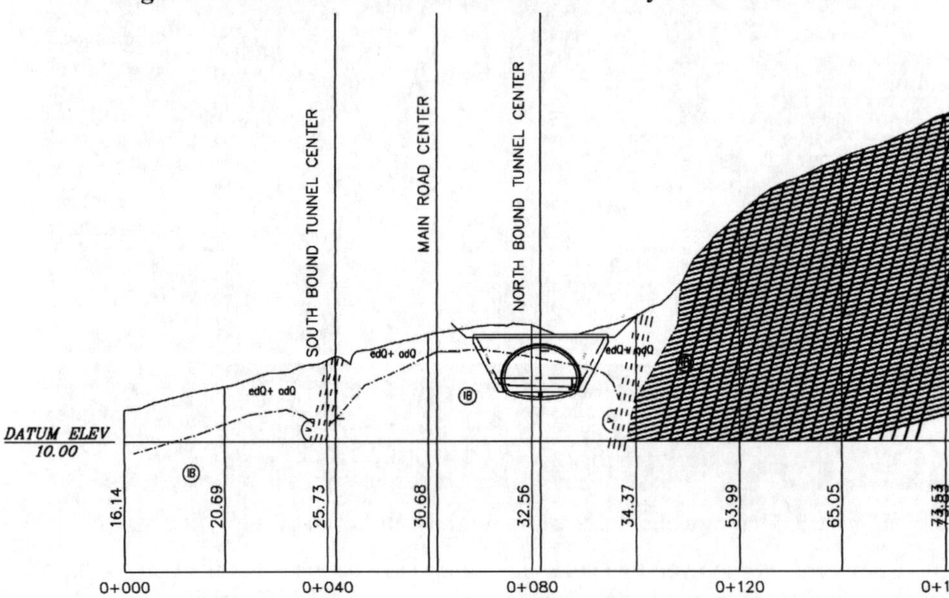

Figure 4. Geological cross section MC02

Geological cross section MC02 at location of the Left Portal is used for stability calculation.

5. Parameters of intact rock and soil to be used for stability analysis of rock slope

- Layer 1: completely weathered sandstone - green grey, brown, stiff sandy clay with thickness of 0.5 to 1.2m (Table 2).

Table 2. Parameters of the layer 1.

Case of analysis	Bulk density γ (kN/m^3)	Internal friction angle φ (degree)	Cohesion c (kPa)
Natural	17.0	19	20
Saturated	20.0	16	14

- Layer 2: Parameters of sandstone mass are determined according to Hoek - Brown methods with RocLab software of Rocscience, as seen in Table 3 and 4.

Table 3. Input parameters of the layer 2 for RocLab software.

Case of analysis	Unconfined compressive strength of intact rock sigci (MPa)	Geological strength index GSI	Intact rock parameter m_i	Disturbance factor D	Type application
Natural	50	65	17	0.8	Slope (15m)
Saturated	25	65	17	0.8	Slope (15m)

Table 4. Output rock mass parameters according to Mohr-Coulomb method determined by RocLab software.

Case of analysis	Bulk density γ (kN/m^3)	Internal friction angle φ (degree)	Cohesion c (kPa)
Natural	25.4	58.70	457
Saturated	25.4	55.08	276

- Bedding layer: physico-mechanical parameters of bedding layer are determined as parameters of hard clayey soil (Table 5).

Table 5. Physico-mechanical parameters of bedding layer.

Case of analysis	Bulk density γ (kN/m3)	Internal friction angle φ (degree)	Cohesion c (kPa)
Natural	17.0	19	36
Saturated	20.0	16	21

Fault system perpendicular to bedding system: parameters of fault system perpendicular to bedding system are determined according to Hoek – Brown method. Its cohesion has been determined by 20% of cohesion of rock mass (according to Hoek – Brown method) (Table 6).

Table 6. Parameters mechanical and physics of fault system perpendicular with bedding system.

Case of analysis	Bulk density γ (kN/m^3)	Internal friction angle φ (degree)	Cohesion c (kPa)
Natural	17.0	33	91.4
Saturated	20.0	29	55.2

6. Cases of calculation

Case 1 (natural construction condition): parameters for calculation used in natural condition.

Case 2 (construction is conducted in raining condition): parameters for calculation used in saturated condition.

Case 3 (exploiting the tunnel in typhoon or flood condition): initial seep flow in layer rock and soil.

Case 4 Rock slope is supported by anchors with tensile capacity of 300kN, spacing of 2m; bond length equals 10% of anchor length.

7. Results of rockslide calculation

Summary of calculation results is shown in Table 7.

Table 7. Calculation results of factor of rock slope stability at geological cross section MC02.

Case of calculation	With circular slipping			With non circular slipping	
	Bishop simplified	Janbu simplified	Ordinary	Spencer	Janbu simplified
Case 1	2.052	2.094	2.388	-	6.628
Case 2	1.783	1.895	1.771	1.914	1.778
Case 3	-	-	-	1.138	1.046
Case 4	1.230	1.265	1.511	-	-

8. Conclusion and recommendation

Calculation results show that rock slope stability factor in all cases ensured the conditions of equilibrium limit. However, it should also satisfy the condition of limiting rock slope stability factor [FS]. According to Vietnamese standard, [FS] it is 1.2 for rock slope. So, the rock slope stability factor in case 3 does not satisfy Vietnamese standard. Rockslide may happen at the Left Portafactor l of the North Tunnel when constructing and exploiting the Tunnel in typhoon or flood conditions.

In Vietnam, anchor is being used to prevent landslides and rockslides at some projects such as Noi Bai – Lao Cai Expressway, Hai Van Tunnel and etc. The anchor method is a highly effective one and of reasonable price. Thus, we propose that anchors should be used to prevent rockslides at the Left Portal of the North Tunnel. Anchor bars have tensile strength of 300kN, spacing of 2m; bond length equals 10% of anchor length. Calculation results in case 4 show that the rock slope stability factor (FS) is greater than in [FS].

Acknowledgments

The investigations were partially supported by the Russian Foundation of Fundamental Research (№ 15-05-00809), as well as by the Ministry of Science and Education of the Russian Federation (as part of the research "Improvement of Constructive and Geoecological Safety, Energy Efficiency of Vessels and Facilities at Continental Shelf of the Far Eastern and Arctic Seas", NIR 543.2014).

References

[1] Doosan Hevy Industries and Construction 2011 *Slope stability Review*. Doc. No DHIC-MS-CMPS-EN-A2 p 255

[2] Joint Venture of Nippon Koei Co., Ltd. and Thai Engineering Consultants Co., Ltd. (2013). *Geotechnical Investigation Report for Pakage 4, Station 1* (Km21+500 - Km26+800) p 589 p

[3] Pham Q T, Phi H T and Nghiem H H 2015 Analysis stability rock slope at left bank of South portal of North tunnel - Da Nang - Quang Ngai expressway *Vietrock 2015 Int. Symp. Series: Rock mechanics for sustainable development* 314–322

[4] Strokova L A 2015 Modeling of tunneling-induced ground surface movement // Scientific and Technical Challenges in the Well Drilling Progress *IOP Conf. Series: Earth Environ. Sci.* **24** (24–27 November 2014), Tomsk, Russia

[5] Strokova L A 2009 Numerical model of surface subsidence during subway tunneling *Soil Mechanics and Foundation Engineering* **46** 3 117–119

[6] Strokova L A 2010 Methods of estimating surface settlement during driving of urban tunnels *Soil Mechanics and Foundation Engineering* **47** 3 92–95

[7] URL: https://www.google.com/?gws_rd=ssl#q=duong+cao+toc+da+nang+-+quang+ngai

Modeling of changing hydrogeological conditions during construction of pier foundations on the Kama river bank

D Purgina[1], L Strokova[2] and K Kuzevanov[3]

[1, 2, 3] Department of Hydrogeology, Engineering Geology and Hydrogeoecology, Institute of Natural Resources, National Research Tomsk Polytechnic University, 30 Lenin Av., Tomsk, 634050, Russia

E-mail: [1]Purgina_darya@mail.ru, [2]strokova@sibmail.com, [3] kki@tpu.ru

Abstract. The article presents the results of hydrogeological studies carried out within the area of the Kama river bank in Perm city. It proposes the hydrodynamic model by means of which a number of forecasting issues have been addressed. The possible scenarios of changes in filtration flow, i.e. water rise before the obstacle and water drop behind the obstacle due to groundwater filtration blockage, have been described [2]. The allowable changes of hydrodynamic conditions within the study area have been outlined.

1. Introduction

The relevance of the study is due to the need to improve flood forecasting methods and corresponding warning techniques. City development, development density, underground utility systems, and changes in natural relief dramatically affect geological environment within the territory under construction. Due to a number of factors (city development, rock compaction caused by overburden engineering load), new perched aquifers form.

The purpose of the current research is to study and provide a long-term forecast of geological, hydrogeological, and hydrodynamic conditions when constructing additional pier foundations for the Kama river bank protection.

Today, flood events appear to become an urgent issue throughout the entire country [9], which, in its turn, brings devastating consequences and can have effects on the economy and environment. This fact urges to carry out a more detailed study aimed at improving flood forecasting methods and protection techniques [7].

2. Materials and methods

The data for the research were taken from “Sibgiprotransput”, which had been collected by the author of the article in 2013…2014 when studying hydrogeological conditions. In the course of this work, 66 wells were drilled (including four slant wells and two horizontal ones), 20 pit-holes were made, and more than 600 samples were collected. In field investigations, static and dynamic zoning was applied. The soils were sampled and tested using rotating method (static loads). To provide reliable estimate of the filtration parameters, the pumping tests within a two-well cluster and five individual wells were carried out. Hydrogeological survey was done, more than 50 observation points being described. In lab conditions, the full scale of physic-mechanical properties of soils was determined. The filtration capacity was defined for conventionally compacted rocks above the water table. To estimate the impact of the newly constructed pier foundations or changes in groundwater recharge conditions, the numerical modeling was mainly used, including solution of differential equation of nonstandard planned filtration based on the finite-difference technique [2]. Surfer and AutoCad were used as pre-processing programs.

3. Results and discussion

In 2013, the Institute "Sibgiprotransput" (Novosibirsk) with direct involvement of the author of the current research carried out geoengineering studies in Perm city. This study was urged by the need to reconstruct the motor over-bridge crossing the railway line Perm 1…Perm 2. During the study period (03.08…24.11.2013), geoengineering and hydrogeological conditions of the area were investigated in detail.

Within the hydrogeological cross-section of the study site and according to its geological structure [6], the following aquifer systems were identified: Quaternary Aquifer System (pore-ground) and Upper Perm Sheshminsky Aquifer System (eP2sI) (fractured-porous), as well as waters found in technogenic deposits. The aquifers are interconnected and sporadically distributed within the study area.

Within the study area, the depth of the groundwater table is 91.50…104.76 m. The groundwater depth ranges within 0.5…12.2 m. The elevation difference is 13.26 m. The filtration flow is directed to the riverbed, with the hydraulic slope being 0.0608 (Figure. 1).

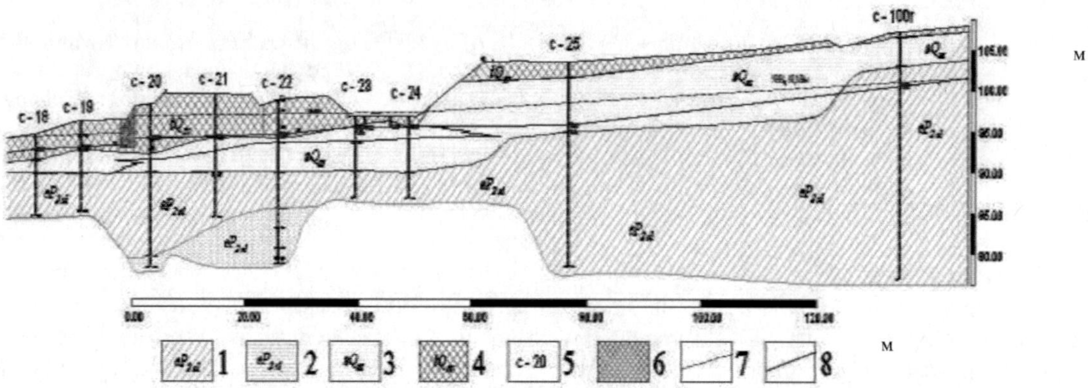

1 – Sheshminsky complex of fractured argillite; 2 - Sheshminsky complex of fractured sandstone; 3 Quaternary alluvial deposits of II and III terraces above-the-floodplain; 4 – technogenic soils 5 – well, its number; 6 – pier foundation; 7 – estimated groundwater level; 8 – groundwater level in October, 2013

Figure 1. Hydrogeological cross-section of the study area.

At the time of the study, the level of the surface water in the Kama river altered within 87.69…87.72 m. The highest values were identified within the motor over-bridge area, i.e. 93.48 m. In that case, a part of the Kama river valley was flooded, the river stages were water backed over a considerable distance [3].

To obtain reliable estimates of aquifer filtration parameters, a number of tests for underground waters flow were carried out. Such tests were an obligatory part for further estimates of the expected changes in hydrogeological conditions (Table 1).

Table 1. Data of pumping tests (individual wells).

Well number	Test interval, from … to m	Pumping parameters			Average filtration coefficient, m/day	Calculated filtration coefficient at a=0.98 (+)	Calculated filtration coefficient at a=0.98 (-)
		Flow rate, l/sec m^3/day	Draw-down м	Specific capacity, l/sec*m m^3/day*m			
C-18	1.9…10.0	0.036	2.05	0.018	0.33	0.39	0.28

	8.1	3.11		1.52			
C-25	8.5…25.0	0.800	6.45	0.124	1.36	1.51	1.21
	16.5	69.12		10.72			
C-56	10.9…15.0	0.180	2.18	0.083	1.57	1.75	1.39
	4.1	15.55		7.13			
C-59	6.6…10.0	0.370	1.20	0.308	5.94	6.28	5.59
	3.4	31.97		26.64			
C-100г	8.8…30.2	0.170	3.40	0.050	1.39	1.57	1.22
	20.4	14.69		4.32			
	Average				*2.12*	*2.30*	*1.94*
	Min.				*5.94*	*6.28*	*5.59*
	Max.				*0.33*	*0.39*	*0.28*

Due to the complex hydrogeological conditions, estimation of groundwater level change induced by rise of surface water level in the Kama river up to 93.936 m (1 % flow duration curve) was carried out by numerical hydrodynamic modeling in the simulation system Processing Modflow [1].

Gauging of filtration model [5] simulating the increase in the Kama river water level involved determination of the conductivity values and filtration coefficients of the underflow and adjacent deposits, respectively.

As the result of model gauging based on the fitting criteria of real and modelled values (well pressure), the problem of inverse filtration was solved. The field of natural head distribution was modeled, by means of which the flood hazard and risk map of the study area was developed (Figure2).

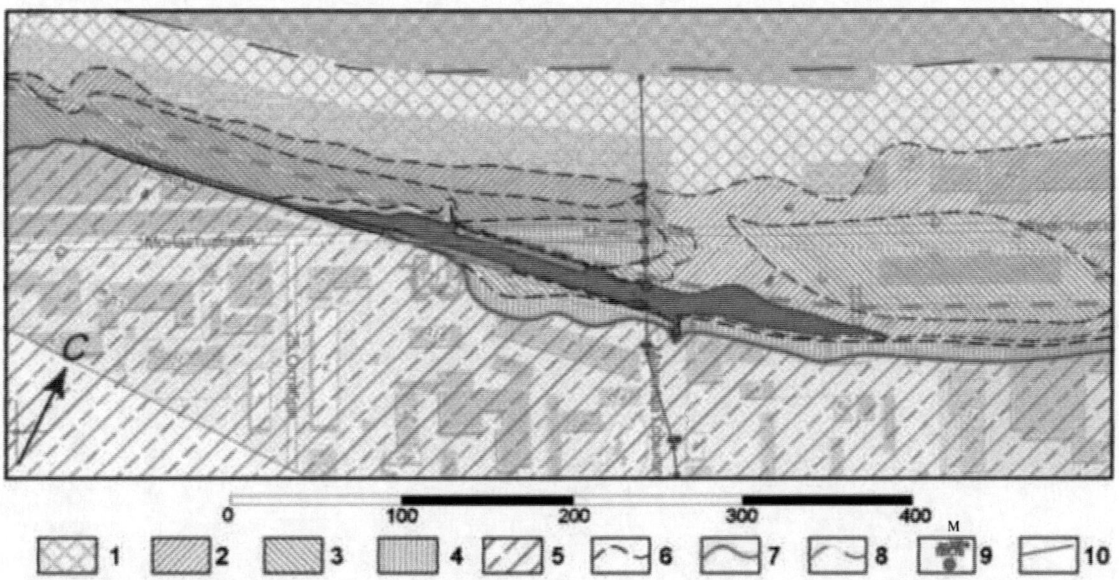

1 – flood subzone; aeration capacity for 2 zone – 0…0.5 m; 3 zone – 0.5…2.0; 4 zone – 2.0…4.0; 5 zone – slow increase in groundwater level; 6 – boundaries of minor flooding zone; 7 – boundaries between zone of seasonal (annual) flooding and zone of long-term increase in groudwater level; 8 – boundary of the Kama river, 26.10.2013; 9 – well; 10 – crosssection line.

Figure 2. Schematic flood hazard and risk map.

Slope protection for further construction of pier foundations could increase the risks of technogenic floods due to the damming effect that can occur during installation of the pier foundation base below the groundwater level [4]. To eliminate such a scenario, it is necessary to define the optimal depth of pier foundation grid installation. Considering complexity of the hydrogeological conditions, it is essential to use the methods of numerical modeling [8] in order to estimate the groundwater level changes during pier foundation base construction. In this case, the conventional hydrodynamic calculations are not efficient as they rarely make it possible to fully consider the complex nature of water-bearing foliation system interacting with the adjacent layers characterized by different hydrogeological conditions [9].

The forecast of groundwater levels involves two design options which are different in the absolute elevation of pier foundation grids. According to the first design, the absolute mark of pier foundation base is 92.0 m (Figure 3), while the second design option suggests 94.5 m.

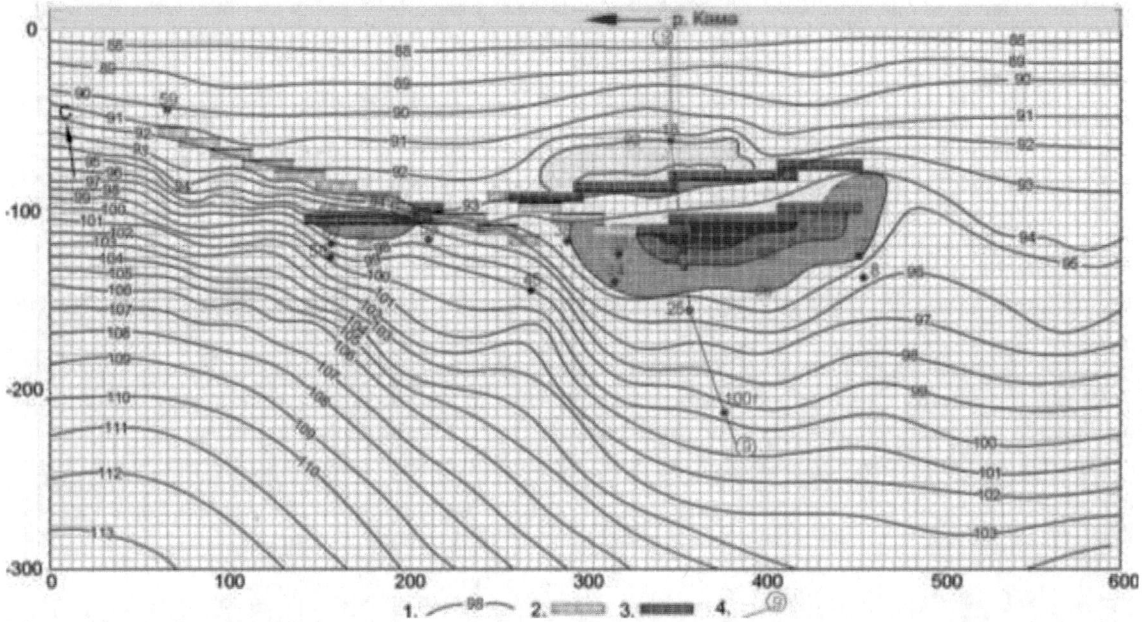

Figure 3. Forecast model of groundwater level changes during pier foundation base embedment up to 92.0 elev.m.: 1 – line of equal heads; 2 – existing pier foundations; 3 – designed pier foundations; 4 – cross-section line.

While the pier foundation base embedded up to the elevation mark 94.5 m, the filtration behavior changes insignificantly. The groundwater level rise around the pier foundations does not exceed 0.3…0.5 m against its standard level restored by means of epignose modeling (Figure 4).

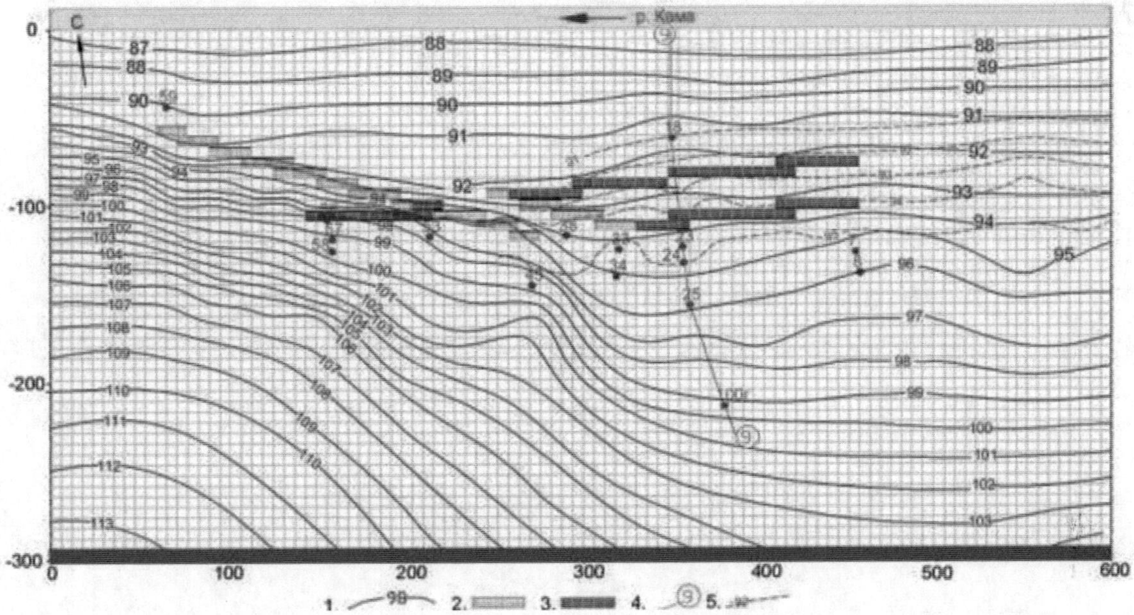

Figure 4. Forecast model of groundwater level changes during pier foundation base embedment up to 94.5 elev.m. 1 – line of equal heads; 2 – existing pier foundations; 3 – designed pier foundations; 4 – cross-section line; 5 – isolines of natural flow.

4. Conclusion

1. The hydrodynamic model to forecast the changes in groundwater levels with regard to maximum possible flood marks (93.94 m) of the surface waters in the Kama river was developed in the software package PMWIN.

2. The flood hazard and risk map of the study area was compiled in accordance with the rules and regulations of geoengineering studies SP 11-105-97 (part 2). The entire area is divided into two zones: zone of seasonal (annual) flooding and zone of long-term increase in groudwater level. The first zone is also divided into four subzones within which groundwater level is projected in line with the maximum level of flood waters in the Kama river. These subzones are identified on the basis of aeration capacity within the four graduation lavels: 0 m; 0.0…0.5 m.; 0.5…2.0 m; 2.0…4.0 m.

3. The forecast of groundwater level change was carried out by means of numerical modeling considering two options of pier foundation base embedment. According to the first design option, the absolute mark of pier foundation base (92.0 m) results in insignificant rise of groundwater level. In the second case (94.5 m), groundwater level rise around the pier foundations does not exceed 0.3…0.5 m against the natural conditions.

References

[1] Chiang Wen-Hsing and Kinzelbach Wolfgang *3D – Groundwater Modeling with PMWIN* (Germany: Springer) p 346

[2] Bukaty M B 2002 Software engineering for the solution of hydrogeological problems *TPU Bulletin Geology, Prospect and Exploration of Mineral Resources in Siberia* Vol 305, **6** 348–365

[3] Burtsev Yu G et al. 1978 *Otchet o rezul'tatakh rabot po izucheniyu dinamiki inzhenerno-geologicheskikh protsessov na beregakh Kamskogo i Votkinskogo vodokhranilishch v 1977-78* (Perm: Sylvenskaya gidrogeologicheskaya partiya) p 212

[4] Gavshina Z P and Dzektser E S 1982 *Usloviya podtopleniya gruntovymi vodami zastraivaemykh territoriy* (Moscow: Stroyizdat) p 116

IOP Conf. Series: Earth and Environmental Science **33** (2016) 012046 doi:10.1088/1755-1315/33/1/012046

[5] Zhernov I E and Pavlovats I N 1976 *Modelirovanie fil'tratsionnykh protsessov* (Kiev: Vysshaya shkola) p 192

[6] Lekhov M 2013 Hydrogeological surveys and modeling for urban construction: critical comments *Engineering Survey* **1** 24-29

[7] Muftakhov A Zh 1975 *Gidrodinamicheskie osnovy prognoza podtopleniya promploshchadok i fil'tratsionnye raschety zashchitnogo drenazha v slozhnykh gidrogeologicheskikh usloviyakh: avtoref. dis. d-ra tekhn. Nauk* (Moscow: VNII VODGEO) p 44

[8] Popov V K, Lukashevich O D and Korobkin V A 2004 *Tekhnogennoe podtoplenie: otsenka, razvitie, preventivnye mery* (Irkutsk) pp 24–27

[9] V D Pokrovsky, E M Dutova, K I Kuzevanov, D S Pokrovsky and N G Nalivaiko Hydrogeological Conditions Changes of Tomsk, Russia URL: http://iopscience.iop.org/article/10.1088/1755-1315/27/1/012031

Change of climate conditions in the forest area of South Yakutia

M V Reshetko[1], Y A Moiseeva[2] and D A Terre[3]

[1,2,3]Tomsk Polytechnic University, 30 Lenin Avenue, Tomsk, 634050, Russia

E-mail: [1]mrechetko@tpu.ru, [2] julchiky@mail.ru, [3]tedina@yandex.ru

Abstract. A statistical analysis of meteorological parameters which influence the intensity of exogenous processes in the forest area of South Yakutia has been performed. The indicated increase in mean annual air and soil temperature rates occurred non-uniformly. Annual precipitation increase was due to precipitation rates in cold or transitional seasons; increase in snow cover depth and decrease in diurnal air temperature range was observed in cold season.

1. Introduction

The negative effects of climate change are considered one of the global problems that mankind is facing today. Planning of effective measures on climate change effect mitigation is of great significance, in particular, ensuring accident-free running of gas transmission lines. It is impossible to consider the climate change influence on intensity of exogenous processes without objective estimate of long-term changes in meteorological parameters, which, in its turn, is the relevance of the research.

The results of the statistical analysis of meteorological parameters for the construction area of the transmission line "Power of Siberia" in the pipeline section Chayanda-Lensk are presented. The research objective is to reveal trends in climate condition change which affect the intensity of exogenous processes in the area under complex natural conditions.

2. Study area characteristics

The study area is characterized by warm summers and severe winters with relatively little snow. The transitional seasons are not long. The spring coming is characterized by great diurnal temperature variations. The maximum temperature range (24°C - 27°C) is registered in March and April. The annual precipitation is 347–454 mm, most precipitation (70–75 %) falling during the warm period. Insignificant snow cover and extremely low temperatures (- 60°C) contribute to permafrost extension. Exogenous processes, freeze-thaw action and karst processes present the most unfavorable conditions for gas line construction. As described in [1], the territory refers to the sporadic permafrost zone. The permafrost occurrences account for 40-80 %.

According to a number of researchers [2-9], the climate conditions of permafrost zones are becoming less favorable; however, in general permafrost thickness does not change. As reported in [4,5], significant changes of meteorological characteristics that affect permafrost conditions took place in Siberia. Therefore, the most serious consequences of climate change can be observed in Siberia.
According to [8], the increase in snow cover and temperature changes resulted in displacement of permafrost borderline at distance up to few tens of kilometers to the north. As noted in [5], in Cisbaikalia and Trans Baikal regions, isotherm specific for the southern border of insular permafrost was displaced up to 500±50 km to the north at the depth of 320 cm in 1976 – 2006. The analysis of

temperature trends in Siberia [10] revealed the areas of warming where temperature exceeded 0.5°C/10 years, i.e. one order of magnitude higher than in the Northern hemisphere. A major cause of warming is the change in atmospheric circulation [10].

3. Materials and methods

The source of information for the current research is data obtained in meteorological stations Lensk and Vitim. They included measurements of surface air temperature, amount of precipitation, and snow cover depth in 1960-2014. In addition, the following information was also considered: the data on soil temperature (320 cm) obtained in meteorological stations Lensk in 1988-2011 and Vitim in 1963-2011[11]; amount of rain, mixed, and solid precipitation in 1960-2010. The accuracy correction of measurement equipment was also made.

The selected series were checked for uniformity, precisely, for the presence of climatological and statistical heterogeneities. As is known, one of the main reasons for climatological heterogeneity is microclimate change as a result of station transfer to different local conditions or change in the station surrounding environment. This may affect meteorological parameter values and can be referred to as climate change improperly.

The investigation procedure started with verification of the null hypothesis about series of homogeneities, randomness, and presence of the observed trend. The homogeneities was checked by means of Student tests, Abbe test, and F-test, while statistical independence and trend presence were verified by Pitman test and the inversion criterion [12]. The conclusions about series fluctuation or nonrandom change were made at 0.05 significance level provided that calculated statistics in the module was above the corresponding threshold or was outside the variable type range. Abbe test, which is used by climatologists to study precipitation and temperature series, implies the condition (1) to be met in case of non-uniform series:

$$1-\frac{1}{\sqrt{n-1}} \le \frac{2A}{B} \le 1+\frac{1}{\sqrt{n-1}} \quad (1)$$

where: $$A = a_1'^2 + a_2'^2 + \ldots + a_n'^2 - \tfrac{1}{2}\left(a_1'^2 + a_n'^2\right) \quad (2)$$

$$B = \left(a_1' - a_2'\right)^2 + \left(a_2' - a_3'\right)^2 + \ldots + \left(a_{n-1}' - a_n'\right)^2 \quad (3)$$

Where a_i' – sample deviation from the sample mean, n – number of samples. This condition being violated, the hypothesis of series uniformity is rejected.

3. Results and discussion

Based on the conducted analysis, statistical heterogeneity of more than one-third of air, soil temperature, and precipitation series has been detected.

The positive trend in mean annual air and soil temperature has been stated. It is worth noting that rise in mean annual air temperature did not occur linearly. During the first half of the study period till 1986-1990 the temperature increased by 0.03°C per year in Lensk and by 0.04°C per year in Vitim. In the later period the rate of increase falls down to 0.02°C per year in Vitim, and no temperature increase is observed in Lensk. Statistically significant increase in average monthly temperature can be observed in transitional period (May-June), being manifested non-uniformly in the considered stations. In Lensk, average monthly air temperature rises approximately by 0.04°C per year in April, May and June, whereas in the meteorological station Vitim (Figure 1) it increases by 0.05-0.07°C per year from March to June (Table 1).

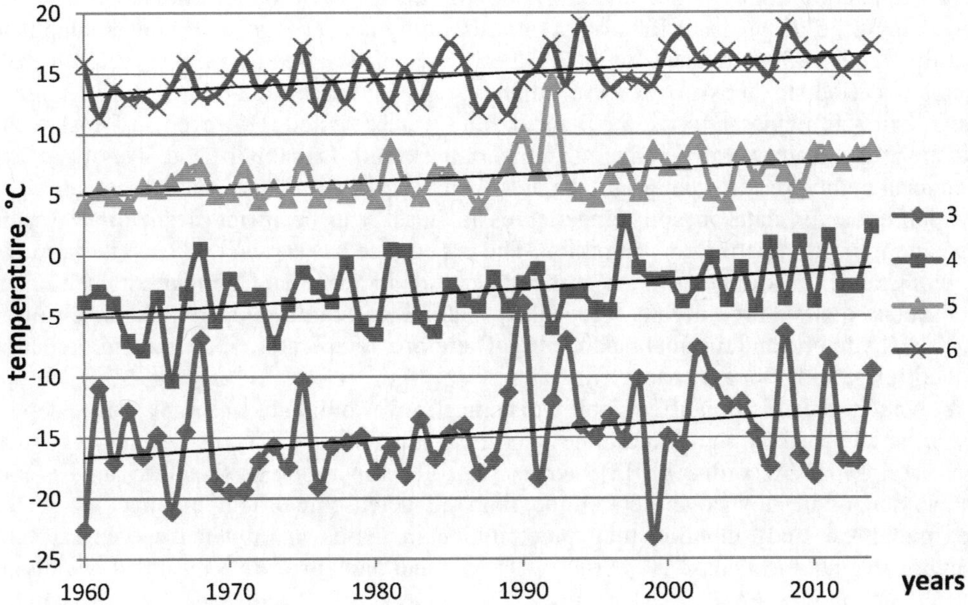

Figure 1. Long-term changes in average monthly air temperature in March (3), April (4), May (5) and June (6). Vitim, 1960-2014.

The statistically significant trend in annual temperature variation was not revealed. Based on the analysis of long-term maximum diurnal temperature range, it has been stated that maximum diurnal temperature variation increases only in May (0.09°C per year) in the meteorological station Lensk, while no trends have been distinguished throughout the remaining period (Table 1). In Vitim the decrease in maximum diurnal temperature range is registered in the cold season from October to April (by 0.05-0.1°C per year) and in July (0.05°C per year).

Table 1. Variations in air temperature (t_{air}), maximum diurnal air temperature range (Δt_{max}) and precipitation amount (ΔX).

	t_{air}, °C/year		ΔX, mm/decade		Δt_{max}, °C/ year	
Month	Lensk	Vitim	Lensk	Vitim	Lensk	Vitim
I	–	–	–	–	–	-0.09
II	–	–	1.5	2.7	–	-0.06
III	–	0.06	1.0	–	–	-0.05
IV	0.05	0.07	–	–	–	-0.1
V	0.04	0.05	–	–	+0.09	–
VI	0.04	0.06	–	–	–	–
VII	–	–	3.8	3.1	–	-0.05
VIII	–	–	–	–	–	–
IX	–	–	4.1	4.1	–	–
X	–	–	–	4.8	–	-0.07
XI	–	–	1.6	2.1	–	-0.05
XII	–	–	–	–	–	-0.09
year	0.02	0.03	16	19		

Currently, according to [4], the lower values in the positive mean annual soil temperature trend compared to air temperature trends are characteristic for the entire cryolithzone of Russia. This tendency persists in Vitim. It has been found that over the study period (1963-2011) mean annul soil temperature at the depth of 320 cm rises by 0.02°C per year, positive trends being revealed over the January-September period. In Lensk over the available study period in 1987-2011 the increase by 0.05°C per year, and soil temperature rise in October-June were detected. However, in Lensk testing did not reveal any statistically valid positive air temperature trend over this period. This peculiarity can result from local anthropogenic changes in the area which surrounds the station, i.e. climatological heterogeneity; subsequently, data on soil temperatures in Lensk can be insignificant for estimating climate change and impact on permafrost condition. The experience has shown that such facts must be considered. For instance, it was determined that temperature increase and permafrost melting, for example in Yakutsk region, are directly caused by natural and anthropogenic damages of the underlying surface, hydraulic and thermal balance [6]. Therefore, temperature increase and permafrost melting do not correlate with climate change [8].

It is well known that precipitation affects soil moistening in warm months and snow cover depth in the cold season, which, in their turn, determine to a greater extent intensity of hot and cold wave penetration into soil depth. According to [2] increase in soil temperature in the study area is more dependent on the increase in snow cover depth rather than the increase in air temperature.

In the area under the study monthly total precipitation increases mainly in the cold season in February, March, November in Lensk; in February, October and November in Vitim; and in July and September in all the stations (table 1). The highest values (up to 4 mm/decade) are registered in September, November. The positive trend in annual total precipitation has been distinguished, precipitation amount increases by 16-19 mm per decade. In Lensk the increase in both annul and monthly total precipitation is greater than in Vitim. Moreover, in the second half of the considered period May total precipitation exceeds three times long-term mean values which are typical for June in Lensk (Figure 2)

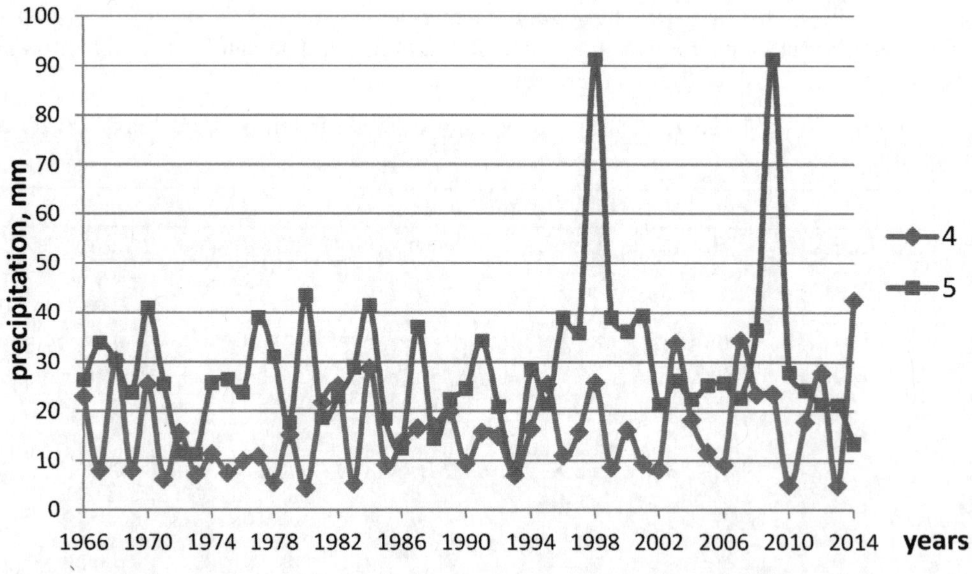

Figure 2. Time variations in monthly precipitation in April (4) and May (5), based on data obtained in the meteorological station Lensk, 1966-2014.

Based on the analysis of long-term variations in monthly total rain, mixed and solid precipitation, it has been detected that the revealed decrease in solid precipitation and increase in mixed precipitation in the transitional periods are due to rise in air temperature. Mixed precipitation amount increases in

April, May, September in Lensk, and in April, October, November in Vitim. Besides, there is a positive trend in rain precipitation in September and October in Vitim. After the period of mid 80-s no solid and mixed precipitation in summer months is observed or it decreases in amounts, whereas at the beginning of the study period solid precipitation amount could be of up to 6.1 mm (Vitim; June,1967) and mixed precipitation could be as high as 9.1 mm (Lensk; June, 1983) a day.

In the analysis of snow cover depth, a positive trend in snow cover changes in January-March was detected in both stations, while in Lensk this trend occurred in April. It should be noted that the most intensive rise was before mid 80-s, after that it slowed down.

The obtained results, compared to forecasts for climate change in the territory of Russia in the XXI century [7, 13], indicated that rate of air temperature rise decreased relative to the expected rate in 2010-2015 [13]. It was confirmed that temperature rises to higher rates in cold seasons than in warm months. The further increase in the average precipitation is due to precipitation amount increase in the cold period [8, 13].

The detected precipitation increase in the cold season will lead to soil temperature rise and intensify permafrost degradation. The intensity of seasonal heaving can be influenced due to decrease in diurnal air temperature range in cold season and the distinguished increase in precipitation in July and September. This trend can intensify mollisoil moistening with water accumulated in summer months. July precipitation increase may intensify landslide and solifluction processes.

4. Conclusion

The conducted research has indicated decrease in mean annual air temperature rates since mid 90-s. Average monthly temperature rises in transitional season (March-June) and this occurs non-uniformly in the considered meteorological stations. Diurnal air temperature range decreases mainly in cold season in Vitim; whereas in Lensk no distinctive changes can be traced.

Mean annual soil temperature increases at the depth of 320 cm both in Vitim and Lensk. Climatological heterogeneity is supposed to be present in data series obtained in Lensk, which is probably due to local anthropogenic changes affecting the station environment. Consequently, data on soil temperature in Lensk can be of no significance for climate change estimates.

In the study area the positive trend in annual total precipitation has been detected. Monthly total precipitation increases mainly in the cold season and in July. The highest values are observed in September, November. In Lensk, the increase in monthly total precipitation dispersion is distinguished in the spring months. Solid precipitation decreases, while mixed precipitation increases in transitional seasons, which is due to air temperature increase.

The positive trend in snow cover depth occurs in January and in April-March. It should be mentioned that rates slow down in the second half of the considered period.

Thus, the conducted analysis of long-term changes in meteorological parameters revealed non-linear climate changes, their non-uniform intensity and orientation observed in data of relatively closely-located meteorological stations. In studying climate change the emphasis should be made on the reliability verification of datasets.

References

[1] Strokova L A, Dutova E M , Ermolaeva A V, Alimova A N and Strelnikova A B 2015 Karst hazard assessment in the design of the main gas pipeline (South Yakutia) *IOP Conf. Ser.: Earth Environ. Sci.* **27** 12032

[2] *Metody otsenki posledstviy izmeneniya klimata dlya fizicheskikh i biologicheskikh sistem* [kollektivnaya monografiya] ed S M Semenov 2012 (M. Rosgidromet)

[3] Pavlov A V and Malkova G V 2009 Melkomasshtabnoe kartografirovanie trendov sovremennykh izmeneniy temperatury gruntov na severe Rossii, Kriosfera Zemli t. XIII 4 32-39

[4] Sherstyukov A B 2007 Temperatura pochvogruntov Rossii na glubinakh do 320 cm v usloviyakh izmenyayushchegosya klimata *Trudy VNIIGMI-MTsD* **173** 72-88

[5] Sherstyukov A B 2009 Izmeneniya klimata i ikh posledstviya v zone mnogoletney merzloty Rossii Obninsk: GU «VNIIGMI-MTsD»

[6] Fedorov A N and Konstantinov P Y 2008 Recent changes in ground temperature and the effect on permafrost landscapes in Central Yakutia *Proceedings of the 9th International Conference on Permafrost, D. L. Kane and K. M. Hinkel (eds)* (June 29-July 3) Fairbanks, Alaska, Institute of Northern Engineering, University of Alaska Fairbanks **1** 433-438

[7] Groisman P Ya, Blyakharchuk T A, Chernokulsky A V et al 2012 Chapter 3 Climate changes in Siberia. *In: Regional environmental changes in Siberia and their global consequences, Springer environmental science and engineering* (Springer, Dordrecht) 57–109

[8] Romanovsky V E, Drozdov D S, Oberman N G, Malkova G V and et al. 2010 Thermal state of permafrost in Russia *Permafr. Periglac. Process.*, 21, 136-155, doi: 10.1002/ppp.683.

[9] Shiklomanov N I 2005 From exploration to systematic investigation: Development of geocryology in 19th- and Early–20th-Century Russia *Physical Geography* **4** 249–263

[10] Gorbatenko V P, Ippolitov I I, Podnebesnykh N V 2007 Atmospheric Circulation Over Western Siberia In 1976-2004 *Russian Meteorology and Hydrology* **32** (5) 301-306

[11] FGBU VNIIGMI-MTsD [Elektronnyy resurs] URL: http://meteo.ru/data

[12] Bendat Dzh and Pirsol A 1989 Prikladnoy analiz sluchaynykh dannykh: Per. s angl (M.:Mir) p 540

[13] Federal Service for Hydrometeorology and Environmental Monitoring (Roshydromet) 2005 Strategic Prediction for the Period of up to 2010-2015 of Climate Change Expected in Russia and Its Impact on Sectors of the Russian National Economy Moscow

Development of a permanent geological environment model of Kazan city aimed to solve various engineering-geological problems (Russia)

N Zharkova[1], A Latypov[2], A Shevelev[3] and I Khuzin[4]

[1, 2, 3, 4]Kremlevskaya str. 4/5, Kazan (Volga District) Federal University, Kazan, Russia

E-mail: [1] kazannad@rambler.ru

Abstract. The article discusses the composition, structure and operation principles of a digital geological environment model for the urban area located in the valley of a large lowland river (the Volga). The model is implemented in ESRI (ArcView and ArcGis) and MapInfo software environments. The basis of the model is the data on the composition and physical and mechanical properties of soils, the information about ground waters and industrial loads. The model has been used to conduct zoning of soil conditions, groundwater aggressivity to the materials of underground structures. Also, the areas of existing and possible exogenous geological processes (flooding, karst, suffusion, erosion, landslides) have been identified. According to the model, it is offered to evaluate the stability of geological environment to human impact using typification on the soil conditions based on the pre-zoning of water content and the degree of drainage. A new monitoring system of dangerous exogenous geological processes has been developed, the impact of exogenous processes on the residential buildings has been estimated, and, also, the analysis and evaluation of geological risks have been performed. According to the data on the composition, density and water saturation of soils, the stability of the ground bases to a dynamic impact has been estimated.

1. Introduction

Kazan is one of the largest industrial centers of Russia, the city with more than one million inhabitants and a millennial history. The city is located in the eastern part of the East European platform, within a complicatedly formed terrace complex of the large lowland river Volga. The upper part of the geological section is represented by alternation of the carbonate and terrigenous Permian rocks; they are erosively covered by sandy-clay sediments of the Pliocene and Quaternary, mostly, alluvial sands, sandy loams, loams and clays.

Complicated geomorphological and geological-hydrogeological conditions, as well as centuries of human impact on the geological environment predetermined the development of flooding processes, karst, suffusion, landslides, river and gully erosion, abrasion, subsidence of soils, chemical pollution of ground waters. Activation of these processes has led to the increase of accidents in industrial and residential assets. Therefore, there is a need for a monitoring system of exogenous geological processes, which would be based on a large-scale multi-layered graphic digital environment model (hereinafter, a GIS-model), corresponding to engineering-geological conditions.

The GIS-model is based on the data on geological and hydrogeological conditions of the city obtained during geological surveys, which have been carried out by various organizations, at scale of 1: 50 000 - 1: 200 000 (from 1961), satellite imageries and topographic maps, at scale of 1: 25 000 and 1: 50 000, data on the composition, physical and mechanical properties of soils (about a thousand wells), and, also, numerous scientific studies. As a part of the GIS-model, a detailed route survey has been conducted to explore geological processes on the territory of the Kazan city, as well as visual evaluation of the technical condition has been given for industrial and civil objects [1].

2. Composition and structure of the permanent model

Under the permanent environment model we understand a unified geoinformation system of the structured and interrelated conditions and factors corresponding to the status of geological space, which is transformed into its cartographic representation. Development of the model has been carried out in ESRI - ArcView and ArcGis environment, as well as in Easy Trace software. Later the model has been transformed into MapInfo environment. The model is a combination of the following blocks (databases):

«Soils&rocks» block is a database of points (1057 points). Primary information required both for modeling and forecasting is fixed to each point in the attribute table of the GIS-model: a type, genesis and age of soils, their physical and mechanical properties, depths of bottom and top, thickness.

«Groundwater» block is also a database of points (about three hundred points). An attribute table of the GIS-model contains information about water table depths and ground water chemical composition.

«Exogenous geological processes» block is a complex database, where a separate database corresponds to one exogenous geological process. The block contains point databases (suffusion, landslides), as well as linear (erosion, abrasion), and polygonal databases (karst, flooding, subsidence of soils). Attribute tables of databases contain information about the process activity. For such processes as karst, waterlogging and subsidence of soils the attribute tables contain information about the probability of activation.

«Technogenic load» block is also a set of databases (linear and polygonal ones). It shows the static load on the geological environment according to weight of buildings and structures, vibration loads from the movement of auto- and electric transport, a barrage effect of buildings, the areas of existing and possible leakage from the water communications (changes in the chemical and hydrogeodynamic regime of ground waters), the dissemination of technogenic soils with indication of their type and thickness.

On the basis of these four blocks (the primary data), as well as with the use of the existing medium-scale geological and hydro-geological maps and a series of engineering-geological routes, we have synthesized the following vector maps (1: 25 000): geomorphological, geological maps of the Quaternary and Pre-Quaternary sediments, a map of ground water chemical composition, and a map of water table depths. As a part of the GIS-model, a map of soil types has been formed. Zoning in compliance with soil conditions has been carried out by approximation of the data according to the nature of soil bedding to a depth of 30 m from the database of the «Soils&rocks» block. Borders control between areas has been carried out using geological maps. Totally, 29 types of soils are identified, which are combined into the following groups: A - various variants of interbedded sandy and clay soils; B - the types similar to group A, but with the presence of soil subsidence differences, C - soils, which contain rocky soils, in addition to disperse, D - soils, which contain peat and /or peaty soils [2].

«Deformation of structures» block is an additional database, which contains the deformed residential buildings, according to visual evaluation of their condition.

The developed GIS-model allows sorting the information, as well as performing statistical analysis, interpolation, spatial operations, forming complex queries to retrieve information from the database, making changes in the database. As a result, it is possible to get a set of new prediction maps, and to display the results in maps.

3. Main results

3.1. Comprehensive evaluation of engineering and geological conditions.

While using the "Master of spatial operations" application, the resulting GIS-model allows us analyzing the factors of engineering and geological conditions formation:

According to data on chemical composition of ground waters, and taking into account the criteria used in the Russian regulatory documents, the areas with various types of groundwater aggressivity towards materials of underground structures have been identified [3].

The areas of carbonate eluvium have been identified, a depth map of eluvial sediments top has been obtained (highly susceptible to dissolution and suffusion), and their thickness has been shown. Later, the information has been refined, additional laboratory investigations of carbonate eluvium have been conducted, which are discussed in a number of studies [1, 4].

New principles of urban areas typification based on the stability of the geological environment to human impact have been substantiated and developed. It is offered to implement zoning according to watering and drainage, at the first stage, and to assess the stability of the geological environment at the level of a soil type, at the second stage (as far as it contains the information about stable or unstable soils under certain drainage and water content). A certain type of technogenic impact has been assigned to each taxon in the attribute table of the database, in relation to which the geological environment is unstable. Also, possible geological processes "have been predicted" (a response to the technogenic impact), as well as the degree of geological environment stability. The data on buildings deformation have been compared with the data obtained at evaluating geological environment stability. This comparison allows revealing a clear accordance "calculated stability - volume and extent of structures deformation", which is not observed while using the other methods of evaluation [5].

The complexity of engineering and geological conditions has been quantified according to the criteria of the Russian regulatory documents. Quantitative analysis has shown that a significant part of the urban area is characterized by the geological conditions of complex (47%) and medium complexity (45%). It is, mainly, determined by hydrological conditions and the extensive development of geological processes. A significant part of the urban area is exposed to the development of waterlogging processes (25%). Also, the areas of karst-suffusion activation (4%), erosion (3%), and slope processes (3%) have been identified.

3.2. Monitoring of dangerous exogenous geological processes on Kazan territory.

Based on analysis of the GIS-model databases new observational sites and areas for the monitoring of exogenous geological processes have been identified (12 sites). The detailed study of the geological and hydrogeological conditions has been conducted within each area, as well as both a program of technical equipping with measuring instruments and a method of observation with time intervals have been developed. Economic feasibility is given for each type of observation [6].

3.3. Analysis and risk evaluation of the geological processes damage on the territory of the historic center of the Kazan city.

The availability of created databases has allowed conducting quantitative evaluation of a geological risk (physical, economic and social) due to the possible activation of exogenous geological processes in the historic part of the city. These risks have been evaluated by the method of IEG RAS involving some economic and statistical data for the city of Moscow. Additional block of the GIS-model has been developed to evaluate the economic risks. This block contains information about the number of storeys in buildings, the type of foundation, the cost and age of buildings. The database on population density at different times of the day has been formed to evaluate social risks. It is observed that the territory of the Kremlin hill is characterized by the maximum geological risk values due to the extensive development of karst-suffusion processes, high population density, and a sufficiently dense development (a physical risk - 1 incident / 10 years, the potential economic losses - $ 23 million, a social risk - 1 pers. / 1000 years) [7].

3.4. Evaluation of ground bases stability to a dynamic impact.

The GIS-model has allowed implementing the idea of city zoning based on the stability of the soil bases to a dynamic impact (sands to a depth of 15m are the most susceptible). Typical responses of sandy soils to the dynamic impact and typification criteria are shown in Table 1.

Table 1. Classification of sandy soils on dynamic instability.

Type of dynamic instability	Porosity coefficient, unit fraction		Water saturation, unit fraction
Postcompaction	pulverescent	>0,80	any
	fine	>0,75	
	gravelly, large and medium size	>0,70	
Decompression	pulverescent and fine	≤0,60	0÷0,5
	gravelly, large and medium size	≤0,55	
Accumulation of shear deformations	pulverescent and fine	≤0,60	0,8÷1
	gravelly, large and medium size	≤0,55	
Liquefaction	any		0,8÷1

As a part of the GIS-model, this idea has been implemented as the following sequence:

- Identification of the areas that contain sands on a depth less than 15 m.
- Typification in accordance with the location of sands in section and in relation to groundwater table.
- "Binding" the data on the density and moisture content of sandy soils by transferring the data from the point objects of the «Soils & rocks» block to polygonal ones (the areas with a certain type of soils).
- Assigning the type of dynamic stability for the areas according to the specified criteria.

Performed zoning has shown that the dynamic instability of soils is the most typical for the complex of low terraces. All the sands located below the groundwater table are potentially liquefiable, regardless of their genesis, density and a particle size. Moreover, shear deformations accumulation and postcompaction are typical for a significant part of sands from the complex of low terraces, while decompression is hardly probable. The majority of sandy soils from high terraces complex is often resistant to the dynamic impact due to the high density and low humidity. Decompression (both for natural and technogenically altered sands) is typical only for some areas [8].

3.5. Geological evaluation of exogenous geological processes impact on the areas of residential houses building (Kazan).

In 2014 the Ministry of the Russian Federation for Civil Defense, Emergencies and Disaster Response compiled the list of the buildings ruining due to influence of exogenous geological processes (581 buildings on the Kazan territory). The list was formed according to the complaints of the citizens, who would like to get new apartments at the expense of the state budget. In order to identify the true causes of buildings and the surrounding area destruction, visual inspection of buildings was carried out, but it was not enough to make a final decision. It was necessary to carry out a full evaluation of engineering and geological conditions, but a very short turnaround time (2 months) and the low cost of the work did not allow us conducting necessary field and laboratory studies. We used the database of «Soils&rocks» block for necessary engineering calculations (slope stability, filled soil postcompaction, etc.). Having been conducted simultaneously, the work allowed us inserting in the GIS-model the information obtained in the course of the survey. As a result of engineering-geological survey aimed at 582 residential buildings, it was found out that exogenous geological processes influenced only 39 buildings, and 12 of them were subject to immediate resettlement.

IOP Conf. Series: Earth and Environmental Science **33** (2016) 012048 doi:10.1088/1755-1315/33/1/012048

3.6. *Updating engineering-geological conditions maps for development of Kazan territory.*

On the basis of the existing GIS-model after the preliminary transformation of vector themes from ESRI environment into MapInfo environment, as a part of preparation for the new General Plan of the city (it is planned to increase the city area to 2025) three new digital engineering geological maps were built: a seismic microzonation (1: 100 000) map, an engineering geological map (1: 25 000), and an aggregate map of the development complexity (1: 25 000). Spatial boundaries of the territory were extended while developing the maps. Engineering and geological conditions were detailed and updated taking into account new regulatory documents. Also, new thematic layers were added in the GIS-model, engineering-geological zoning was performed again, and recommendations on measures for land development were made. GenPlan Institute of Moscow is currently planning a new residential and commercial development taking into account the newly created GIS-model maps. A fragment of one map is shown in Figure 1.

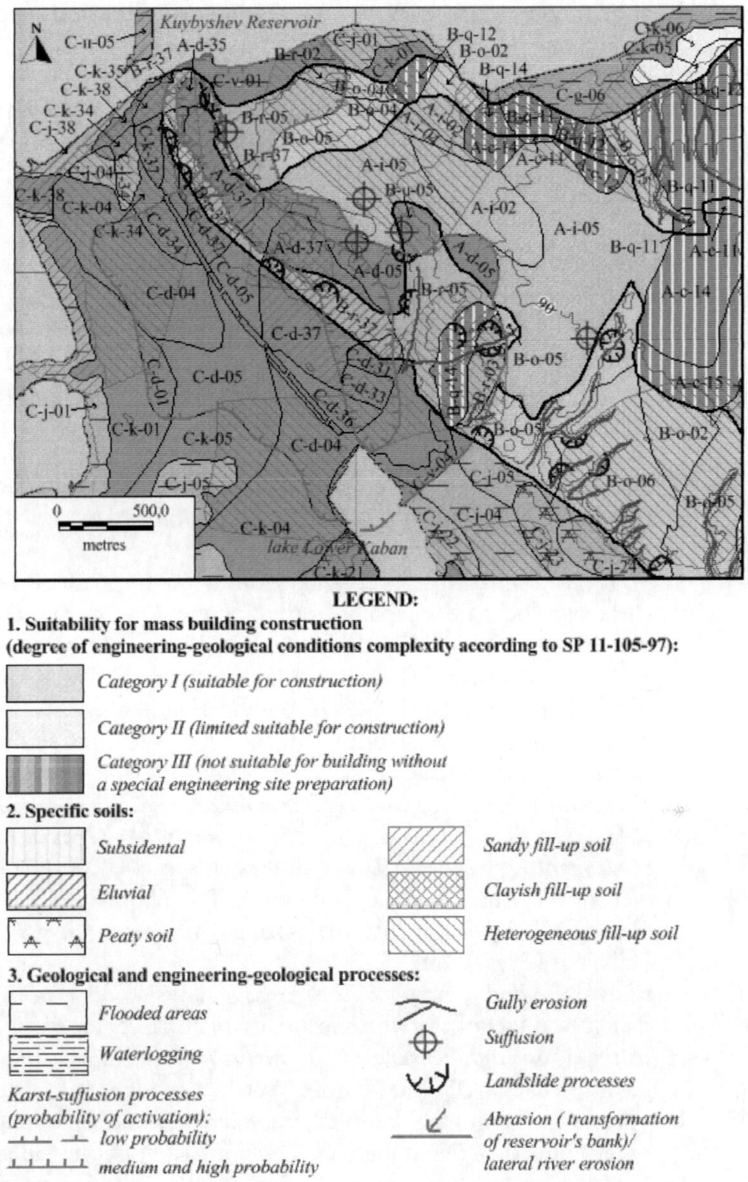

Figure 1. A fragment of an aggregate map of Kazan city development complexity. Indexes within polygons (e.g., C-j-05): A - watersheds (high terraces), B - slopes, C - lowlands (lower terraces); d, k, j and other uppercase letters of the alphabet - a certain set of exogenous geological processes; 01, 02, 03, and other two-digit numbers - the type of soil stratum.

4. Conclusion

The offered digital model of Kazan city geological environment meets the modern requirements for a complex model. This model has a structured database of Kazan hydrogeological and engineering-geological objects with the ability to search for, sort out, and analyze information. The possibility of hydrological conditions modeling and further forecasting, as well as convenient form of results representation are present there. The above mentioned examples show that the offered model of the geological environment can be successfully used to solve various applied and scientific problems.

Acknowledgments
The work is performed at the expense of grants under the state support Kazan (Volga-region) Federal University in order to improve its competitiveness among the world's leading research and educational centers.

References

[1] Shevelev A I, Zharkov N I, Huzin I A, Polyakov S I and et al. 2012 *Hydrogeological and geotechnical conditions of Kazan city* (Kazan: K(VR) FU p 236

[2] Zharkova N, Zharkov I and Cherniychuk G 2014 Engineering geological characteristics of artificial soils in Kazan city (Russia) *Int. Multidiscip. Sci.GeoConf Surveying Geology and Mining Ecology Management SGEM Albena Bulgaria (17-26 June 2014) Vol.* 2 (Sofia: STEF92 Technology Ltd) 323–334

[3] Zharkova N and Khuzin I 2012 Groundwater aggressiveness as an important factor of engineering-geological conditions in the Kazan city (Russia) *Int. Multidiscip. 12th Sci. GeoConf SGEM Albena Bulgaria (17-26 June 2014) Vol* 1 (Sofia: STEF92 Technology Ltd) 117–126

[4] Latypov A I, Zharkova N I and Mouraviev F A 2013 Dispersed weathering products of carbonate rock: Features and formation conditions from the construction's point of view (by the example of Kazan, Russia) *Global View of Engineering Geology and the Environment Beijing China* (London: Taylor & Francis Group) 891-896

[5] Zharkova N I and Huzin I A 2008 Engineering geological basics of typification the territory of Kazan city on the stability of the geological environment to human impact *Sergeev Readings. International Year of Planet Earth: Tasks for Environmental Geoscience, Engineering Geology and Hydrogeology Issue 10. Annual session of the Scientific Council RAS on the problems of Environmental Geoscience, Engineering Geology and Hydrogeology, (March 20-21, 2008)* (Moscow: GEOS) 413–418

[6] Shevelev A I, Zharkov N I, Bubnov Y P, Latypov A I, Huzin I A and Galeev R K 2014 Geological Environment Monitoring of Kazan city *J Georesources 3* (Kazan: Kazan University Press) 3–8

[7] Zharkova N I and Garayeva A N 2013 Analysis and risk assessment of geological processes damage on the territory of Kazan city historic center *Collection of scientific works dedicated to the 155th anniversary of the birth of Academician P.A.Tutkovsky: modern problems of geology* (Kiev - Olevsk: National Science and Natural History Museum of NASU) 217–223

[8] Latypov A I, Zharkova N I and Cherniychuk G A 2013 Zoning of Kazan city on stability of soil bases to the dynamic impact *Geotechnics 1* (Moscow: Geomarketing) 42–48

Application of groundwater aggressiveness assessment method for estimation of the karst process at main gas pipeline construction

A V Ermolaeva

National Research Tomsk Polytechnic University, 30 Lenin Ave., Tomsk, Russia

E-mail: alyona7@inbox.ru

Abstract. Main pipelines maintenance is connected with hazard engineering and geological working conditions. The article deals with the use of groundwater aggressiveness assessment method to estimate the karst processes development during the construction of main gas pipelines. The possibility of using this method is analyzed on the example of the initial section of the designed gas pipeline "Power of Siberia" (section "Chayanda-Lensk"). The calculation of the nonequilibrium index Ca was made in accordance with the geotechnical survey data. The dependencies between the geomorphological features of the terrain and the natural waters aggressiveness were determined.

1. Introduction

At present, the key criteria for transportation systems are industrial and environmental safety. At the moment, the Russian pipeline system – is one of the largest transportation systems in the world. Typically, the main gas pipelines maintenance occurs in the complex of dynamic, engineering-geological and nature-climate conditions. In recent years, there is an increase in gas pipelines accident rate due to geological hazard. Particular difficulties arise at the construction of the pipeline facilities in the karst areas due to the complexity of process development identifying and forecasting.

One of the key problems in the karst areas assessment study is the degree of karst process hazard zoning. The TSN 22-304-06 gives the following definition of karst hazard assessment - karst hazard is expressed in terms of probability of karst displacement occurrence for a given period of time (for example, the service life of the construction in the area of site location), which may cause unacceptable strain of the construction. A common integral measure of karst hazard for a long-term period was the average annual doline occurrence intensity (doline occurrence/km^2 a year) proposed by ZA Makeev (1948) and also applied in TSN 22-204-06. According to the results of the analysis conducted both in Russia and abroad, the following features that characterize the karst hazard could be highlighted:

1. Presence of surface karst forms;
2. Dissolving ability of natural water;
3. Presence of zone of the unconsolidated soil, recorded according to geoelectrical and velocity abnormality;
4. Intensive vertical filtration;
5. Presence (absence) of aquaclude which covers soluble rocks.

Among these features, the one should be noted that characterizes the future of karst processes development through space and time: the aggressiveness of natural waters. Since this feature describes the "behavior" and the intensity of the process in whole, the assessment on its base will be promising not only at the design stage, but also during the main gas pipeline maintenance. Under natural conditions, the system "water - rock" has equilibrium-nonequilibrium character. In this case, water, irrespectively of depth and speed, is always in nonequilibrium with some minerals by dissolving them, but is in equilibrium with others by generating them. Great contribution to the study of this process was made by the American researchers R M Garrels and Ch L Kraist.

Any form of interaction reaction could be expressed by the following schematic equation: $aA + bB \leftrightarrow cC + dD$, where A, B, C, B – stoichiometric coefficients of initial substances and reaction product. Thermodynamic equilibrium ratio of this process is defined by the equation: $K_a = \frac{a^*(C)^c \cdot a^*(D)^d}{a^*(A)^a \cdot a^*(B)^b}$. Equilibrium ratio characterizes the correlation of initial substances thermodynamic activity and materials in a state of ohmic equilibrium, in accordance with mass action law. Water aggressiveness index is used for natural water aggressiveness assessment, which characterizes the ability of water to transform solid substance in to the solution: $A = \lg \frac{K_a}{P_i \cdot a_i^{v_i}}$, where $P_i \cdot a_i^{v_i} = \frac{aC^c \cdot aD^d}{aB^b \cdot aA^a}$. The index of aggressiveness is often used for underground water dissolving ability comparison. The higher is the index A the faster is the substance dissolving process. To identify the character of the process (dissolving/sedimentation), the nonequilibrium degree index is used: $\theta = \frac{\Pi_i \cdot a_i^{v_i}}{K_a}$. The nonequilibrium degree characterizes undersaturation (θ < 0 – dissolving process occurs) or oversaturation (θ > 0 – mineral formation process occurs). Thus, application of the above mentioned method could help to estimate the probability of the process development along the main gas pipeline sections, which are predisposed to the development of karst.

2. Experimental design

The above mentioned method was applied to assess the influence of natural waters aggressiveness along the initial section of the designed gas pipeline "Power of Siberia" – Chayandinskoye oil and gas condensate field - Lensk, length – 160 km. The laboratory tests of water composition sampled through the period from 2010 to 2012 were used as the initial data for calculation.

Ground waters at the present site are characterized by sporadic distribution. Waters confined to the deluvial deposits are thin and low-yield, tapping at a depth of 3-5 m, according to the composition – bicarbonate, magnesium-calcium, sodium-magnesium-calcium. Aquaclude is presented by rocky and half-rocky soil, randomly multi-year soils.

Waters confined to deluvial-proluvial deposits are associated with over-permafrost and permafrost taliks. The calculation was done on the basis of 23 chemical water analyses, sampled throughout the studied area, available from the results of engineering surveys.

The assessment of carbonate rocks equilibrium was carried out with the help of HydroGeo software complex [1], designed by M B Bukaty (1999).

The following was taken as a system of calculation: water and basic ions such as H+, Na+, Mg^{2+}, Ca^{2+}, OH-, Cl^-, $(HCO_3)^-$, $(SO_4)^{2-}$; associate: $(CO_3)^{2-}$, H_2CO_3, $(CO)^{2+}$, CO_2, $NaHCO_3$, Na_2CO_3, $(NaCO_3)^-$, Na_2SO_4, $(NaSO4)^-$, $NaHSO_4$, NaCl, NaOH, $Mg(HCO_3)_2$, $(MgHCO_3)^+$, $(Mg(CO_3)_2)^{2-}$, $MgCO_3$, $MgSO_4$, $(Mg(SO_4)_2)^{2-}$, $(MgHSO_4)^+$, $Mg(HSO_4)^2$, $MgCl^+$, $MgCl_2$, $Mg(OH)_2$, $MgOH^+$, H_2SO_4, $(HSO_4)^-$, HCl, $(CaHCO_3)^+$, $Ca(HCO_3)_2$, $(Ca(CO_3)_2)^{2-}$, $CaCO_3$, $(Ca(SO_4)_2)^{2-}$, $CaSO_4$, $Ca(HSO_4)_2$, $(CaHSO_4)^+$, $CaCl^+$, $CaCl_2$, $CaOH^+$, $Ca(OH)_2$; the following minerals $MgCO_3$ (corr.) – magnesite (corr.), $MgCO_3$ - magnesite, $CaCO_3$ – calcite, $CaCO_3$ (IV) - calcite - IV, $CaCO_3$ a - aragonite,

$CaMg(CO_3)_2$ н – dolomite (disordered.), $CaMg(CO_3)_2$ – dolomite (ordered), $CaMg_3(CO_3)_4$ - huntite, $CaMg(CO_3)_2$ - dolomite, $NaHCO_3$ – nahcolite. Results of nonequilibrium index calculation (Ca) are given in Table 1. It is obvious that in most cases rock dissolving process will occur.

Table 1. Results of nonequilibrium index calculation.

№ Sampling point (well/site)	**Salinity, g/dm^3**	**Nonequilibrium index θ (Ca)**	**Absolute marks of wells (height, m)**	**pH**
101	0.153	-1.6	247.89	7.38
103	0.219	-0.4	243.27	7.6
211	0.791	0.99	461.74	7.35
156	0.618	-1.1	244.31	6.59
6943	0.356	0.26	357.2	7.2
6969	0.225	1.6	447.5	8
17	1.13	0.13	387.88	7
35	0.8	0.42	439.33	7.04
36	0.6	0.072	434.87	6.99
38	0.6	0.1	433.75	7
42	0.6	0.23	423.31	7
43	0.7	0.78	424.03	7.13
54a	0.4	-0.035	344.71	7.22
74	0.4	-1.7	416.50	6.54
85	0.4	-1.4	396.50	6.71
1034	0.548	0.2	347.45	7.2
1352	0.425	-0.54	414.95	7.01
2152	0.589	-0.15	458.3	6.99
2207	0.605	0.27	277.5	7.04
2211	0.395	0.13	276.5	7.4
2225	0.401	-1.3	276.08	6.74
2255	0.22	-1.6	253.67	7.08
2278	0.4	-1.1	255.87	7.33

According to the calculation results of calcite equilibrium to the studied waters, correlation dependence was derived, which allowed assessing the possibility of karst process origination in the studied area. Relation of nonequilibrium index CaCO3 and water salinity value was plotted as a diagram in accordance with the obtained results [2]. Analysis of the data showed that waters with a salinity value of more than 0.66 g / L are saturated to calcite. Thus, these areas of water sampling are free from karst process origination or the karst process will be of passive character. During the analysis of nonequilibrium index and pH relation, it was revealed that the waters in the studied area are weakly alkaline and the pH ranges from 6.5 to 7.8 [3, 4, 5]. In the sampling points where the pH value was more than 7.35, waters are likely to be rich in calcite, thus there is no threat of karst process origination.

For a more accurate prediction of karst hazard it is advisable to identify the relationship between the value of nonequilibrium index and terrain features in the area of pipeline construction . To reveal this dependence, the relationship between the wells absolute marks (from which water samples were taken) and the value of nonequilibrium index was analyzed [6, 7].

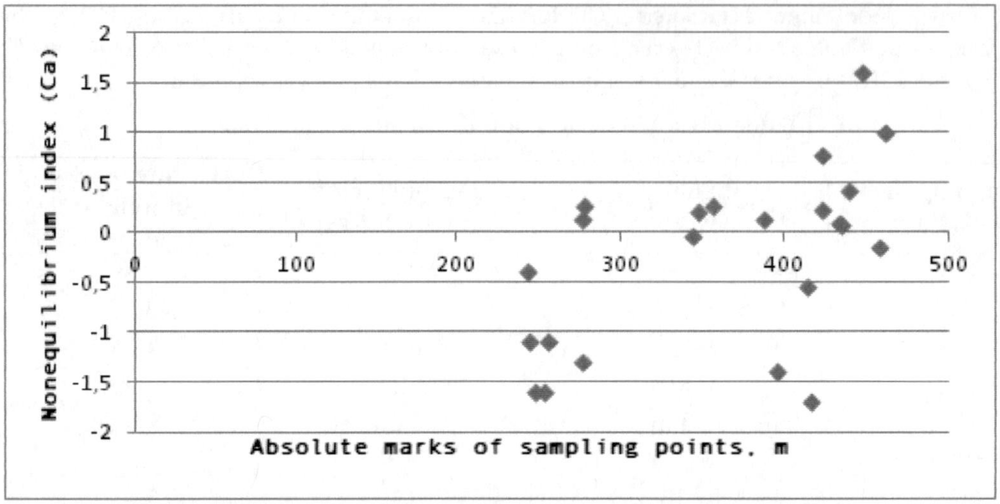

Figure 1. Variation of nonequilibrium index (Ca) regardless the absolute marks.

3. Summary

The correspondence between the geomorphological features of the territory and the groundwater aggressiveness could be noted as a result of the research.

The researched method is promising for the study of geoengineering conditions during the design of linear structures in karst areas [8, 9]. Thus, the samples collected from surface streams and wells confined to the bottoms of the local depressions relate to the negative values of the nonequilibrium index relative to calcite. Origination of less water saturation zones relative to calcite is determined by the increase of water exchange intensity. [10]

Consequently, on the ground of the geoengineering research results within the area of the initial section construction [11-13] of the main gas pipeline "Power of Siberia", the influence of the chemical composition of groundwater on the development of karst process was studied. It seems appropriate to perform such hydrogeochemical modeling across the whole gas pipeline.

References

[1] Bukaty M B 1997 Development of the software in the field of oil and gas hydrogeology *J Investigation and protection of a subsoil* **2** 37 (in Russian)

[2] Garrels R M and Christ C L 1968 *Solutions, minerals, equilibria* (Moscow: Mir) p 368 (in Russian)

[3] Gvozdetsky N A 1981 *Karst* (Moscow: Myisl) p 214 (in Russian)

[4] Gvozdetsky N A 1988 *Karst landscapes* (Moscow: Moscow State University Press) p 112 (in Russian)

[5] Zverev V P 2007 *Underground waters of crust and geological processes* (Moscow: Scientific world) p 256 (in Russian)

[6] Katayev V N 2001 *Methods and practice of comparative and evaluation karst division into districts* (Perm: Perm State University Press) p 85 (in Russian)

[7] Makeev Z A 1948 Principles of engineering-geological division into districts *Karstovedeniye* **4** 43 (in Russian)

[8] Maksimovich G A 1963 *Karst bases* (Perm: Perm State University Press) **1** p 443 (in Russian)

[9] Strokova L A, Dutova E M, Ermolaeva A V, Alimova A N and Strelnikova A B 2015 Karst hazard assessment in the design of the main gas pipeline (South Yakutia) *IOP Conf. Ser.: Earth Environ. Sci.* **27** 12032

[10] Teterin E A and Strokova L A 2015 Methods for karst hazard forecast and pipeline protection in South Yakutia *IOP Conf. Ser.: Earth Environ. Sci.* **27** 12033

[11] Waltham T, Bell F and Culshaw M 2005 *Sinkholes and Subsidence. Karst and Cavernous Rocks*

in Engineering and Construction (Berlin: Springer) p 404
[12] Antaki G 2003 *Piping and Pipeline Engineering: Design, Construction, Maintenance, Integrity, and Repair* (CRC Press) p 564
[13] McAllister E 2009 *Pipeline Rules of Thumb Handbook*: A Manual of Quick, Accurate Solutions to Everyday Pipeline Engineering Problems (Gulf Professional Publishing) p 762

Sedimentological sand grain orientation in oil-producing U1 layer Kazan oil-gas-condensate field (Tomsk Oblast)

L Krasnoshchekova[1], D Cherdansteva[2] and I Vologdina[3]

[1] Department of Geology and Mineral Exploration, Institute of Natural Resources, Tomsk
Polytechnic University, 30 Lenin Ave., Tomsk, Russia
[2] Department of Oil and Gas Industrial Facilities Design Construct, Institute of Natural Resources, Tomsk Polytechnic University, 30 Lenin Ave., Tomsk, Russia
[3] Department of Petrography, Tomsk State University, 36 Lenin Ave., Tomsk, Russia

E-mail: [1]krasnl@tpu.ru, [2]dannia@tpu.ru

Abstract. The paper describes the results in identifying the prevalent directions of elongated quartz grains being the major oil-producing layer component in sand reservoirs. Studying the orientation of quartz to its grain shapes in paleogeographical oriented core samples made it possible to identify the hydrodynamic reservoir regimes and facies type. The spatial confinement of pore spaces and cataclasis fractures in grain material to the prevalent elongated quartz grain directions was defined.

1. Introduction
Kazan oil-gas-condensate field is located in the central east of Western Siberia lowland within Tomsk Oblast. The cross-section embraces sand- argillaceous sediments of Mesozoic-Cenozoic sedimentary mantle embedded in metamorphically altered Palaeozoic folded basement rocks. Geologically, Upper Jurassic U1 Vasugan suite is the major oil-gas producing layer which can be observed in all well logs. It occupies a significant area of the Kazan complex.

2. Research methods
The research target includes core samples from oil-producing rocks of U1 layer, Kazan field wells. The general core sample interpretation embraced more than 130 meters. Paleomagnetically-oriented 25 rock samples were selected from two field wells. Paleomagnetic method involves the direction reconstruction of specific rock samples on the paleogeographic pole. The standard orientation method for viscous components was applied in determining the orientation of the core sample itself. Elongated quartz grains, base rock-forming components in investigated terrigenous reservoirs, were measured in preliminary prepared oriented thin sections. Generation of oriented sedimentation structures of quartz grains could determine the sedimentation facies conditions of sedimentary rocks and trace the development of pore spaces relative to debris in these rocks.

3. Results and discussion
U_1 layer sediments embrace medium/fine-grained sandstones and aleurolite-clay rocks. The lithological and mineral petrographic features of above-mentioned rocks were previously studied [1].

Macrofacies classification of Jurassic sediments [2] was applied to identify the facies environments in investigated sediments. Upper Jurassic terrigenous layer material could have accumulated in open active shallow-water basins and/or within off-coast areas [3].

Clastic material accumulates within shallow waters, which, in its turn, migrates with longshore currents from outboard coast towards its back land. In this case, well-sorted rocks with low-inclined and cross-bedding and banded structure develop. Simultaneously, homogeneous horizontally-bedded sediments accumulate mainly in distant coastlines, predominately in coastal-marine environments bounded by river lowlands.

Morphological analysis included the measurement of elongated grains in oriented thin sections relevant to specified direction (paleogeographic north pole). This involved the measurement of strike azimuth of elongated quartz grains. The minimum number of accurate and precise quartz grain orientation measurements in a rock embrace 100-150 grains in one thin section. The measurement of elongated grain orientation in photomicrographic thin sections was also conducted. Obtained photomicrographic data was processed in ArcGIS and Grapher 9 programs. Plotted rose diagrams defined the prevailing development direction of elongated quartz grains in a rock. Comparably, in both cases, the rose diagrams reflected that the microscopic measurement of thin sections and photomicrographic processing were practically identical.

The proposed computerized measurement method and calculations of grain orientation simplify the processing of enormous numerical data.

The measurement procedure and further calculation is as follows:

1. Oriented thin sections are photographed, processed and produced photos are glued into one photo via CorelDRAWX5. When photographing thin sections under the polarized POLAM 312 microscope, a line was selected on the thin section orienting to north paleogeographic direction; and along this line a series overlapping photos were compiled.

The photos were glued together via CorelDraw to produce a continuous image and this, in its turn, makes it more convenient to perform TIF file in ArcMap.

2. Prevailing elongated quartz grain orientation directions were drawn up in ArcMap (figure 1, A). Produced TIF file was uploaded in ArcMap, then the file shape was designed, coordinates of which were imported from the uploaded image. The direction of elongated quartz grains was drawn up and saved in the produced file shape.

3. According to designated directions the grain azimuths were recalculated in ArcInfo. To measure the linear grain orientation in ArcInfo and to calculate the azimuth of produced lines relative to the north direction, the results of half a circumference at an angle of 0 to 180° were processed. After statistic processing produced results were imported into circular diagram.

4. Rose diagrams of identified azimuths to visualize the produced data were plotted in StereoSN and Grapher 9 (Figure 1, B).

Grapher 9 was used to visualize and compare the elongated quartz grain orientation with other petrostructural data (optical axis). Based on produced data rose diagrams and prevalent elongated quartz grain line were plotted.

Based on the morphological analysis of sandstones the most common orientation types of quartz grains were classified (Figure 2).

Type I (predominate)- orientation in two directions of elongated clastic quartz material, where one is dominating. Biomodal orientation of clastics can be found in coastal (inward) zones of water basins affected by wave actions and/or sediment reworking by two currents sorting the grains in different directions. For example, backwash, including wind-induced and longshore currents, approaches the land at an angle and most grains are oriented either subnormal or subparallel to the coastline. In this case, distinct orientation of clastic material is predominate.

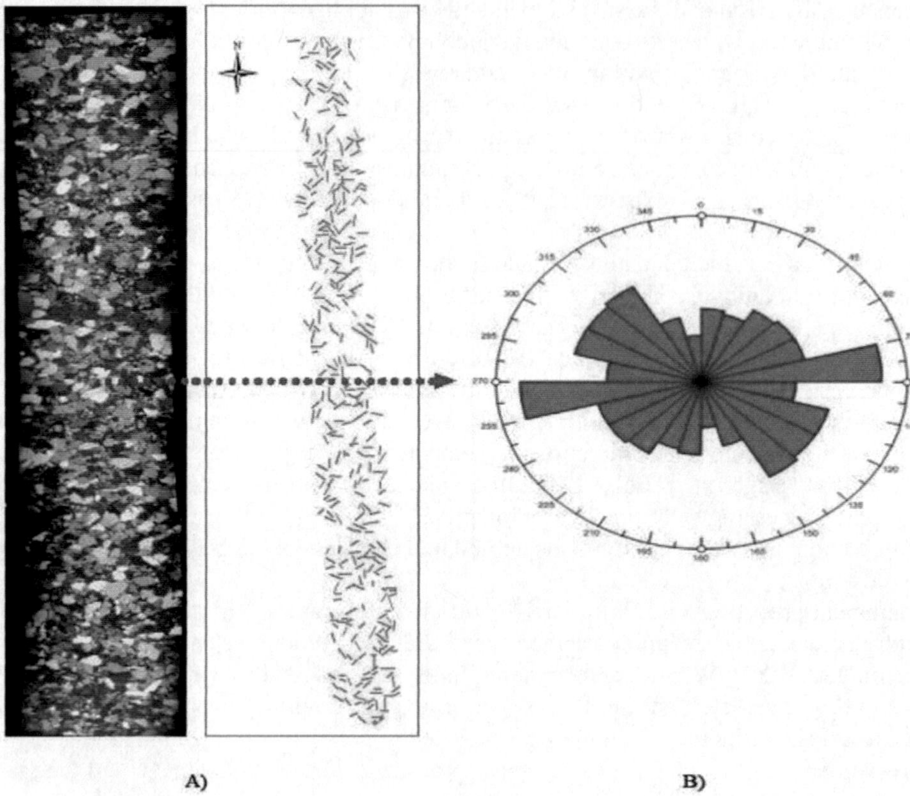

Figure 1. A) Prevalent orientation directions of quartz grains via ArcMap by photomicrographic oriented thin section (sample 2); B) Rose diagrams of prevailing orientation directions of quartz grains plotted in Grapher 9 (sample 2).

Type 2 (less common) – one maximum longitudinal grain orientation and insignificant cross orientation of elongated quartz. This orientation can occur in rapid stream flows where grain point end is downward to the current. In this case, the prevalent quartz clastic arrangement is accordant to the current.

Type 3 (insignificant) - polymodal development of elongated grains in rocks. Such orientation could occur in sediment detachment, similar to turbidite type. In non-compacted sediments with significant plastic clay matter terrigenous grains could be mixed and turned under conditions of external factors.

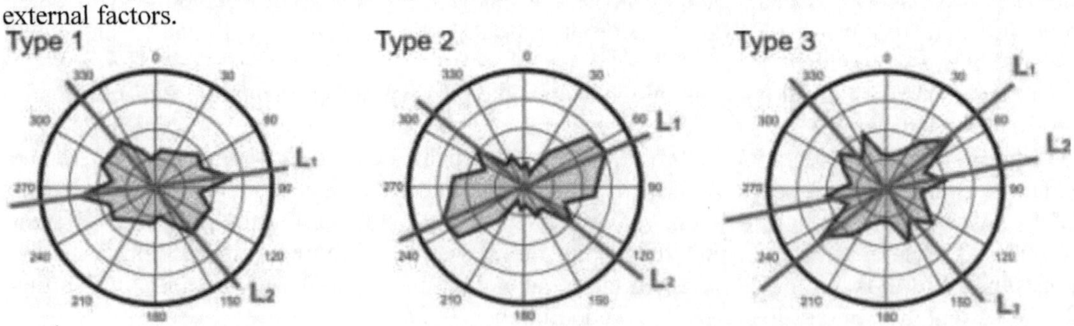

Figure 2. Rose diagrams of oriented elongated quartz grains in sandstones of U_1 Kazan field. L_1, L_2, L_3 – prevalent directions of elongated quartz grains in rocks.

4. Conclusion

Analysis of elongated quartz grain orientation in studied sedimentary rocks reflected two predominately distinct orientation directions of these elongated grains being close to N-E and N-W. Facies and dynamic conditions of sedimentation is governed by the shelf water environment, i.e. wave behavior and shoreline currents. Wave velocity is moderate to strong. Morphological quartz analysis data identified the location of the coastline within the field, i.e north-east. The development of pores and pore channels in sandstones coincides with the orientation of elongated quartz grains. There is also an insignificant number of cataclasis cracks in terrigenous sandstones involving simultaneous development of additional breaking directions in the grains. This prevalent quartz orientation in N-E and N-W directions can be found in rocks of Tomsk Oblast oilfield [3]. In this case, maximum porosity values can be found along these predominately oriented elongated grains.

Acknowledgments

The investigations were partially supported by the Russian Foundation of Fundamental Research (№ 15-05-00809), as well as by the Ministry of Science and Education of the Russian Federation (as part of the research "Improvement of Constructive and Geoecological Safety, Energy Efficiency of Vessels and Facilities at Continental Shelf of the Far Eastern and Arctic Seas", NIR 543.2014).

References

[1] Krasnoshchekova L et al 2015 Oriented core application in texture analysis of J_1 formation in Kazan oil-gas condensate field (Tomsk Oblast) *IOP Conf. Ser.: Earth Environ. Sci.* **27** 012005

[2] Alekseev V 2007 *Atlas fatsiy yurskih terrigennyih otlozheniy* (uglenosnyie tolschi Severnoy Evrazii) Ekaterinburg Izd-vo UGGU 209

[3] Korovin M O *et al* 2015 Oil extraction optimization case: horizontal permeability heterogeneity evaluation and application *IOP Conf. Ser.: Earth Environ. Sci.* **24** 012043

Theoretical principles of petroleum hydrogeology of the West Siberian megabasin (WSMB)

V Matusevich[1], V Popov[2], L Kovyatkina[3] and G Pozdeeva[4]

[1, 3] Tyumen State Oil and Gas University, Tyumen, 38 Volodarsky Ave., Russia
[4] Tomsk Polytechnic University, Tomsk, 30 Lenin Ave., Russia

E-mail: [1]vladmichtyumen@mail.ru, [2]pvk@tpu.ru, [3]vyru@mail.ru, [4]galyunya-tom@yandex.ru

Abstract. Comprehensive study of the chemical and gas composition, temperatures, levels, pressure of deep underground water in deep wells is associated with the beginning of the systematic development of the oil and gas potential in Western Siberia and the first discovery of large deposits here. The development of new branches of hydrogeology is due to the fact of more and more available data. Thus, fundamental understandings of the WSMB hydrogeological conditions are being translated into new theories. Geodynamically, the WSMB structure was revised and based on hydrogeological data, regional and local prediction of oil and gas occurrence exploration criteria were developed. Based on the dispersion halo water-dissolved substance theory, exploration methodology of "neglected" deposits were formulated, conceptual issues of technogenic changes of oil and gas hydrogeosphere areas were being developed.

1. Introduction

Comprehensive study of the chemical and gas composition, temperatures, levels, pressure of deep underground water in deep wells is associated with the beginning of the systematic development of the oil and gas potential in Western Siberia and the first discovery of large deposits here. The development of new branches hydrogeology is due to the fact of more and more available data. Thus, fundamental understanding of the WSMB hydrogeological conditions are being translated into new theories. Geodynamically, the WSMB structure was revised and based on hydrogeological data, regional and local prediction of oil and gas occurrence exploration criteria were developed. Based on the dispersion halo water-dissolved substance theory, exploration methodology of "neglected" deposits were formulated, conceptual issues of technogenic changes of oil and gas hydrogeosphere areas were being developed.

2. Materials and methods

Fundamental concepts of petroleum hydrogeology are reflected in the works of A.A. Kartsev, V.N. Kortsenshteyn, G.M. Suharev, L.M. Zorkin, M. I. Subbota, L.N. Kapchenko, S.B. Vagin, V.M. Shvets, E..A. Bars, Yu.P. Gattenberger, A.R. Ahundov, A.M. Nikanorov and many others. Thanks to V.A. Sulin works in hydrogeochemistry (1935, 1948), petroleum hydrogeology was developed as a separate branch.

As petroleum hydrogeology developed, numerous regional scientific schools were being established, including the Siberian School. The representatives of the latter are A. Antsiferov, R.N.

Abdrashitova, M.B. Bukaty, A.I. Vozhov, V.P. Danilova, A.A. Dzyuba, L.A. Kovyatkina, A.R. Kurchik, O.I. Leuhina, V.M. Matusevich, D.A. Novikov, E.V. Pinneker, A.G. Plavnik, R.G. Prokopeva, B.P. Stavitsky, N.F. Chistyakova, S.L. Shvartsev, O.V. Shiganova and others. This article mainly highlights the contribution of Tyumen and Tomsk research school representatives to the petroleum hydrogeology theory

3. Research target

Comprehensive study of the chemical and gas composition, temperatures, levels, pressure and flow rates, organic matter and microflora of underground water in deep wells is associated with the beginning of the systematic development of the oil and gas potential in Western Siberia and the first discovery of large deposits here. The groundwater impact on the deposit formation and destruction and its energy is widely discussed, as well as interaction impact of formation and injection waters, their impact on rock porosity and permeability, their colmation. Researchers are considering technogenic hydrogeosphere transformation.

4. Background information

The forerunner of petroleum hydrogeology in Western Siberia is the outstanding scientist-hydrogeologist A.M. Ovchinnikov. In 1959 at the first hydrogeological meeting he outlined hydrogeochemistry and paleohydrogeology significance in mineral exploration and introduced the water drive system classification. In the 60's A.A. Kartsev published the first summary papers on petroleum hydrogeology.

As late as the 50's of the last century a number of works were devoted to the WSMB hydrocarbon potential assessment based on published hydrogeological data, including O.V. Ravdonikas, N.M. Kruglikova, V.F. Nikonova, A.A. Rozina, B.V. Mavritskiy. The hydrogeochemical indicators of hydrocarbon potential are regarded in details by N.N. Rostovstev. The most turbulent period of petroleum hydrogeology development is observed in the 60-80's as well as the industrial boost.

A.E. Kontorovich (1963), B.P. Stavitsky (1964), V.M. Matusevich (1966) summarized the chemical and gas composition of the groundwater trace elements in oil and gas regions of Western Siberia. Due to V.M. Matusevich large-scale research data in both ion-salt composition, and trace elements and water-dissolved organic matter of the deep horizons were obtained. The monograph "Groundwater Geochemistry of West Siberian Basin" presents unique data on water reduction-oxidation potential and experimental study of diffusion-osmotic processes.

Monograph, published in 1970, "Hydrogeology of the USSR. West Siberian Plain. Volume 16" with a series of maps, is still an indispensible guide on issues related to groundwater megabasin for scientists and specialists.

In the 80's fluid geodynamic concepts in the formation and development of WSMB water drive systems are becoming more and more popular among followers. Comparable to the artesian concept of the formation of deep groundwater horizons, V.M. Matusevich presented expelled water concept, explaining the composition and salinity difference, inversion and anomaly in WSMB.

A significant step in furthering above-mentioned concept was N.B. Vassoevich stages of petroleum generation. According to the theory of organic formation and petroleum accumulation, high trace element concentrations and organic substance in groundwater determine the main oil generation zone (MOGZ). It has been established that MOGZ has "floating" position in the cross section, its greatest depth is characteristic of the northern regions (5.5 to 4.5 km or more), and southward MOGZ intervals are mainly within the range of 2-3 km. This can be explained by the following factors: insufficient rock heating, thermal flux instability due to climatic variations in the Late Quaternary. The concept base and applying hydrogeothermal data to predict WSMB oil and gas occurrence via A.R. Kurchikov mathematical model has been presented in his monographs and scientific papers (co-authors; I.I. Nesterov, S.B. Stavitsky and others). A.R. Kurchikov monograph "Hydrogeothermal criteria of oil and gas occurrence" (1992) is of current issue today.

Accordingly to underground water geochemistry, petroleum generation and carbonification stages are regarded as organic matter transformation througout lithogenesis (V.M. Matusevich, V.K. Popov). In meso-catagenesis, not only organic matter conversion, but also groundwater interaction between host rocks and organic matter actively occur. In this case groundwater is significantly enriched by these components.

In the 80-s V.M. Matusevich revised hydrogeological stratification of WSMB cross section and identified three superposed hydrogeological basins with seven-level complexes. Later they were enhanced by the eighth level was identified as result of separating the Triassic and Paleozoic periods based on the deep well drilling data (V.M. Matusevich, 1984; V.M. Matusevich, L.A. Kovyatkina, 2013, 2014). According to this concept, sub-sequential member – West Siberian megabasin (WSMB) was identified, being a complex system of independent hydrogeological Cenozoic, Mesozoic and Paleozoic basins. Seven hydrogeological complexes within the whole megabasin cross section are distinguished; Oligocene and Quaternary, Turonian-Paleogene, Aptian-Albian-Cenomanian, Neocomian, Upper Jurassic, Lower-Middle Jurassic and Triassic-Paleozoic. The exploration maturity of these hydrogeological complexes is extremely variable; up to the present day the Paleozoic-Triassic hydrogeological complex is characterized only by fragmentary information obtained from single deep and ulterdeep wells.

Further oil and gas well exploration reveal new and new data on the composition and structure of deep horizons in sedimentary mantle and foundation of the West Siberian geosynecline basement. Drilling and well testing, as well as 2D- and 3D-geophysical studies are aimed to obtain information about oil and gas reservoir content and their hydrocarbon potential. This data interpretation makes it possible to identify aquiferous properties, formation conditions of underground waters and their composition, regional prediction of oil and gas occurrence, including Pre-Jurassic WSMB basement.

Tomsk scientists have made a significant contribution in the common concept of hydrogeology and petroleum hydrogeology of WSMB. For example, S.L. Shvartsev, his colleagues and students have implemented new and new ideas into the concept. S.L. Shvartsev equilibrium-non-equilibrium concept "water-rock-gas-organic-substance" system is like "a fresh grasp of air" in the existing concept of fresh groundwater formation in the hypergenesis zone as well as catagenetic processes in the deep crust horizons.

S.L. Shvartsev innovative idea of the formation of granite layer in the continental crust highlights the fact that crystallization of minerals in granitoids is a global evolutionary transformation, process of basalt which could not develop without water, controlling the equilibrium-non-equilibrium state of the thermodynamic multicomponent system "water-rock". The main factor of this system evolution is associated with the continuous dissolution of aluminosilicate calcium, magnesium, iron and formation of alumino silicate sodium, potassium, quartz within filtrated waters. "The main rock transformation process is influenced by water from the continents to form granite". In the deep water oceanic conditions the interaction of water with rocks results in rock serpentinization, and according to S.L. Shvartsev, this can explain the fact the oceanic floor is composed of not granite but of basalt.

This fact can also prove the origin of igneous rock vugs in the Pre-Jurassic WSMB basal complex and, consequently, being filled with fluids and further formation underground water horizons and oil deposits.

The groundwater technogenesis ideas of E.V. Pinneker, B.I. Pisarsky, Yu.P. Gattenberger, A.A. Kartsev became the basis for their testing and development under the conditions of Western Siberia, being the largest worldwide oil and gas complex. WSMB technogenic aspects attract Russian scientists. Our research in this field is the subject of discussion at scientific conferences and in papers from the 90-s to the present day. Some issues are directly related to petroleum hydrogeology. In 1997, the classification of WSMB technogenetic hydrogeological systems was presented and from this moment on the issues concerning reservoir and injection water into actions, their impact on rock reservoir properties, their colmation are still being investigated. Technogenic field as a factor in the formation of the geological environment, geological environment classification of oil and gas deposits, technogenesis zoning are reflected in recent scientific works (2010-2015). The latest works present the concept of the stage development of natural and technogenic hydrogeological systems, applicable to all types of systems, including oil and gas fields [7].

5. Conclusion

Oil-field development, underground water intake exploitation for maintaining reservoir pressure (MRP), subsurface disposal of waste and associated water excess in the oil fields, provision of production field industries and the population with high quality drinking water necessitate the investigation and theoretical justification of technogenic geological environment transformation and their impact on the environmental conditions of oil and gas regions.

References

[1] Abukova L A and Vagin S B 2005 Osnovnye jetapy razvitija neftegazovoj gidrogeologii (The main stages of development of hydrogeology). *Fundamental'nye problemy neftegazovoj gidrogeologii. Mezhdunarodnaja konferencija, posvjashhennaja 80-letiju A A Karceva* (Moscow) 251 URL: http://ipng.ru/publications/labs/

[2] Karcev A A 1963 *Gidrogeologija neftjanyh i gazovyh mestorozhdenij* (Moscow: Gostoptehizdat) 354

[3] Matusevich V M, Ryl'kov A V and Ushatinskij I N 2005 *Geofljuidal'nye sistemy i problemy neftegazonosnosti Zapadno-Sibirskogo megabassejna* (Tyumen: TyumGNGU Publ.) p 225

[4] Matusevich V M and Popov V K 1982 Stadijnost' nefteobrazovanija s pozicij geohimii podzemnyh vod *Organicheskaja geohimija vod i poiskovaja geohimija* (Moscow: Nauka) 79 – 83

[5] Tyumenskaja sverhglubokaja skvazhina *Rezul'taty burenija i issledovanija: Sbornik nauchnyh dokladov* (Perm: KamNIIKIGS) p 376

[6] Shvarcev S L 2012 Pochemu net granitnogo sloja v okeanicheskoj kore*? Geologicheskaja jevoljucija vzaimodejstvija vody s gornymi porodami: materialy Vserossijskoj konferencii s uchastiem inostrannyh uchenyh* (Tomsk: NTL) 495

[7] Matusevich V M and Kovjatkina L A 2015 Koncepcija stadijnogo razvitija prirodno-tehnogennyh gidrogeologicheskih sistem Zapadno-Sibirskogo megabassejna.(The concept of stage development of natural and man-made hydrogeological systems of West Siberian megabasin) *Sovremennye problemy nauki i obrazovanija* **1** 1844. URL: http://elibrary.ru/item.asp?id=23658416

Electromagnetic emission in mineral and rock dehydration

V Salnikov[1], V Popov[2] and D Terre[3]

[1,2] Department of Geology and Land Management, Institute of Natural Resources, National Research Tomsk Polytechnic University, Tomsk, Lenin Avenue, 30, Russia
[3] Department of Foreign Languages, Institute of Natural Resources, National Research Tomsk Polytechnic University, Tomsk, Lenin Avenue, 30, Russia

E-mail: [1]salnikov-40@mail.ru

Abstract. The article considers regularities of radiofrequency electromagnetic radiation from minerals and rocks, with samples being heated in a vacuum to 20° C- 1000° C. The examples of electromagnetic emission correlation with electric conductivity, thermoluminescence and thermographic analysis during physic-chemical processes resulting from diagenesis, catagenesis and metagenesis have been provided.

1. Introduction

In 1970 a geological and physical exploration team was organized at the department of Solid State Physics, Electrophysical Faculty (SSP EPF), Tomsk Polytechnical University. Salnikov V.N., teaching assistant of SSP department, was appointed to head it. Based on variable anomalies of natural field, the first measurements of the earth's impulse electromagnetic field were conducted under the supervision of A.A. Vorobyev in the Altai and ore mine "Kommunar" in summer surveying season, 1971 [1].

Consequently, the research objective was set up – "Which physico-chemical processes, apart from fracturing as a result of rock decomposition due to earthquakes, can cause electromagnetic emission in the Earth's shells?" In 1971 the plan of laboratory investigations, aimed to detect electromagnetic emission caused by physico-chemical processes which occur due to rock and mineral heating, was proposed. In Laboratory of Dielectric and Semiconductor Physics V.N. Salnikov and his students made a circuit for electrical conductivity operating set-up to measure electromagnetic radiation. It is measured upon heating naturally occurring and man-made dielectrics in a vacuum at temperatures of 20 °C -1000 °C simultaneously with conductivity and separately. The procedure and methods of mineral and rock properties investigations are provided in [2]. The experimental results in measuring electric conductivity of muscovite and rock samples being heated and cooled in temperature interval of 20 °C-1000 °C have been obtained. Variation in electric conductivity timed to occurrence of impulse electromagnetic radiation at radio-frequencies between 247,5 kHz and 35,8 mHz has been detected. The changes in electrical conductivity and electromagnetic emission maxima correspond to vaporization temperature of weakly-bound water (about 100° C), electrical activity of mica surface centers (20-450 ° C), temperatures of muscovite crystallization in hydrothermal process (480-160 ° C), phase transitions 550-590 °C and constitutional water release (780-800-1000 ° C)

In addition to three known water types in crystals - constitutional water, crystallization water and adsorbed water, N.P. Ermakov (1972) distinguishes the fourth type, trapped or mother water, which is geometrically captured in mineral pores. Removal of this water by means of heating in inclusion

IOP Conf. Series: Earth and Environmental Science **33** (2016) 012052 doi:10.1088/1755-1315/33/1/012052

disintegration (dehydration) leads to significant reduction in mineral specific gravity and considerable loss at calcination [3]. It has been stated that electromagnetic emission in heated samples of quartz, orthoclase, mica and rocks which contain these minerals can be detected due to separation of mineral-forming solutions. This separation brings about volume charge, polarization, current magnitude impulse change in minerals in the intervals of mineral and ore formation temperatures (Figure 1, 2, 3).

2. Materials and methods

Self-organization effects resulting in the formation of time-ordered varied structures can occur during certain nonlinear chemical reactions in the lithosphere due to metasomatism, metamorphism and hydrothermal process. Thereby, the speed of nonlinear reaction product formation depends on reactant concentrations. Of particular interest in analyzing hydrogeological consequences of catagenesis of sedimentary rocks is mesocatagenesis (MC1-MC3). It is a so-called "thermal furnace", where considerable transformations of mineral and organic matter including water solutions take place at the temperature of 80-160 ° C [4].

In studying physico-chemical conditions of the considered catagenesis grade the question arises concerning the source of oxygen necessary for oxidation of organic matter, for instance, in the reactions:

$\rightarrow SO$

$CH_4 + 2O \rightarrow H_2O + CH_2O \rightarrow HCO$ (3.1)

H_2S

$CH_4 + 4O \rightarrow CO_2 + H_2O$ (3.2)

Silicon is of great significance for the studied physico-chemical processes. Our attention was attracted by the fact that in underground waters of oil-gas bearing basins adjacent to oil deposits there is a great deficiency of silicon, whereas country rocks are subjected to silicification. During regional catagenesis of Permian and Carboniferous deposits in Kuzbass secondary quartz occurs at MC1-MC2 stage and is abundant at later catagenesis stages:

$[SiO_4]_n^{4-} \rightarrow nSiO_2 + 2nO^{2-}$ (3.3)

As reactions 3.1-3.3 proceed simultaneously and involve the common element – oxygen, and its release in the reaction 3.3 contributes to oxidation of dispersed organic matter, it is possible to deal with coupled reactions when there is interdependence of mineral and organic matter transformations (equation 3.1, 3.2). Silicon-oxygen tetrahedron turns into bent SiO_2 molecule according to rock silicification mechanism, consequently forming geochemical barriers for hydrocarbon gases. During catagenesis in silicate rocks, exoelectron emission accompanied by acoustic emission is brought about. For the described catagenesis conditions hydrocarbon-silicate interaction is the main factor. In petroleum geochemistry, it has been shown that mineral catalysts are of great importance for thermodynamic destruction of organic matter. Their activity is due to ions of aluminium, iron, silicon and acid interlayer water in minerals (Khant, 1982; Tikhomirov, 1978; Dinisenko, 1974).

In energy terms hydrocarbon-silicate interaction reactions can occur since Si-O and O-H dissociation energy is equal to 224 and 462 kJ/mole, respectively; and corresponds to energy level of hydrocarbon transformation:

– CH_3—H (453 kJ/mole),
– C_2H_5—H (407,4 kJ/mole),
– CH_3—R (252 kJ/mole).

IOP Conf. Series: Earth and Environmental Science **33** (2016) 012052 doi:10.1088/1755-1315/33/1/012052

Physico-chemical analysis shows that between hydrocarbons and rock-forming silicates spontaneously proceed reactions which are accompanied by heat release over the range of 480.8 kJ/mole and are similar to that given below:

$2[SiO_4]^{4-} + CH_4 \rightarrow 2SiO_2 + 2H_2O + CO_2$

In these reactions, oxygen is intensively removed up to 10...20 %. One ton of hydrocarbons can affect nearly 10 m^3 of rocks (Dinisenko, 1974). In hydrocarbon migration paths this effect causes leaching of rocks in Kuzbass. The deposited silicon interacts easily with alkali-hydroxide-water solutions resulting in secondary silicate formation at the later stages of catagenesis.

3. Results and discussion

In studying electric conductivity-temperature dependence, the experiments showed that exponential electric conductivity–temperature relationship in complex compounds such as minerals and rocks is not consistent (Figure 1). What causes inconsistency of electric conductivity-temperature dependence of minerals and rocks?

Deviation of mineral and rock conductivity from exponential law due to change of element number (n) and temperature (T) of exothermic and endothermic reactions can be caused by the following reasons:

– Vaporization of weakly-bound water and semi-bound water (dewatering) at temperature of 0-420 ° C;

–Dehydration at higher temperatures (zeolite and constitution water separation) $KAl_2[AlSi_3O_{10}]\times[FOH]_2 \rightarrow AlSi_3O_5$;
muscovite mullite

– Disintegration of gas-liquid inclusions in minerals (decrepitation);

–Oxidation-reduction reactions involving exothermic and endothermic effects;

–Piezo-effect which determines dielectric polarization;

–Polymorphous transformations, for example, ά-β inversion at 573° C;

–Natural radiation which causes energy accumulation in mineral defects and release of stored energy upon heating by light, heat, radio frequency impulse electromagnetic radiation (Schottky and Fresnel defects);

–Artificial radiation exposure to charged particles, gamma-rays, x-rays and other types of rays that result in electretic state;

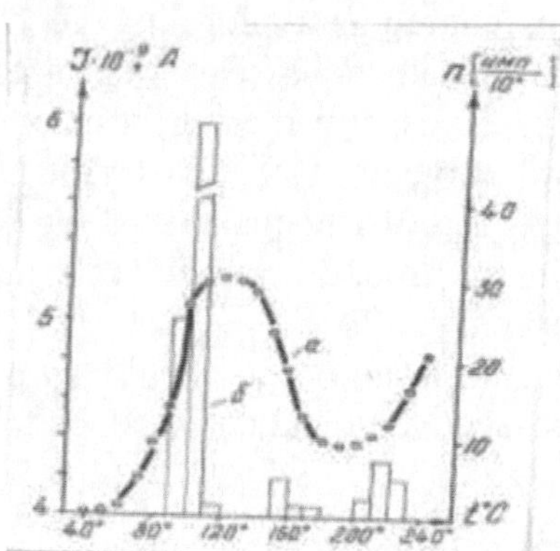

Figure 1. Temperature dependence of electromagnetic impulse intensity (a) and current rate change (b) in the interval of weakly-bound water (granite, 715-M sample, Shindinskiy Khrebet).

–Formation of new phase in heterogeneous dielectric system matrix, formation and decomposition of solid solutions, for instance, NaCl + KCl + t = solid solution.

Let us consider modeling of conductivity anomalies in σ = f (T°) or lgσ = f (–) dependence. While dielectric being heated, conductivity anomalies occur both in impurity and intrinsic range which is determined by basic ions of the crystal lattice. It should be reminded that electric conductivity-temperature dependence is analytically expressed as follows:

–— , where A —— —— —

On heating dielectric in which there are no phase transitions of the first and the second type, charge (q) – constant; ion mobility (lattice parameter) δ and their quantity (n) change insignificantly with temperature increase (T); particle vibration frequency (ν) does not practically change until melting point.

In chemical reactions, we deal with self-sustained oscillations which are determined by reacting substance amount over time, i.e. reaction intensity increases with increase in temperature or time, and then reaching extremum, it decreases. Temperature changes depending on heat absorption or heat release reaction effects. Redistribution of heat in samples, supplemented by furnace effect or on the contrary, its absorption by the system, traces a curve, passing over maximum or minimum. Therefore, in accordance with whatever physic-chemical process (related to phase transition of the first type) is, it is advisable to substitute n or T1 in a pre-exponential variable for normal law of distribution or Gaussian law in electrical conductivity formula (Rytov, 1978).

It is concerned with the case of modeling, whereas in reality it can be much more complicated and does not follow Gaussian law at the moment of phase transition. The mathematical scheme of Bernourlli problem contains a number of specific issues which may represent model situations for defining electric conductivity at the period of I type phase transition occurrence: shot effect (Schottky effect) in a vacuum; random walk; intensity fluctuations.

Nine modifications of electric conductivity temperature dependence deviation from exponential law were first presented by V.N. Salnikov in [2].

Formation and decomposition of defect complexes are always accompanied by neutralization of definite amount of ΔQx charges in a dielectric sample being heated.

Hence, f potential between unlike defects and their complexes drops by Δf, additional charge ΔQx is required for its restoration. Intensity of internal electrical field between defect complexes reaches 106 V/sm.

For a single neutralization event:

—; where —

C_x – capacitance of two mica plates or crystal faces, or a defect sample; S – plate area; d – distance between plates; ε – specific inductive capacity of medium; E_0 – dielectric constant

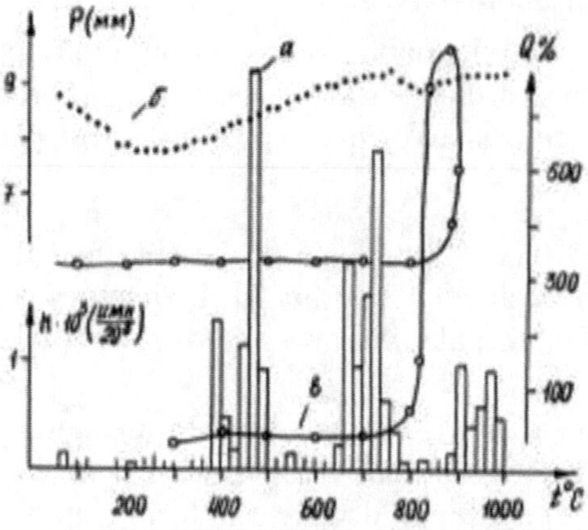

Figure 2. Temperature dependence of electromagnetic pulse (EMP) intensity of Sogdiodon muscovite mica: a – EMP intensity; b – vacuum change in a system while EMP measuring (according to V.N. Salnikov); c – heaving of mica as a result of zeolite and constitutional water loss according to the research done by E.K. Lashava, 1948.

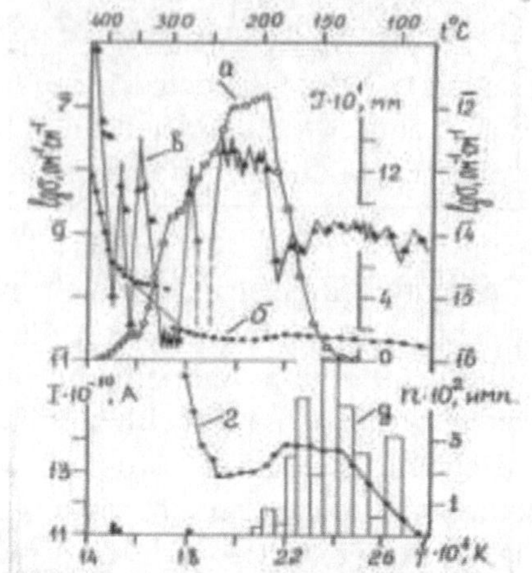

Figure 3. Correlation of electric conductivity with thermoluminescence of microcline samples: a – thermoluminescence; b – electric conductivity dependence at high temperatures; c – electric conductivity dependence on cooling, after sample being heated to 800°C; d – conduction current change upon heating; д – electromagnetic pulse intensity at 0,15-30MHz frequency band.

For example, in muscovite during neutralization immediate increase in Cx capacitance occurs between electrodes which host dielectric (Figure 2). Mica heaving in course of constitutional water separation, and its splitting results in decrease in Cx capacitance between mica plates, while f potential and electrical field intensity (E) increases.It is supposed that intensity of internal field (EIF), produced by defect-adsorbed charges alters according to formula:

—

E_0 – internal field intensity, produced by adsorbed charges in defect spacing or pore counter-surfaces at the moment of emission (charge) termination; t – time of charge (emission or quantum diffusion); τ – time constant of defect-adsorbed charge accumulation, i.e. Maxwellian dielectric relaxation time.

Dielectric relaxation time is dependent on volume resistivity ρ_v and dielectric capacity ε:

or —

The formula shows that the more volume charge relaxation time (i.e. time of its retention by dielectric) is, the less electrical conductivity is, and the bigger dielectric capacity which is the ratio of capacitance of a capacitor filled with dielectric C to capacitance of a capacitor C_0, which is filled with vacuum between plates. Consequently, the formula is written as: —

Volume charge relaxation may take place both due to slow diffusion resulting from conduction and fast relaxation which is accompanied by electrical breakdown, cold emission induced emission, electromagnetic signal generation.

4. Summary and conclusion

Based on experiments in electrical conductivity, electromagnetic radiation and results of differential thermal analysis (DTA) of carbonaceous rocks of Yangoyakhinskoe ore deposit and Paleozoic deposits of gas-oil fields in Tomsk Oblast, we have detected the stages of decarbonatization which can be used to determine mineralization and diagenesis conditions. In DTA curves of limestones it is possible to observe one minimum at 960°C, which is due to $CaCO_3$ decomposition to CaO and CO_2. In case of dolomite two minima can be observed: the first is related to decomposition of solid solution $CaMg(CO_3)_2$ to $CaCO_3$ and $MgCO_3$, which in turn, decomposes to MgO and CO_2 (780 °C); the second occurs due to $CaCO_3$ decomposition to CaO and CO_2 (Figure 4).

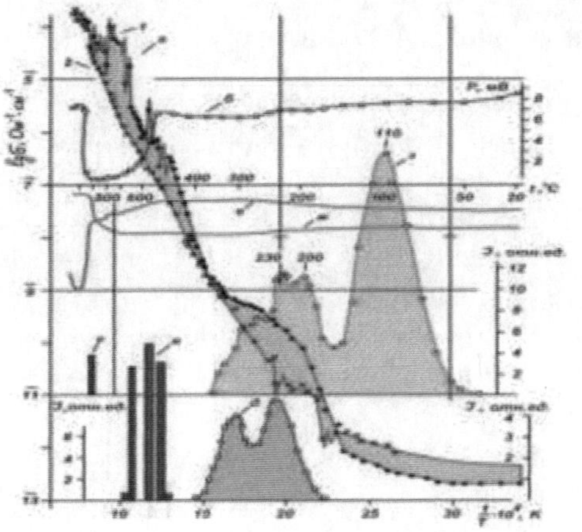

Figure 4. Conductivity-temperature dependence of limestone sample (a): 1) heating; 2) cooling; vacuum change in a system (b); electromagnetic impulses while heating (c); electromagnetic impulses in cooling (d); natural thermoluminescence (e); differential thermal analysis (f); differential thermal analysis on weight loss (g); gamma-thermoluminescence (sample 1052) (h).

Application of electro-physical methods can define kinetics of solid solution decomposition more precisely and introduce new method to determine temperature of geological processes. The integral analysis of electro-physical properties of carbonaceous rocks, different stages and lithogenesis phases can decipher the history of rock formation and transformation. Detection of evolution regularities of carbonaceous substance exposed to thermal and electrical field using electro-physical methods can be applied to differentiation and correlation of Paleozoic carbonaceous deposits of gas-oil fields.

References

[1] Salnikov V N 2008 *Fizika tverdogo tela na sluzhbe geologii Istoriya stanovleniya Sibirskoj geologicheskoj shkoly i geologicheskikh issledovanij* II 402-410

[2] Zavertkin S D, Salnikov V N and Arefyev K P 2010 *Electromagnetic emission at phase transitions in minerals and dielectric materials*

[3] Salnikov V N, Gankina L N 1978 Nablyudenie electromagnitnykh impulsov v nagrevaemykh obraztsakh gornykh porod vsledstvie otdeleniya mineraloobrazuyushikh rastvorov *Izvestiya VUZov «Geologiya i pazvedka»* **5** 61-68

[4] Popov V K 1998 *Geologiya podzemnykh vod Yugo-vostochnogo obramleniya Zapadno-Sibirskogo megabassejna*: Avtoreferat diss.doct.geol-min. nauk

Reflectivity and microhardness of sulfide minerals as genetic information source (case study: pyrite and arsenopyrite)

E A Sinkina[1], M V Korovkin[2], O V Savinova[3] and A A Makarova[5]

[1, 3, 4, 5]Tomsk Polytechnic University, 30 Lenin Avenue, Tomsk, 634050, Russia
[2]Tomsk State University, 36 Lenin Avenue, Tomsk, 634050, Russia

E-mail: [1]vea1986@mail.ru

Abstract. Reflectivity and microhardness of pyrite and arsenopyrite of black shale gold-ore deposits in Chertovo Koryto (Patom upland) were studied. It was found that sulfides of different generations are characterized by different values of above-mentioned parameters which is associated mechanical and isomorphic impurities.

1. Introduction

Reflectivity is the main quantitative and diagnostic constant of ore minerals which is closely connected with mineral composition and structure. Reflectivity can be measured along one wavelength and across a visible spectrum. The study of reflection the spectrum is more informative, as it indicates the changing reflectivity in relation to the wavelength itself. Both mechanical and isomorphic impurities, as well as ultrafine structure of decomposed solid solutions affect the spectral reflectance curve shape. One and the same element of different accompanying elements could increase or decrease the reflectance. For example, reflectance increases to increasing iron content in oxides [1, 2].

Microhardness parameter is successfully used as a typomorphic indicator for ore (sulfides) and non-metallic (grenades, fluorite, topaz and others) minerals. Based on this parameter minerals are classified to generations and the influence of certain impurities on the microhardness value is also determined [1, 3, 4, 5].

The present paper deals with the study results of reflectivity and microhardness of pyrite and arsenopyrite, being the main ore minerals in Chertovo Koryto gold-ore deposit.

This deposit is located in the north of Patom upland within the Big Patom river basin. Thick (up to 150 m) ore deposit was formed in the early Proterozoic carbonaceous terrigenous slates of Mikhailovskoe suite. The suite embraces metasomatic rocks of beresite-propylite formation with vein-disseminated sulfide-quartz mineralization. Sulfides occur rarely in quartz veins and veinlets. Pyrite, arsenopyrite and pyrrhotite are predominate. Some galena, sphalerite, chalcopyrite, cobaltite, native plumbum, ulmanita, tellurobismuthit, valleriite impurities can be found in the quartz sulfide complexes. Free gold dominates in quartz. A more detailed description of this deposit is given in [6, 7].

2. Research technique

Pyrite and arsenopyrite monomineral samples were selected to study the reflectivity of sulfides. A portion of samples was abraded into powder for the atomic emission analysis in order to determine the impurity element content. The remaining pyrite and arsenopyrite grains were crushed into pieces and carefully polished to obtain more reliable reflectivity values. Polished sections were also used. Microscope spectrophotometer MSFU-K was applied in measuring the reflectivity. The spectral reflectance curves were recorded in the range from 400 to 750 nm at the sampling rate of 4.0 nm. More than 100 measurements were performed.

The device PMT-ZM with ocular photoelectric micrometer (FOM-2-16) attachment was used in studying microhardness. The indenter was the diamond pyramid Vickers. 30-40 shots were conducted, each of which involved 5 diagonal measurements and further calculation of the mean value. More than 600 shots were performed in 20 samples. Analysis of measurement results was performed on the basis of S. I. Lebedeva method [8] which makes it possible to evaluate the results more objectively. The aim of this method is to find the most probable mineral microhardness value via variation curves. In this case, the microhardness of minerals (kgH/mm^2) was determined on different face sections and the variation magnitude and width of microhardness interval were determined.

Further, measurement occurrence frequency (P, %) was calculated in specified intervals. Finally, variation curves were designed for each sample and the most probable values were identified.

3. Research findings

Based on previous crystallomorphology studies, analysis of polished sections and thermal emf it was established that there are 4 generations of pyrite and 2 generations of arsenopyrite in Chertovo Koryto deposit [9].

Pyrite

Pyrite I – widely spread in the carboniferous area of the metasomatic halo as porphyroblasts (up to 1 cm).

Pyrite II – well- developed within the rock mass as chain clusters of cubic crystals of up to 1 cm with the distinct edges and accompanied by thin quartz-carbonate veinlets.

Pyrite III—well- developed as granular aggregates in quartz veins and embracing a variety of morphological crystal types and different thermal emf parameters: from pyrite I and II generations.

Pyrite IV – formed during the replacement of pyrrhotite, located in quartz and are marcasite-pyrite aggregates. It is developed on the peripheral part of pyrrhotite or along its the cracks.

The first three generations of pyrite could be applied in study of reflectivity and microhardness.

Arsenopyrite

According to mineralographic results 2 generations of arsenopyrite were revealed. Arsenopyrite I found in rocks as short prismatic porphyroblasts. Arsenopyrite II found in quartz veins and veinlets as granular aggregates. The cataclastic texture of arsenopyrite II revealed in the polished sections indicates the fact the mineral grains were crushed and then quartz cemented. Both generations are suitable for studying reflectivity.

As minerals have different optical properties, the approach in studying the reflectivity was different. Optically, pyrite is an isotropic mineral, although there is evidence that in nature anisotropic pyrite exists. The studied pyrite showed no evidence of reflectivity anisotropy. In this case, arbitrary cross-sections were examined. In the case of arsenopyrite it was reverse. Visually, it was very difficult to identify the maximum darkening time of the grain, as it can be more than 5 degrees. To determine this moment visually it was necessary to perform the measurements from the original position of the grain rotating at +5 and -5 degrees. All graphic measurements were compared with reference samples. Reference samples were curves for «pure» pyrites and arsenopyrites imported from http://www.webmineral.com/.

The spectral reflectance curve for pyrite can be seen in Fig. 1. The spectral reflectance curve of the pyrite III (Figure 1, a) differs from the " reference sample" one which is probably due to the high content of Co and Ni in comparison with pyrite I and II (Table 1). According to the previous studies [9, 10] it was found that the typical electronic conductivity of pyrite III is associated with isomorphic Co and Ni impurities. The differences in the spectral reflectance curve of pyrite III and pyrite reference sample are expressed by the reflectivity increase from 53,4 to 57.2% at the wavelength of 540 nm. The spectral reflectance curves of pyrites I and II (Figure 1, b) coincide and they are slightly different from the reference sample curve at 500...600 nm, which could be explained by the presence of cobalt and nickel. Pyrites have different microhardness parameter (Table 1) which is associated with the presence of Co and Ni. Pyrite III involves increased Co and Ni content, but low microhardness.

Table 1. Morphological parameters of different pyrite generations.

Minerals	Mean value H, kgH/mm^2	The range of values of the thermal emf, mV	Content Co, g/t (average content)	Content Ni, g/t (average content)
Pyrite I (5 samples)	1500	+(46…64)	23.0-82.1 (44.3)	93.4-268.5 (151.6)
Pyrite II (3 samples)	1580	+(49…58)	26.2-79.3 (31.5)	175.5-290.0 (233.8)
Pyrite III (4 samples)	1400	+(25...67) −(60...67)	150-200 (177.8)	370.7-514.6 (421.6)

The obtained spectral reflectance curves for arsenopyrite are presented in Figure 2. The reflectance spectra of arsenopyrites are non-parallel with very weak dispersion and intersect at 530-580 nm depending on the section. The spectral reflectance curve of arsenopyrite I isn't different from the sample curves (Figure 2, a) and the spectral reflectance curve of arsenopyrite II have the same shape, but they are located below the ordinate axis (Figure 2, b), i.e. the mineral has higher reflectivity values at particular wave length. This can be related to the presence of mechanical impurities of galena in arsenopyrite, which was revealed by using the electron microscope. According to the references the spectral reflectance curve for galena is typical, which is lower at the ordinate axis than for arsenopyrite and at the wavelength of 540 nm the reflectivity is about 44 % [1, 11].

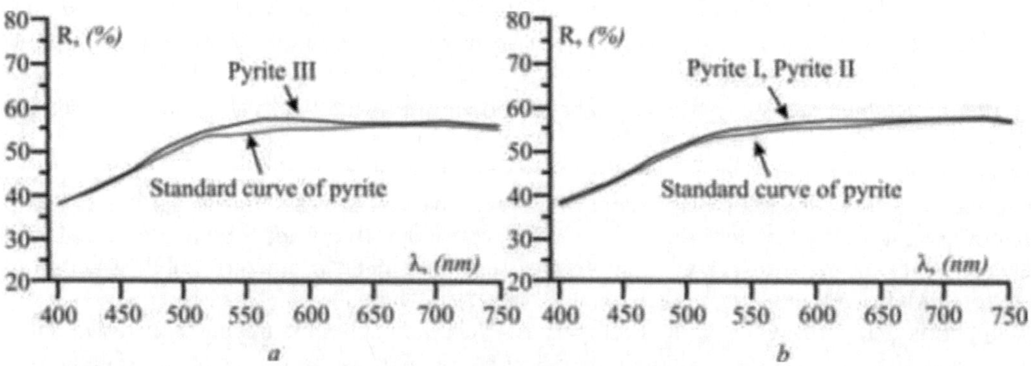

Figure 1. The spectral reflectance curves of pyrites in Chertovo Koryto deposit.

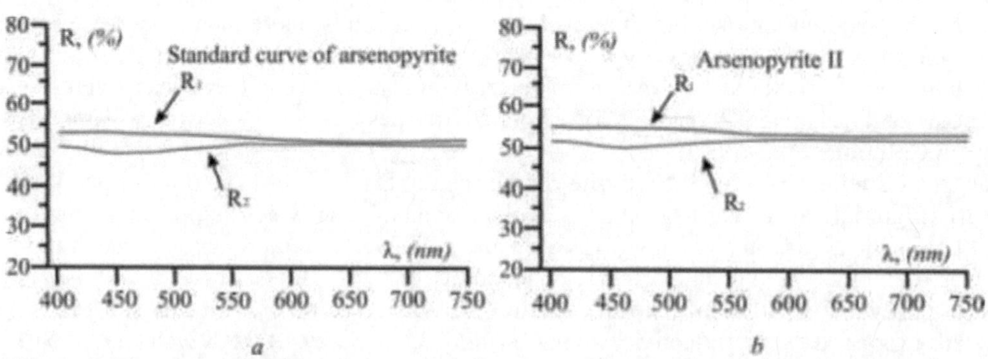

Figure 2. The spectral reflectance curves of arsenopyrites in Chertovo Koryto deposit.

The research showed that arsenopyrite in terrigenous rocks as porphyroblasts, has a higher microhardness, changing in the interval 1070 1270...kgH/mm^2 with the mean value of 1200 kgH/mm^2. Arsenopyrite, occurring in thin quartz veins and veinlets, is characterized by a lower value ranging from 800...1000 kgf/mm^2, with a mean value of 950 kg/mm^2. The graph of the variational curves is presented in Figure 3. During the study of trace elements in arsenopyrites, it was established that the average content of Co, Ni, and Pb in different generations is approximately the same, in contrast to the content of the noble elements (in arsenopyrite II - increased content of Ag and Au) (Table 2).

Table 2. Average content elements in arsenopyrites.

Element	Arsenopyrites I	Arsenopyrites II	Type of analysis
Co, g/t	400	500	Microprobe
Ni, g/t	800	700	
Pb, g/t	1500	1300	Atomic emission
Au, g/t	17	30	Atomic absorption
Ag, g/t	3	7	

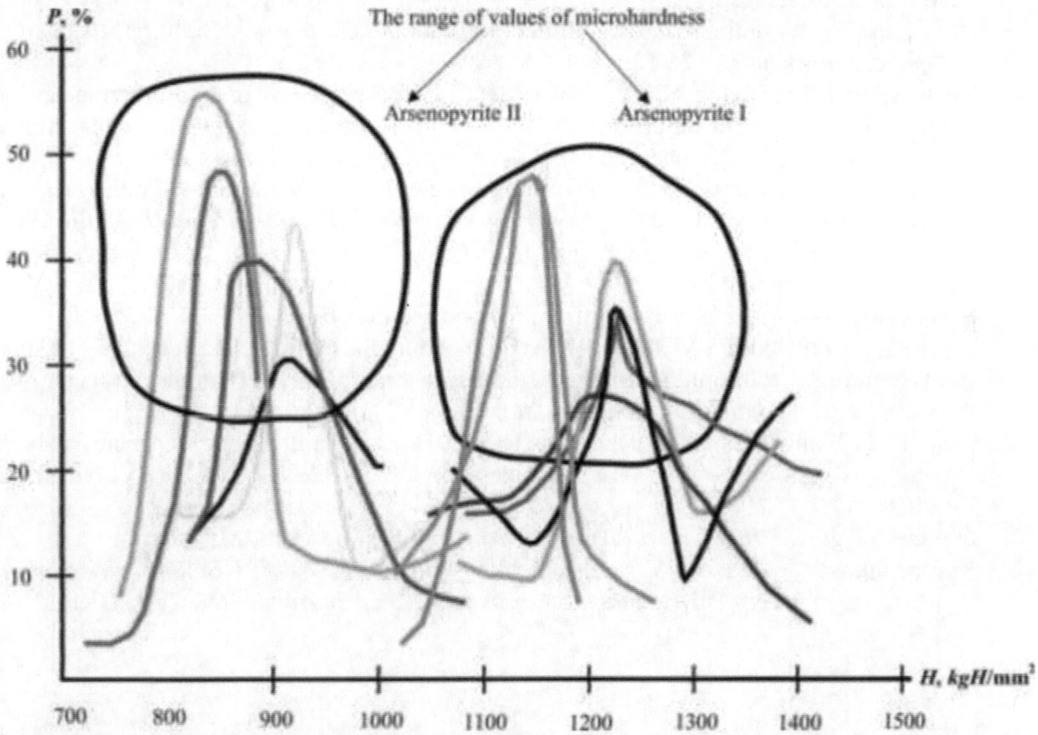

Figure 3.Variational curves of arsenopyrite microhardness (R shares of occurrence, H - value of microhardness, colors denote different samples).

4. Summary and conclusion

From the above mentioned studies it follows that both mechanical and isomorphic impurities contained in the mineral can influence the reflectivity. Many scientists have come to similar conclusions in the study of physical properties of minerals [2, 3]. Reflectivity can be a good diagnostic parameter, but it is limited by many factors (polishing, clean grains, isomorphic impurities etc.).

The study of microhardness showed that pyrite and arsenopyrite of different generations have different values of microhardness, due to the presence of isomorphic and mechanical impurities and the conditions of formation of minerals as well. In particular, Co and Ni as isomorphic impurities was found in pyrite. These isomorphic impurities decrease the value of microhardness. The presence of e-type conductivity in pyrite III suggests that Co and Ni are included in the crystal lattice of the mineral. This was confirmed by the decrease of the value of microhardness of up to 1450 kgH/mm^2, because the isomorphic inclusion of Co and Ni reduces the hardness due to the weakening of the bonds in the mineral structure to the higher electropositivity of nickel and cobalt in comparison with iron [12].

It is presupposed that the change of microhardness of arsenopyrite was due to the high Au content and intense tectonic conditions of the formation.

References

[1] Shumskaja N I and Ljahnickaja V D 1993 *Mineragarficheskie issledovanija rud s primeneniem mikroskopov-spektrofotometrov* Metodicheskoe rukovodstvo (Izd-vo VSEGEI)

[2] Belozerceva N V, Silichev M K and Avvakumov S V 1998 Vzaimosvjaz' fizicheskih svojstv i sostava arsenopiritov zolotorudnyh mestorozhdenij v terrigennyh otlozhenijah *Mineral'nye aspekty metallogenii Jakutii: Mater. dokl. Sessii* (Jakutsk)

[3] Mozgova N N 1966 Ob otrazhatel'noj sposobnosti, mikrotverdosti i termoJeDS galenita v svjazi s izomorfnymi primesjami serebra i vismuta *Geologija rudnyh mestorozhdenij* 3

[4] Nakhla F M 1956 The hardness of metallic minerals in polished section *Econ. Geol.* **51** 8

[5] Desbarats A J, Parsons M B, Percival J B, Beauchemin S, John Kwong Y T 2011 Geochemistry of mine waters draining a low-sulfide, gold-quartz vein deposit, Bralorne, British Columbia *Applied Geochemistry* **26** 12

[6] Kucherenko I and Zhang Yuxuan 2014 Metallogenic problems of hydrothermal gold deposit formation: facts and arguments *Scientific and Technical Challenges in the Well Drilling Progress* (24–27 November, Tomsk Polytechnic University, Tomsk) **24**

[7] Gavrilov R Ju, Khmelev M V and Abramova R N 2014 Volumetric geochemical model of Chortovo Koryto gold-ore deposit (Patom range) *XVIII International Scientific Symposium in Honour of Academician M.A. Usov. Problems of Geology and Subsurface Development* (7–11 April, Tomsk) **21**

[8] Lebedeva S I 1977 *Mikrotverdost' mineralov* pod red. S A Jushko (M.: Nedra)

[9] Vagina E A and Rudmin M A 2010 Kristallomorfologija i termojelektricheskie svojstva pirita i arsenopirita v zolotorudnom mestorozhdenii Chertovo koryto (Patomskoe nagor'e) *Izvestija Tomskogo politehnicheskogo universiteta* **1**

[10] Ding Y C, Xiang A P, He X J and Hu H F 2011 Structural, elastic constants, hardness, and optical properties of pyrite-type dinitrides (CN_2, SiN_2, GeN_2) *Physica B: Condensed Matter* **406** 8

[11] *Opredelitel' glavnejshih mineralov rud v otrazhennom svete* 1978 (M.: Nedra)

[12] Korobejnikov A F, Narseev V A, Pshenichkin A Ja and Revjakin P S 1993 *Pirity zolotorudnyh mestorozhdenij (svojstva, zonal'nost', prakticheskoe primenenie)* (M.: TsNIGRI)

Modeling the damming effect of pile foundations (Tomsk city)

K I Kuzevanov[1], D S Pokrovsky[2], V D Pokrovsky[3] and K K Kuzevanov[4]

[1, 3, 4,5] Tomsk Polytechnic University, 30 Lenin Avenue, Tomsk, 634050, Russia
[2]Tomsk State University of Architecture and Building, 2 Solyanaya sq., Tomsk, 634003, Russia

E-mail: [1]kki@tpu.ru, [2]pokrovskiy@gmail.com, [3]vdp@tpu.ru, [4]kuzevanovkk@mail.ru

Abstract. The authors have considered the impact of pile foundations on the structure of filtration flows in the conditions of urban development. Hydrodynamic simulation methods have shown that a groundwater level rise might occur due to the damming effect that can be created by pile fields in semipermeable rocks. This phenomenon can intensify anthropogenic waterlogging processes in urbanized territories.

1. Introduction

The interest in studying the damming effect of pile foundations in urbanized territories is caused by the need to forecast changes in hydrogeological conditions under the influence of the construction development of the geological environment. The underground hydrosphere reacts to man-induced impact most dynamically. Directional groundwater level rise, i.e. waterlogging, occurs as the result of distortions in the water regime and balance of territories. This complex hydrogeological and engineering-geological process distorts the required conditions of construction and operation of specific structures and demands the application of protective measures. Waterlogging caused by different reasons is registered in many cities of the world in Brazil, Great Britain, Germany, Egypt, India, China, USA, France, etc. Tomsk is not an exception in this respect. We have been studying the problem of changes in the hydrogeological conditions in the territory of this Siberian city for a long time [1- 4].

It is found that the hydrogeological conditions of the city territory are predetermined by the peculiarities of a geological structure. There are two structural levels in the cross section. The foundation is composed by the compact dislocated fractured rocks of the Paleozoic era that is covered by the loose sand-clay deposits of the Meso-Cenozoic age. The role of the separating layer between them is played by the clay weathering mantle of the Cretaceous-Paleogene age characterized by varying thickness and composed by seat clays. The dense rocks of the foundation contain fracture, predominantly head waters that are partially used for the utility and drinking water supply of the city. The loose rocks of the sedimentary mantle are characterized by almost horizontal bedding, in compliance with which one can identify aquifers on the basis of the lithostratigraphic principle. The special conditions of groundwater occurrence are typical of the complex of alluvial deposits of the rather well-developed hydrographic network. The groundwater of the upper part of the hydrogeological section is exposed to the strongest anthropogenic impact and, in its turn, has a significant influence on the living environment of the city. The groundwater is recharged locally due to the infiltration of precipitated waters and leaks from water-bearing utility systems [1, 2, 3, 5, 6].

2. Hydrodynamic simulation

The need to apply mathematical simulation methods to study changes in the hydrogeological conditions under the influence of anthropogenic waterlogging processes in local areas is explained by the complexity of schematizing and assigning boundary conditions within the framework of analytical solutions to forecast geofiltration problems [7,8,9]. Within the territory of the city, waterlogging development schemes are difficult to forecast due to the damming effect of pile foundations [7].

As the interaction between groundwater and an artificial barrier implies the distortion of flow lines and forming a complex shape of filtration flow, it is expedient to use numerical methods of hydrogeological condition simulation to assess changes in levels. In this case, it is offered to perform simulation with a multiple-path approach for a local area having a standard size and typical conditions of a filtration section when locating pile foundations of different shapes within the boundaries of the area under study [6, 7].

We have used the ***Processing Modflow*** software complex as a tool for simulating local models. The peculiarity of this programming environment is a simplified user interface that has only the essential set of capabilities to prepare initial data with due account for the immediate setting of filtration parameters and boundary conditions for the space of the finite-difference grid. Along with that, the software complex uses ***modflow96,*** which is the core of the most popular hydrodynamic simulation programs, such as ***Groundwater Modeling System*** or ***Visual Modflow***.

For local models, we have used a square finite-difference grid with the permanent size of 21×21 computational cells. The same size of the cells (10×10 m) is selected, on the one hand, to minimize the number of computational cells and, on the other hand, to have a possibility to approximate the shape of file foundations satisfactorily [8,9].

All local models have a three-layer structure to the depth of 21 m by analogy with the structure of typical filtration sections. However, the model implies the more exhaustive division of the water-saturated and semipermeable layers to set water-bearing utility systems at the depth of 3 m and piles having the length of 9 m from the surface. Thus, to simulate the conditions of interaction between a typical filtration section and a pile foundation, it is offered to use a five-layer numerical model where layers 1 and 2, 3 and 4 pairwise have similar filtration parameters, as they represent the same lithologic layer of the cross section (Figure1).

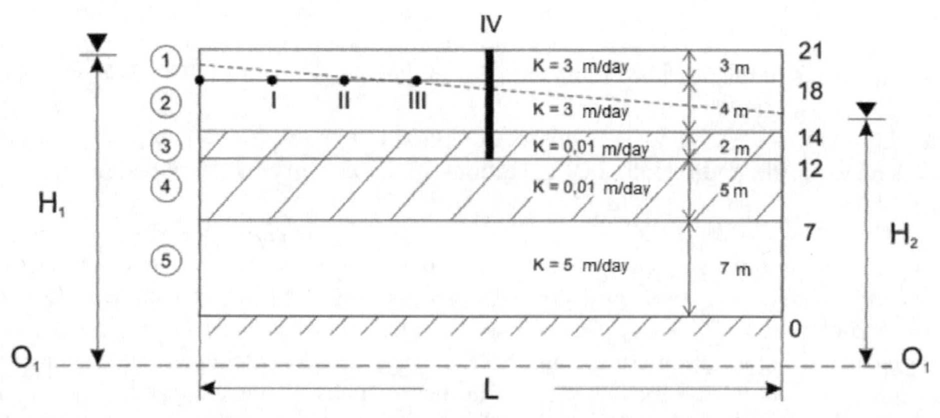

Figure 1. Boundary conditions in the numerical model of a typical filtration section:
1-5 – number of layers in the numerical model; I-III – configuration of the linear source of additional recharge at different distances from the pile field; IV – pile field acting as an impermeable groundwater cutoff wall; H1 and H2 – heads on the boundary of the filtration area; L – filtration area length (square numerical model size); ▼ – groundwater level.

Constant heads are set on the opposite boundaries of the filtration area, filtration heterogeneity is restricted by the different degrees of permeability of the layers, spatial heterogeneity is absent.

To register levels, we have set observation wells to ensure that there is a possibility to compare groundwater levels in different layers of the numerical model and in terms of the direction of a filtration flow (Figure 1). The observation wells are located on a single common beam pairwise in each layer of the numerical model. Their location is selected in such a way that two control points are before the filtration barrier on the side of the recharge area, and two more points of observation are behind the pile field on the side of the discharge area. Such an approach allows comparing groundwater levels in the conditions of natural and distorted filtration flows for similar points. In this case, it is noteworthy that the quantitative measure of anthropogenic impact might be the difference between the reference level surface of groundwater in the current simulation option and the basic surface taken as the natural field of heads obtained in the first simulation option without the impact of the damming effect.

Initial heads are determined on the model as the result of solving a stationary problem for the filtration area without regard to urban development (piles are absent).

The calculated values of the heads are used for the comparative assessment of the influence of the geometrical configuration of impermeable barriers on the structure of the filtration flow.

3. Hydrogeological conditions

To predict the development of anthropogenic temporary waters, we have schematized the upper portion of the cross section to the averaged depth of the area of interaction between aquifers and engineering systems and the geological environment. We have determined typical cross sections and mapped them in the area of the city [3, 4, 7]. In the area of the city, we have localized filtration areas with the single-, two- and three-layer structure of filtration sections that differently react to the supply of an additional infiltration recharge. The analysis of the spatial arrangement of these types of filtration sections has allowed zoning the territory of the city and qualitatively characterizing it in terms of the degree of potential waterlogging.

We have presented the materials of hydrogeological studies conducted in different years in the form of a multilayer electronic map. It gives a possibility to use them within the framework of comprehensive analysis with additional updated information. We have suggested a concept of a municipal geographic information system for Tomsk. It allows using the address of an object to determine the qualitative and quantitative characteristics of the hydrogeological environment (type of filtration section, additional infiltration recharge determined on the basis of the averaged value of per-capita water consumption for this portion of the territory, depth of temporary waters and groundwater). It allows promptly assessing the degree of the potential waterlogging of a specific object.

Forecasting the influence exerted by the pile foundations of buildings on the structure of filtration flows is essential for the territory of Tomsk. First, the analysis of hydrogeological conditions shows that the upper portion of the cross section composed by the bedded formation of semipermeable loose deposits tends to develop anthropogenic waterlogging in case of excessive additional infiltration recharge. Second, overmoistening is caused by the non-ideal conditions of the relatively old water-bearing utility systems. Third, in the conditions of the current urban development, the prevailing type of civil buildings foundation is pile foundation with the length of piles usually exceeding 9 meters that forms practically impermeable barriers to natural filtration flows. The maximum effect of this impact on groundwater is expected in transit zones.

The analysis of the developed areas of the city territory with high-rise modern buildings that drastically differ from the districts with private houses allows systematizing the planned configuration of pile foundations for the quantitative assessment of the damming effect. We suggest six main shape-based types of pile foundations in plan view in terms of their impact on natural filtration flows (Figure 1).

Due to the limited range of the typical shapes of pile foundations in plan view there is a whole family of computational models in case of different orientations of foundation contours in relation to the direction of the filtration flow (Figure 2). Their diversity grows with the joint inter-quarter arrangement of single-type or different-type foundations.

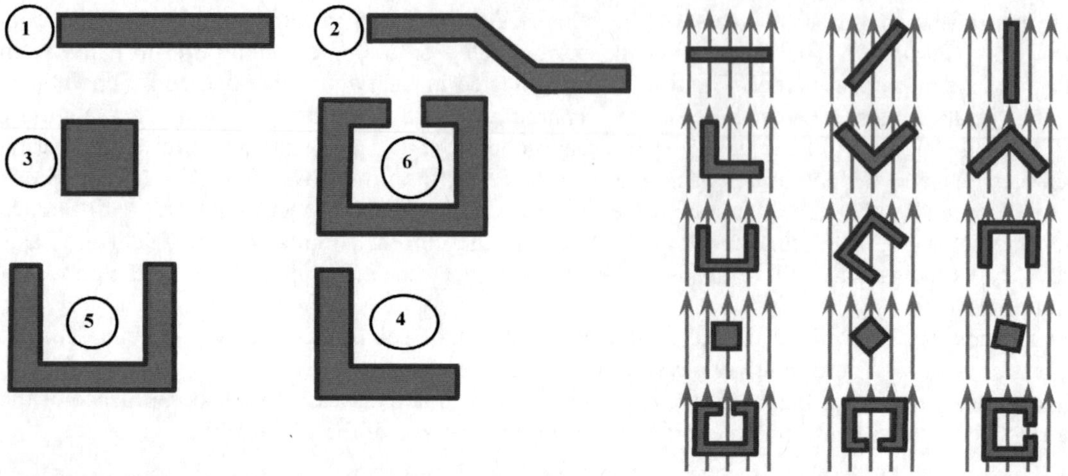

Figure 2. Shape-based types of pile foundations in plan view: 1 – linear; 2 – linear with peculiarities; 3 – isometric continuous; 4 – L-shaped; 5 – U-shaped; 6 – closed U-shaped foundation and their orientations vs. the direction of a natural filtration flow.

4. Results and discussion

For illustrative purposes, Figure 3 presents the results of the numerical simulation of the influence exerted by specific types of pile foundations.

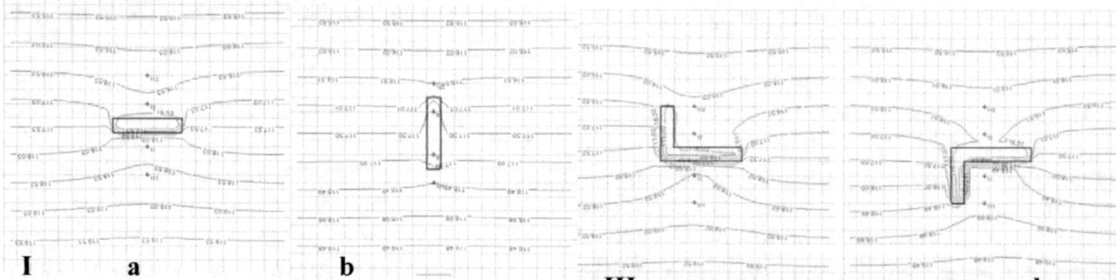

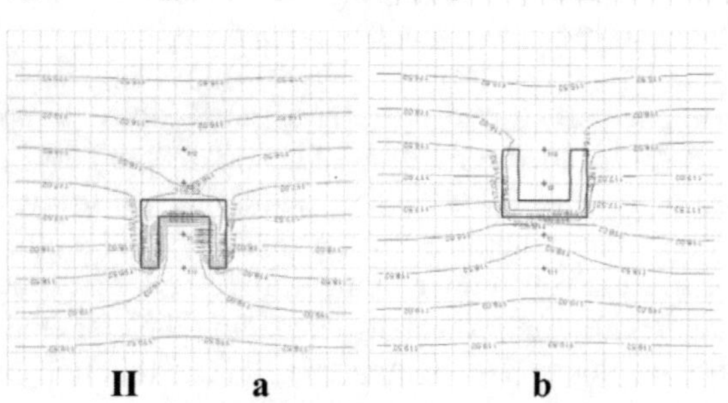

Figure 3. Solving stationary problems of assessing the impact of different types of pile foundations on the structure of a filtration flow: I – linear type (*a* – long side against the flow; *b* – short side against the flow; II – U-shaped foundation (*a* – protrusion against the flow; *b* – protrusion along the flow); III – L-shaped foundation (*a* – protrusion against the flow; *b* – protrusion along the flow).

To be definite, the model arrangement of the hydroisohypses is shown for the second layer of the model located above the surface of the separating semipermeable horizon.

It is possible to perform the quantitative assessment of damming effect manifestation by comparing the structural plans of hydroisohypses for a distorted filtration flow in different typical conditions or by comparing heads for observation wells (Table 1). To obtain more credible comparative assessment results, we have used the fixed spatial arrangement of observation wells for all simulation options, including the reproduction of a natural head field.

Table 1. Model values of heads in observation wells.

Well No.	No piles	Shaped-based types of pile foundations in plan view							
		linear		U-shaped		L-shaped			
		Length against the flow	Length along the flow	Protrusion against the flow	Protrusion along the flow	Protrusion along the flow	Protrusion against the flow	Vertex against the flow	Vertex along the flow
11	117.87	118.40	116.96	119.08	118.49	118.54	118.77	117.80	119.20
12	117.13	116.60	116.96	116.51	115.92	116.23	116.46	116.00	116.46
21	117.86	118.39	116.12	119.07	118.48	118.53	118.76	117.36	119.20
22	117.12	116.60	115.94	116.51	115.92	116.23	116.45	116.00	116.40
31	113.58	113.82	113.91	114.08	113.86	113.89	113.97	113.87	114.02
32	113.20	112.87	113.90	112.82	112.43	112.64	112.79	112.49	112.81
41	111.58	111.58	111.58	111.58	111.58	111.58	111.58	111.58	111.58
42	111.27	111.27	111.27	111.27	111.27	111.27	111.26	111.26	111.26
51	111.15	111.15	111.15	111.15	111.15	111.15	111.15	111.15	111.15
52	110.86	110.86	110.86	110.86	110.86	110.86	110.86	110.86	110.86
111	118.37	118.65	118.50	119.09	118.72	118.76	118.90	118.85	119.22
112	116.63	116.35	116.50	116.28	115.91	116.10	116.24	115.94	116.22
121	118.36	118.64	118.69	119.08	118.72	118.75	118.89	118.84	119.21
122	116.63	116.35	116.30	116.27	115.91	116.10	116.24	115.94	116.18
131	113.63	113.76	113.65	113.90	113.79	113.81	113.85	113.82	113.87
132	112.74	112.57	112.59	112.53	112.26	112.40	112.51	112.29	112.51
141	111.73	111.73	111.73	111.74	111.74	111.74	111.74	111.74	111.73
142	111.02	111.01	111.02	111.01	111.01	111.01	111.01	111.01	111.00
151	111.34	111.34	111.34	111.34	111.34	111.34	111.34	111.34	111.34
152	110.6	110.66	110.66	110.66	110.66	110.66	110.66	110.66	110.66

The results of the comparative assessment of groundwater levels for observation wells are given in Table 2.

Table 2. Stratigraphic throw of groundwater levels on different sides of a filtration barrier (for the second model layer).

Layer	Well No.	Shaped-based types of pile foundations in plan view							
		linear		U-shaped		L-shaped			
		Length against the flow	Length along the flow	Protrusion against the flow	Protrusion along the flow	Protrusio n along the flow	Protrusion against the flow	Vertex against the flow	Vertex along the flow
2	21	0.53	-1.75	1.20	0.62	0.67	0.89	-0.50	1.33
	22	-0.53	-1.18	-0.62	-1.20	-0.89	-0.67	-1.13	-0.73
	121	0.28	0.33	0.72	0.36	0.39	0.54	0.48	0.86
	122	-0.28	-0.33	-0.36	-0.72	-0.53	-0.39	-0.69	-0.45
Stratigraphic throw		1,06	1.06	0.66	1.82	1.82	1.56	1.56	1.61

Note: the color marking shows cells used for calculating the stratigraphic throw with due account for the pile foundation shape.

The simulation results allow making substantiated conclusions about the peculiarities and differences of changes in hydrogeological conditions that accompany construction of buildings on different types of pile foundations.

In the conditions of a three-layer filtration hydrogeological section, when the depth of the pile field does not exceed the bounds of semipermeable separating layer, no changes in the groundwater level occur in the lower aquifer. This fact is testified to by the absence of difference in heads in the conditions of natural and artificial flows for all types of pile foundations (wells Nos. 51, 52, 151, 152).

The peculiarity of damming effect is not only an increase in the groundwater level from the side of oncoming flow, but also a simultaneous drop in the levels behind the impermeable barrier vs. natural conditions (Table 2). Almost in all the cases under study, the amplitude of such changes is over one meter with the maximum value of over two meters. Such types as U-shaped and L-shaped foundations have the strongest impact on the filtration flow among all shapes of pile foundations under study. They can cause the growth in the level by 1.2 and 1.3 m, respectively, even in permeable rocks with the filtration coefficient of 3 m/day. The simulation results show that a predicted level rise can be decreased by almost 50% down to 0.62 m and 0.67 m, respectively, only by choosing the optimal orientation of the pile field vs. the direction of filtration flow (Table 2).

5. Conclusion

The simulation materials allow concluding that filtration barriers with a rectilinear axis in plan view have the minimum impact on the structure of filtration flow. The maximum growth in groundwater levels can be caused by filtration barriers with one and two changes in the direction of their longitudinal axis at the angle of 90 degrees forming a semi-closed shape.

These changes can only give a general idea about the possibilities of predicting changes in groundwater levels. We should note that we can significantly change the assessments of groundwater level rise by taking additional recharge into consideration. However, these examples are quite sufficient to substantiate the need for taking into account the regional direction of filtration flows at all stages of area planning for urban quarters.

References

[1] Pokrovsky V D, Dutova E M, Nikitenkov A N, Kuzevanov K I and Pokrovsky D S 2015 Methodological approaches to the assessment of the degree of drainability of the urbanized territories *Modern problems of science and education* 1 URL:http://www.science-education.ru/125-20130

[2] Pokrovskiy V, Pokrovskiy D, Dutova E and Nikitenkov A 2014 The research underflooding processes of architecture monuments on the territory of tomsk with using gis-technology. *Paper presented at the International Multidisciplinary Scientific GeoConference Surveying Geology and Mining Ecology Management, SGEM*, 935 – 941.

[3] Pokrovsky V, Pokrovsky D, Dutova E, Nikitenkov A and Nazarov A 2014 Degree of areal drainage assessment using digital elevation models *Paper presented at the IOP Conf. Ser.: Earth Environ. Sci.* **21**(1)

[4] Pokrovsky V, Dutova E, Kuzevanov K, Pokrovsky D and Nalivaiko N 2015 Hydrogeological Conditions Changes of Tomsk, Russia *IOP Conf. Ser.: Earth Environ. Sci.* 27 012031 (http://iopscience.iop.org/1755-1315/27/1/012031)

[5] Adam A M, Novoselov A L, Chepurnykh N 2000 *In. Environmental problems of Russian regions. Newscast. The Tomsk region.* No. 6

[6] Olhovchenkov V E 2015 *Hazardous natural and anthropogenic processes on the territory of Tomsk and their influence on the stability of natural-technical systems* (Tomsk) p141.

[7] Sologaiev V I 2002 *Filtrational calculations and computer modeling at protection against flooding in city construction: Monograph* (Omsk: Publishing house SibADI) p 416

[8] Nazarenko I A 2010 *GIS in hydrogeological studies* (on the example of JSC "AVTOVAZ"). *Geography and Geology. Hydrology and water resources*. SSC RF research Institute "VODGEO", Moscow. [Electronic resource]. URL: http://www.rusnauka.com/10_NPE_2010/Geographia/62643.doc.htm. free

[9] Bukaty M B 2002 *Development of software for solving hydrogeological problems* // News of TPU Geology prospecting and exploration of mineral resources of Siberia (Tomsk: Izd. TPU) T 305 **6** 348 – 365

Oil and gas company policy regarding the concept of sustainable development (water resources)

E G Matyugina [1], O V Pogharnitskaya [2], L S Grinkevich [3], D S Belozerova [4] and A B Strelnikova [5]

[1, 2, 4, 5] National Research Tomsk Polytechnic University, 30 Lenina Ave., Tomsk, 634050, Russia
[3] National Research Tomsk State University, 36 Lenina Ave., Tomsk, 634050, Russia

E-mail: [1] emk512542@mail.ru, [2] pov@tpu.ru

Abstract. The paper considers oil and gas companies implementing the strategy of sustainable development. Being vital to the national economy, oil and gas companies have a significant impact on the environment. Having analyzed the statistical data, the authors state that the leading Russian oil and gas companies contribute to the industry dynamics and conduct eco-friendly production practices. The environmental component is reported to be integrated in production, HR, information and other company policies, which results in "greening" both economic cooperation and place of production. The authors report the inverse relation between production dynamics and significance of the impact on water resources.

1. Introduction
Intensive development of the productive forces, limited resources, and carrying capacity insufficient to mitigate the human impact have boosted the choice and implementation of measures for sustainable development. Though the concept of sustainable development is assumed to support all national industries, this approach is particularly important for oil and gas producing regions due to the tradition of using fossil fuels in goods production.

2. Materials and Methods
Under sustainable development we mean a way of life which ensures expanded production, however, particular attention should be paid to the interaction between production and environment. Intensive development of Russian oil and gas companies shown in table 1 [1] makes them key factors for dynamic development of national economy, i.e. its sustainability in satisfying society needs.

Table 1. State and perspectives of leading national companies.

Company	Production (compared to the same period in 2013)	Development perspectives
Rosneft	Over the past 9 months of 2014: oil recovery – 159.9 million tons (+12.5%), gas recovery – 40.9 billion m^3 (+56.6%)	Production increase by 30% by 2020; twice as much by 2034.

Lukoil	Over the past 9 months of 2014: oil recovery – 72 million tons (+6.6%), gas recovery – 14.7 billion m³	The company is going to expand its business. through cooperation with Gazprom (2014–2024 General Agreement on Strategic Partnership).
Gazprom	oil recovery – 26.3 million tons; gas recovery – 320.3 billion m³	To increase the global market share, Gazprom is implementing Vladivostok LNG project (up to 15%) and constructing LNG production complex in Tomsk oblast with the capacity of 20–23 tons per year.
NOVATEK	Over the past 9 months of 2014: gas recovery – 45.86 billion m³ (+1.76%), oil recovery (including gas condensate) – 4275 million tons (+21.8%)	Government support (Yamal LNG).
Bashneft	Over 2014: oil recovery – 17.8 million tons (+10.8%)	Four GCF development licenses.

However, oil and gas complex has a negative environmental impact on water and land resources, as well as flora and fauna, at all production stages – geological and seismic survey, exploratory drilling, field facilities construction, field development and oil transportation, site restitution. Water resources contamination takes place throughout field development, production, and well abandonment stages. For instance, according to the research data on the fields of Tumen' oblast, there are 30 thousand exploratory wells (most of them drilled more than 15 years ago) collapsing due to long downtime and lack of maintenance, which, in its turn, results in seal failures and oil, gas and water showings. The total square area of oil-contaminated and saline lands and water surfaces is from 20 to 30 thousand ha [2].

Following the sustainable development strategy, oil and gas companies need to develop an environmental policy and then integrate it into the production one. However, environmental regulations of exploration and production are created earlier and become the requirements, which a company should meet to start a business. This indicates the prevalence of the environmental component over the production one. In compliance with the Federal Mineral Resources Act [3], the major requirements to sustainable use and conservation of mineral resources are "prevention of subsurface pollution during subsurface works, especially those connected with underground storage of oil, gas, and other substances and materials; hazardous and processing waste burial; waste water discharge, etc.". For example, NOVATEK conducted the environment research while constructing and commissioning the Ust-Luga Fractionation and Transshipment Complex processing stable gas condensate, with particular attention being paid to Luzhskaya Guba, the Gulf of Finland, and its water conservation zone [4].

The company environmental policy is not developed in isolation of production, quite the opposite, the environmental component is integrated into all activities performed. For example, in 2014, Gazprom conducted eco-friendly HR policy: 9012 employers of Gazprom Group attended environmental training courses (with 4669 persons trained for environmental management system (EMS)), including 7290 employers of OAO Gazprom (with 4580 persons trained for EMS) and 1306 employers of Gazprom Neft (with 54 persons trained for EMS) [8].

In order to meet environmental regulations, the companies not only improve current production, but also identify the strategic priorities (Table 2).

Table 2. Implementation of the sustainable development policy by national leading oil and gas companies [4–8].

Company	Activities
NOVATEK	In 2013, Bureau Veritas Certification audited EMS, industrial safety, occupational safety and health management system to determine whether they conform to the international standards ISO 14001:2004, OHSAS 18001:2007. In 2013, the costs for water resources utilization and conservation were 26436

	thousand rubles (145,3% increase in comparison with 2012).
Gazprom	The Company declared 2014 to be the Year of Environmental Culture and Industrial Safety. According to certification audit of 2013, the integrated EMS complies with the international standard ISO 14001:2004. Waste water collecting and treatment account for 26.1%; the company conducted campaigns on water reclamation to remove residential waste and submerged trees.
Bashneft	Circulation water supply system and water recycling are operational in the oi fields; at the petrochemical plants, after two-stage filtration, about 65% of water is returned into the technological process. The water-body management system includes 1391 water monitoring centers, with 70% monitoring the surface water quality and 30% – underground fresh water. There is also a large-scale project aimed to modernize the biological treatment plants of the Bashneft-Ufaneftekhim branch.
Lukoil	The program for 2014–2018 presumes additional treatment of 6.8 billion m³ of sewage water and a decrease in water consumption by 7.6 billion m³. To ensure a sustainable use of water resources and prevent water contamination, the company takes actions as follows: pipeline failures prevention and repair, constructing the system for produced water disposal; pipeline water crossing inspection and maintenance; slop water treatment system, etc.

Eco-friendly policy conducted by oil and gas companies stipulates establishment of special departments and services responsible for environmental monitoring, control, and implementation of the required production systems (table 3) [8,9,10].

Table 3. Changes in company structure.

Company	Activities
Gazprom	Establishment of environmental inspectorate responsible for meeting environmental regulations, adoption of environmental content for corporate code of conduct; EMS internal audit; meeting corporate requirements to sustainable use of mineral resources; cooperation with the government environmental protection agencies to improve company environmental activities. In 2014, there were 597 inspections in 54 company subsidiaries: 411 – scheduled inspections examining for the compliance with environmental regulations, including 135 EMS audits, and 186 themes inspections examining for the compliance with environmental regulations regarding capital construction .
Lukoil	EMS has been implemented for offshore field construction. In relation to the offshore field located in the Baltic sea, hydrodynamic data have been studied, bottom flows have been measured, etc.; special types of ships are involved in environmental monitoring; two double-bottomed observatories, a hydrophysical survey station, two hydrometeostations and three weather stations are operational on the offshore platform, at Kurshskaya Kosa, in Klaipeda, Baltiysk and Pionersky. It is for the first time when the company established an autonomous seismic station in this region.
Rosneft	The department of environmental safety and technologies is responsible for environmental risks assessment, internal environmental audit and project expertise. The department is in charge of supporting the program on Environmental Protection and Biodiversity Preservation in Exploration and Production of Mineral Resources in the Arctic Continental Shelf of the Russian Federation.
Bashneft	PJSOK Bashneft has in place HSE policy and is currently implementing the group's HSE strategy. The Group has implemented a Standard on Cooperation with Contractors in the Sphere of Health, Safety and Environment. Compliance of potential contractors with environmental requirements set out in the Standard is a necessary prerequisite for entering into

IOP Conf. Series: Earth and Environmental Science **33** (2016) 012055 doi:10.1088/1755-1315/33/1/012055

	contracts with them (organization of waste collection, storage and accumulation sites and the transfer of waste for recycling; making payments for negative environmental impact; compliance of water consumption and discharge processes with environmental legislation).
NOVATEK	The company has implemented HSE policy and operates an Integrated Health, Safety and Environment Management System (IMS). Before carrying out activities involving environmental impact, the company, through media, informs the society about the data on environmental risks assessment, hold public consultations, establishes community liaison offices, present the information on the company's website. The company representatives are among the members of the consulting group contributing to research and conservation of the Atlantic walrus in the Barents sea south-east and contiguous zones; they also make a part of the study team considering environmental impact of Sabetta dockyard construction, Obskaya guba.

It is noteworthy that the above mentioned companies improving their production regarding environmental safety and sustainable development set environmental requirements for their partner (for example, contractors), inform the society about an environmental impact (actual or anticipated), and consider public opinion. Firstly, this indicates "greening" not only economic cooperation, but also the place of production, since all participants are involved in this process. Secondly, this defines the external environmental policy (including the information one). Moreover, making project solutions with due regard to environmental risk assessment signifies "greening" investment policy.

The companies currently promote environment-conscious policies, which leads to the fact that production growth and environmental impact are associated with opposite dynamics (for example, Gazprom and Lukoil data [7, 8, 11], Figure 1).

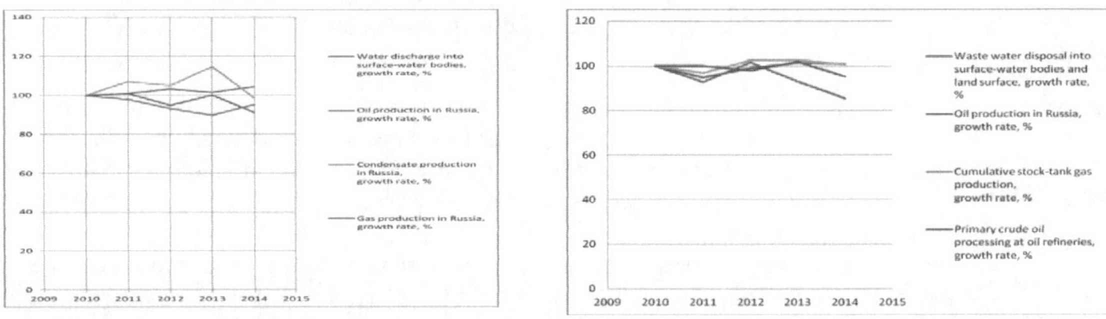

1 2

Figure1. Comparison of business capacity and water resources use track records (1 – Gazprom data, 2 – Lukoil data).

As for OAO Gazprom, water discharge into surface-water bodies decreased (by 22 % over the study period) due to shrinkage of water required for process needs in the companies of Gazprom Energy Holding. Moreover, Gazprom Group took actions on sustainable use of water resources and preventing water contamination. If we turn to Lukoil, the drop in waste water disposal into surface-water bodies (25 % over the study period) results from high-tech and resource-efficient equipment implementation at the sites. Table 4 gives more information on water resources use.

Table 4. Comparison of business capacity and water resources use track records.

	2010	2011	2012	2013	2014
			Lukoil		
Waste water disposal into surface-water bodies and land surface, million m^3	354.9	329.3	334.6	311.1	265.5
including meeting quality standard, million m^3/	326.1/ 91.9	296.7/ 90.1	302.2/ 90.3	281.9/ 90.6	243.0/ 91.5

percentage of total amount, %					
Oil production, thousand tons	89 767	84 966	84 234	85 481	86 328
Cumulative stock-tank gas production, million m^3	13 599	13 174	13 553	13 916	14 031
Primary crude oil processing at oil refineries, million tons	45.21	45.28	44.43	45.25	45.28
Gazprom					
Water obtained, million m^3/ growth rate, %	6 259.00/ 100	5 793.00 / 92.6	5 462.45/ 94.3	5 130.18/ 93.9	4 895.38/ 95.4
including water from natural sources, million m^3/ / percentage of total amount, %	6 015.73/ 96.1	5 572.42 96.2	5 212.95/ 95.4	4 890.63/ 95.33	4 410.68/ 90.1
Water used for production, million m^3/ percentage of total amount, %	5 982.12/ 95.6	5 550.79/ 95.8	5 209.31/ 95.4	4 919.51/ 95.9	4 506.18/ 92.0
Water discharge into surface-water bodies, million m^3	5 364.05	5 257.71	4 892.96	4 389.91	4 179.09
including meeting and treated to quality standard, million m^3/ percentage of total amount, %	5 321.36/ 99.2	5 096.23/ 96.9	4 691.55/ 95.9	4 227.86/ 96.3	3 991.59/ 95.5
Gaz production in Russia, billion m^3	508.6	513.2	487.0	487.4	443.9
Condensate production in Russia, million tons	11.3	12.1	12.8	14.7	14.5
Oil production in Russia, million tons	32.0	32.3	33.3	33.8	35.3

According to the data presented, the companies are reducing their water use with the production growth; also, the percentage of water meeting the quality standards and water treated is almost 100 %.

3. Conclusion

In conclusion, the oil and gas companies lay foundation for sustainable development of national economy, however, this does not result from their particular status of energy suppliers but is due to the consistent measures taken to mitigate the negative environmental impact. The environmental policy being integrated in company management system underlies all company activities (HR, investment, information, etc.), stipulates "greening" economic cooperation, involving contractors, public, and local authorities.

References

[1] *Development of oil and gas complex in Russia over 2014 (as of 02.02.2015)* 2015 (Moscow, National Rating Agency) p 37 URL:http://www.ra-national.ru/sites /default/files/other/neftegaz_2014.pdf.

[2] Soromotin A V 2005 Environmental issues of oil and gas production in West Siberia *UT Research Journal* **3** 137–145

[3] *The Federal law "On subsoil" dated 21.02.1992, № 2395-1*. ConsultantPlus. URL:https://www.consultant.ru/popular/nedr/66_1.html
[4] *OAO NOVATEK Sustainability report 2010–2013* [official website] URL: http://www.novatek.ru/
[5] *OAO Gazprom Sustainability report 2012–2013* [official website] URL: http://www.gazprom.ru/
[6] *PJSOC Bashneft Sustainability report 2010–2014* [official website] URL: http://www.bashneft.ru/
[7] *OAO Lukoil Sustainability report 2014* [official website] URL: http://www.lukoil.ru/
[8] *OAO Gazprom Environmental report 2014* [official website] URL: http://www.gazprom.ru/
[9] Tsibulnikova M R, Kupriyanova O S and Strelnikova A B 2015 Economic assessment of environmental impact in the course of oil field development and production *IOP Conf. Ser.: Earth Environ. Sci.* **27** 012076
[10] Matugina E G, Glyzina T S, Kolbysheva Yu V, Klyuchnikov A S, Vusovich O V. Compensation for oil pollution damage *IOP Conf. Ser.: Earth Environ. Sci.* **27** 012070
[11] Khattu A A, Solodovnikov A Yu 2015 Hydrochemical survey of new zone surface waters as an environmental indicator *J. Oil industry* **4** 104–108

Performance assessment of groundwater drainage basin in small-sized mountain rivers based on DTM

A Nikitenkov[1], E Dutova[2] and D Pokrovsky[3]

[1,2] Tomsk Polytechnic University, Tomsk, Russia
[3] Tomsk State University of Architecture and Building, Tomsk, Russia

E-mail:[1]corestone@mail.ru, [2]dutova@sibmail.com, [3]dsp@sibmail.com

Abstract. The paper describes the performance assessment methods for groundwater drainage basin based on the analysis of morphometric indices of small-sized mountain river drainage basins. DTM was applied to analyze morphometric indices of drainage basins.

1. Introduction

Performance assessment of groundwater drainage basin is commonly used to solving the problems associated with mineral resource management: water supply provision due to groundwater utilization, water inflow calculations in mining, water-base ecological problem- solving. Existing drainage assessment methods are predominately based on either hydrometric data analysis obtained from systematic observations (methods based on the separation of runoff hydrographs) or synchronous measurements of river drainage conducted during independent groundwater regimes, i.e. when the river drainage develops due to separation of groundwater flow [4]. Obtaining such data within field experiments requires significant material expenses. However, theoretical approach in estimating the groundwater drainage values has been attracting the attention of numerous scientists [[3], [5]]. In this respect, the methods assessing groundwater drainage is based on mapping involving minimum volume of field experimental data. Today this is of specific interest.

The investigated area is located in northern Kuznetsk Alatau. The authors proposed a method in studying the mountain groundwater drainage basins within this area. Such areas are considered to be a problematic site due to poor hydrogeological exploration.

Solution of this problem involves factual material harvesting, performance assessment of groundwater drainage, analysis identifying their environmental factors and morphometric dependency via DTM.

2. Materials and methods

Factual data, being background experimental data, includes meteorological data and information on average monthly river discharge from 10 hydrometric stations (Roshydromet) within an observation period (1030-2014), as well as terrain altitude survey data produced within the framework of SRTM program [6].

The investigation involves two basic trends- hydrological and morphometric, which, in its turn, embraces data analysis phase and presentation of obtained results.

Hydrogeological investigation included spatial water balance assessment (spatial variability of rainfall and evaporation), as well as assessing drainage via the method of separation of runoff hydrographs.

The location of hydrometric stations and basic water balance elements are depicted in Figure 1. The water balance assessment of major drainage basins within this area was based on the rainfall-evaporation data from Roshydromet.

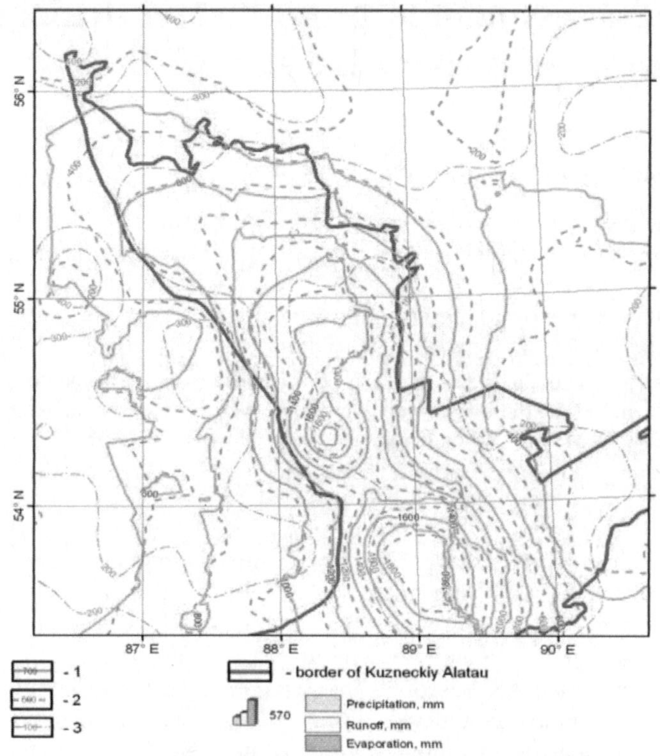

1 – precipitation; 2 – runoff; 3 – evaporation.

Figure 1. Map of water balance elements.

Rainfall supplying groundwater recharge is one of the basic factors in generating groundwater runoff and their variability within the investigated area ranges from 400 to 1800mm/year. The rainfall variability within this area is distinguished by pronounced vertical zonation, characterizing regularly upbuilding of layers with increasing absolute relief elevation. Low rainfall occurs in areas with absolute elevation of more than 700m. Atmosphere circulation influences the distribution of rainfall throughout the territory. Local distribution of rainfall volume is predominately determined by the landscape features. Slope exposure is of significant importance in this process, especially in the winter months.

Evaporation plays a prominent role in the water balance of this area. Based on Roshydrometa observation network data it has been established that the investigation area has one and the same specific feature typical for all mountain areas-regular evaporation discharge distribution pattern i.e. decreasing discharge to height. It varies from 200-300mm/year in basins and foothills to 50-100mm/year in investigated mountain area; whereas the most significant evaporation discharge decrease (averagely, up to 100mm/year) is observed in the height interval from 200-300 to 1000m.

Separation of runoff hydrograph method for rivers is based on the classical approach proposed by B.I. Kulelin [4]. Separation of runoff hydrograph differentiates surface and underground gathering zones. However, based on the different research data [3], the groundwater recharge in significantly large rivers embraces ground and deep water zones (where ground water runoff increases due to the inflow of deep groundwater). Practically, this method is implemented by Visual Basic for Application. A more detailed description of runoff rate assessment can be found in previous research papers [1].

Morphometric survey based on DTM data includes data acquisition and further analysis of relevant characteristic features of river networks and drainage basins. Obtained numerical morphometric parameter values are applied in identifying the characteristic groundwater runoff- numerical relief characteristic relation.

GIS Ilwis [10] was applied in the investigated area. The results involved morphometric parameter values of rivers and drainage basins, including 12 thousand river drainage divide ratios within the investigated area. Based on this data it was possible to conduct an integrated analysis of the morphometric river network parameter values, plotting datum surface and other morphometric survey.

Data analysis implementation of river valleys and river drainage basins and determination of their morphometric parameters was performed via GIS ArcGIS 9.3 and Ilwis 3.5 on digital topographic base and comparable to map scale 1:50000.

Morphometric drainage basin parameters were obtained by standard hydrological analysis methods of DTM within the framework of Ilwis. The following morphometric parameters were obtained- area size and perimeter, minimum and maximum relief elevation within drainage basin, above and average basin altitude, as well as length of major streams and their slope. All in all, more than 2500 small river drainage basins located within 10 base drainage basins were described. Detailed morphometric survey was described in [2]. The final investigation stage (after analyzing climatic, morphometic and hydrogeologic data) included statistical analysis to identify the significant factors determining the formation of groundwater runoff within the investigated area.

Based on the multiple linear regression (MLR) method a simulation model was designed which numerically describes the relation of the modification of groundwater runoff rate to total relevant governing factors.

3. Results and Discussion

The major morphometric factors determining the hydrogeological features of the investigated area are the geomorphological conditions and hyposographical location of river drainage basins. The most distinctive one is the total and surface runoff rate – hyposographical elevation of drainage basin relation. Hydrogeologically, the role of rivers in the investigated area is only the draining of different igneous, sedimentary and metamorphic rocks of various ages.

Numerous domestic and foreign scientist have also described similar approaches in identifying the interrelation between drainage basin parameters with river runoff parameters [7, 8, 9]. We as authors, propose a more detailed and integrated approach in assessing the factors determining the formation of groundwater runoff.

Interrelated bound between the relief parameters and modification of flow has been determined by correlation analysis (Table 1). All modification parameters of flow show a positive relation to datum surface slope, while negative – drainage basin area.

Table 1. Pair correlation coefficients: modification of flow to morphometric drainage basin parameters.

Drainage module	Area	Perimeter	Drain length	Relief elevation			Draining zone capacity	Slope
				Minimum	Maximum	Average		
$M_{riv.}$	**–0.61**	–0.50	–0.46	**–0.75**	–0.23	**–0.52**	–0.06	**0.65**
M_{surf}	**–0.60**	–0.49	–0.46	**–0.79**	–0.26	**–0.58**	–0.08	**0.61**
$M_{und.}$	**–0.58**	–0.49	–0.45	**–0.52**	–0.07	–0.23	0.06	**0.76**
$M_{deep\ und.}$	**–0.53**	–0.47	–0.48	–0.41	–0.16	–0.20	–0.07	**0.71**

IOP Conf. Series: Earth and Environmental Science **33** (2016) 012056 doi:10.1088/1755-1315/33/1/012056

Modification of groundwater runoff rate shows the most close relation to the slopes (pair correlation coefficients 0.71–0.76).

$M_{riv.}$–river runoff module; M_{surf} - surface river runoff component module; $M_{und.}$– underground river runoff component module; $M_{deep\ und.}$ – deep fraction in underground river runoff component module. Significant correlation coefficients were identified for 0.10 significance level.

Parameters characterizing the formation of underground run-off with the involvement of multiple linear regression (MLR) method could be combined into one equation. A series of similar regression equations, describing the underground run-off dependency in different draining levels to governing morphometric parameters was obtained on the basis of data from detailed investigated areas (Tsentralninsk granitoid massif). An extended equation, including both morphometric features of the area and climatic and hydrogeological factors was applied in the case of Kuznetsk Altai:

$M = -2.24 - 3.26PDOL + 0.000113DLP - 0.00128S - 0.00804H_{max} - 0.00721dH + 34.0UKL + 0.00582Os$, where PDOL – valley sequence; DLP – drain length m.; Hmin – height of emerging spring, m; Hmax – height of accumulation area, m; dH – thickness of draining zone, m; HIR – width of stream, m; UKL – slope; S – area of drainage basin, km^2; Os – rainfall mm., at R2=0.71; N=46

Based on obtained dependency detailed areal assessment of subsurface water flow rate was conducted within the limits of elementary drainage basins by morphometric parameter substitution. In this case groundwater run-off level for investigated areas could be assessed, as well those hydrometric stations which have not been included.

Based on the generalization of obtained values, numerical run-off parameters were calculated, which could characterize the water exchange intensity at different draining levels of groundwater by relevant river ordering. Based on obtained data of subsurface water flow rate by IDW method and ArcGIS the following map was plotted, reflecting the distribution of subsurface water flow rate (Figure 2).

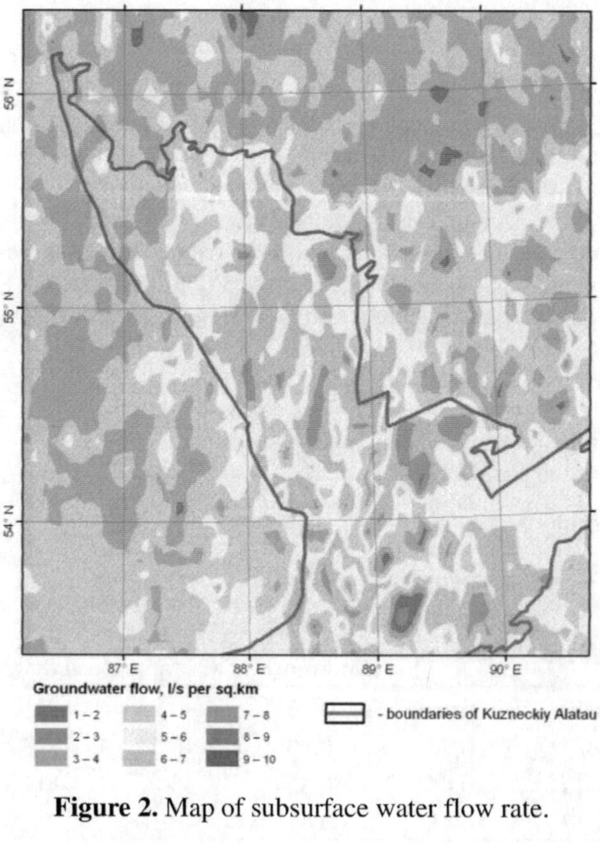

Figure 2. Map of subsurface water flow rate.

IOP Conf. Series: Earth and Environmental Science **33** (2016) 012056 doi:10.1088/1755-1315/33/1/012056

Obtained results show the changes of subsurface water flow rate to drain depth: from valley of 2nd ordering to valley of 6th ordering the water exchange slows down characterizing decreasing run-off rate from 12 liter/sec.·km^2 to 5-6 liter/sec.·km^2 within the most elevated areas of investigated territory.

4. Conclusions

The main investigation results are the integrated methodological approaches in obtaining numerical parameters of groundwater run-off for small-sized river drainage basins of mountain regions. The investigation methods are based on an integrated approach applied in classical, hydrological and hydrogeological surveys in combination with up-dated processing equipment and analysis of spatial geodata. As a result of data processing map documents were compiled providing a detailed description of the spatial distribution of different morphometric parameters within the investigated area and further analysis of the quantitative distribution of groundwater resources.

The proposed method could be the base for detailed investigation of the quality and quantity parameters of water resources in mountain regions. This, in its turn, could optimize water resources management and water consumption, predict the possible quality changes of water resources after different technogenic effects and further hydrogeochemical exoloration of mineral deposits. Investigation results are illustrated as map material.

References

[1] Nikitenkov A and Dutova E 2010 Rechnoj stok i morfometricheskie parametry vodosborov severnoj chasti Kuzneckogo Alatau *Izvestiya Tomsk Polytechnic University* **316** (1) 142 – 147

[2] Nikitenko A, Dutova E and Pokrovski D 2013 Kartograficheskie postroenija i ocenka morfometricheskih parametrov vodosborov gorno-skladchatyh territorij po dannym sputnikovoj s"emki (SRTM) (na primere severnoj chasti Kuzneckogo Alatau) *Vestnik Tomsk Architecture-Building University* **1** 223 – 231

[3] Komlev A 2002 *Zakonomernosti formirovanija i metody raschetov rechnogo stoka* (Pub. House of Perm University) p 163

[4] Kudelin B 1960 *Principy regional'noj ocenki estestvennyh resursov podzemnyh vod (*Moscow MSU) p 344

[5] Kirukhin V and Shvets 2010 Podzemnyj stok – kljuchevaja problema regional'noj gidrogeologii *University Izvestiya Geology and exploration* **5** 42 – 47

[6] Pokrovsky V D and et al. 2014 Degree of Areal Drainage Assessment Using Digital Elevation Models *IOP Conf. Ser.: Earth Environ. Sci.* **21** [Electronic resource]

[7] *Mapping water drainage with SRTM elevation data in Python* URL: http://stevendkay.wordpress.com/2009/09/18/mapping-water-drainage-with-srtm-elevation-data-in-python Accessed date 15.05.2015

[8] James Hamski et al. *Estimation of Hydraulic Parameters and Discharge of the Brahmaputra River from the Shuttle Radar Topography Mission Digital Elevation Model* URL: http://bprc.osu.edu/water/presentations/ASLO_Hamski_Poster.pdf Accessed date 15.05.2010

[9] *The Han River watershed management initiative for the South-to-North Water Transfer project (Middle Route) of China* URL: http://www.springerlink.com/content/g7227j6277346g4n/

[10] *ILWIS 3 TRAINING COURSE FOR USER-FRIENDLY GIS* – Chapter 10. Using Digital Elevation Models URL: http://www.birdlist.org/downloads/ilwis/ilwis3_chap10.pdf Accessed date 15.05.2015

Ecologization of water-land property matters on the territory of the Tom lower course

V K Popov[1], M V Kozina[2], Yu Yu Levak[3] and E V Shvagrukova[3]

[1, 2, 3] National Research Tomsk Polytechnic University, 634050, Lenin Ave., 30, Tomsk, Russia

E-mail: [1]pvk@tpu.ru, [2]marijamkozina@gmail.com

Abstract. In the present paper the water-land property complex is considered as a strategic resource of the city development. The formulated question is expounded through the example of water-land property complex usage on the territory of the Tom lower course for land-use planning and developing the systems of water recourses management and land tenure. Consequences of liquid radioactive waste (LRW) landfilling are investigated in terms of arable farming. Also, forming a water budget of the soils spread on the area of the Tomsk underground water supply cone of depression and its role in the development of agricultural industry are studied. The main aspect of the analysis is the incorporation of social, economic, and ecological requirements for the system of life-supporting branches of municipal economy and social services. As far as the system of land tax payments plays an important role in land property complex management, the common issues and tendencies are specified in the paper. These problems are concerned with the inadequate incorporation of an ecological constituent in the methods of cadastral valuation of lands, as well as the situation of the narrow area of its results usage in the Russian Federation. Natural factors (hydrological, territorial, geological (geomorphologic) territory conditions) are combined by the authors into a special group. These factors should be reflected in the results of cadastral valuation. Also, in order to protect the interests of water consumers, it is offered to establish the Water Consumers Association based on the international experience of such countries as Spain and Uzbekistan.

1. Introduction

In the middle of the current decade the transition toward a market economy has completed in Russia. The transition of the Russian economy from an export raw type to an innovative social-oriented type of development is connected with the formation of a new mechanism for social development based on the balanced social equity and national competitiveness. This approach requires implementing the complex of interrelated transformations.

The changes in the approach to regional planning (formation of the integrated system of strategic and operative planning at all levels) belong also to the above-mentioned type of transformations.

Priority guidelines of strategic policy are as follows: development of life-supporting branches of municipal economy and social services; development of land and property markets, involving simplification of land rehabilitation procedures; formation of the institute of mass real property valuation and further administration of real property full-value taxation.

Implementation of priority guidelines in municipal structures is influenced not only by the presence of raw and financial resources, but also by the level of land resources reactivation and their infrastructure supportability. Therefore, one of the main aims of land policy in the city is formulated as an efficient land and property use, which is able to provide a breakthrough in solving social and economic problems of the city.

2. Research object

It is possible to consider the stated issue through the example of the urbanized water-collecting area of the Tom lower course consisting of Tomsk, Seversk, Tomsk and Shegarsky regions lands.

Development of the analyzed urbanized territory depends, primarily, on the quality of land resources and anthropogenic activities operating upon it.

A distinguished feature of the considered urbanized territory is its location. The territory is situated in the South-East part of the West Siberian artesian basin, where all aquifer systems are in the condition of intensive water exchange, and they contain infiltration waters having a similar hydrogeochemical profile [1, 2]. Only in the northern part of an interstream area the waters of chalk and Paleozoic deposits are identified in the zone of delayed water exchange. 45 settlements are situated on the territory of the Ob-Tom interstream area: 38 settlements come under 6 administrative country districts of Tomsk region, 5 of them form a part of Shegarsky region and 1 is a part of Kozhevnikovsky region. Besides, Nizhniy Sklad settlement goes for Tomsk city boundary, administratively. The considered territory develops intensively. All the main processes of human life activities are concentrated there as well as the objects of land and property complex with various functions, which need to be in the ecological balance with nature in order to increase the life quality of population.

Engineering systems of water resources utilization and military-industrial complexes are operated jointly on this territory (Figure 1).

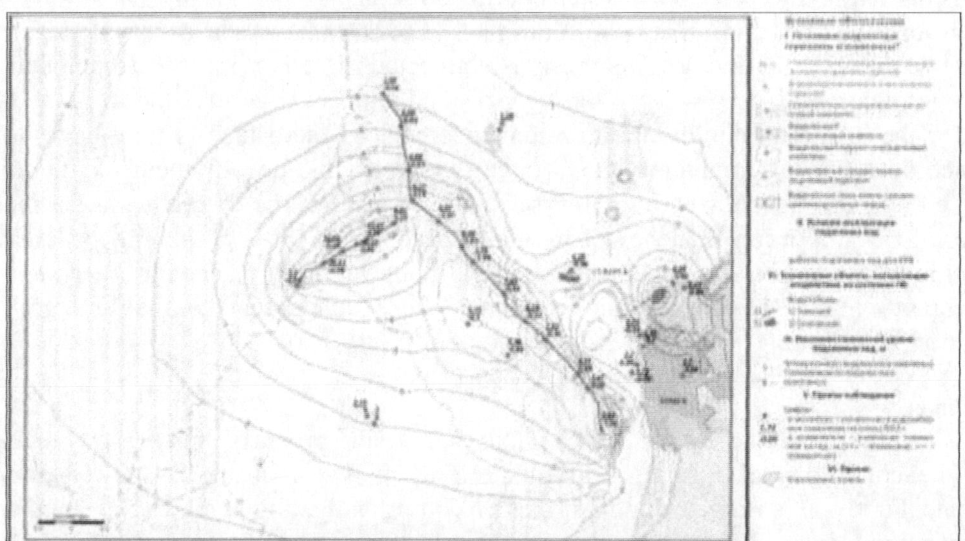

Figure 1. Schematic map of ground waters operational conditions in the influence zone of Tomsk water supply of ground waters (LLC «Tomskvodokanal») in 2013.

Underground water supply facilities complex occupies the territory equal to 5 Ha. There are 198 wells in operation, 93 of them are used in full-time service, and others are suspended. Well depth ranges from 80 to 198 meters, average daily water transport in the city is 136 000 cubic meters. In the city water is transported via two water pipelines. Tomsk water supply of ground waters consists of 2 linear series of water wells. Subsurface sites are situated on the territory of Tomsk region and Tomsk itself.

3. Problems

The interests of four main water consumers (Siberian Integrated Chemical Plant, Tomsk municipal enterprise of water supply, Seversk municipal enterprise of water supply and the general public) have come into conflict on the examined territory. As a result of these four water consumers' activities, two acute problems have arisen. The first problem is as follows: formation of a depression cone as a result of long water supply operation on the Ob-Tom interstream area and formation of a depression cone on the territory of Seversk water supply. For such cities as Tomsk, Seversk and surrounding rural areas the most important negative effects are hydrodynamic and hydrochemical changes in the characteristics of hydrosphere, phreatic decline, groundwater depletion and ground water pollution, changes of ground water run-off, occurring as a result of land settlement and transformation of natural landscape.

On the one hand, development of the depression cone causes water quality deterioration as a result of water exchange intensification due to the rates increase in inter-reservoir communication and waste water taking up [1]. On the other hand, soil drainage occurs, and, consequently, cropping capacity decreases requiring additional costs for irrigation and land improvement. As a result, agricultural complex bears losses as a missed profit.

The second problem is a subsequence of the first one: the waters of chalk (cretaceous) complex, which come in contact with buried liquid radioactive waste (LRW), may transfer into paleogenic complex. As a result of complicated hydrodynamics, the waters of chalk deposits discharge into paleogenic water-bearing stratum.

Therefore, we may assert that the joint impact of various factors is always stronger than the separate impact of each factor. This thesis is connected directly with the arrangement of land and property complex usage on the territory of the Tom lower course due to its multi-aspect value for the local community as far as enhancement of these factors may have a negative influence on the local population. Consequently, sustainable development of the considered urbanized territory depends, to a greater or lesser degree, on the correctness of a managerial decision. Life-supporting branches of municipal economy, particularly, the system of water consumption can not be developed separately without taking into account the system of land and property matters, economy and ecology. These constituents are an integral part of population's well-being.

4. Methods

The basis of the modern methods in the sphere of land and property matters state management under the conditions of forming and functioning market industry is the system of land payments. Firstly, it is implemented as taxes providing budgeting at all levels. Regulation of a land-value tax rate allows government directing, stimulating or, otherwise, hindering development of separate branches and territories. Land payments are very big and take the second place in the structure of own financial returns of business entities. As of 2013 they include 17 % - 35 % of all city returns in Tomsk [3]. The results of government cadastral valuation of lands are used as a taxation base in Russia connected with a land-value tax and other payments for land. These results draw on the classification of lands according to their designated purpose.

5. Conclusion

Using as an example the cadastral valuation of Tomsk and Tomsk region lands, it is possible to make the following conclusion: the modern methods of cadastral valuation for land payments rating are imperfect. The main reason for a current situation lies in the initial choice of incorrect methodology: land taxation base is established by the authorities, who are responsible for the development of the given direction. International and Russian practices connected directly with land resources valuation are insufficient, what generates a need for detailed scientific and methodical elaboration of single elements in the above-mentioned valuation procedure. First of all, it is concerned with the development of basic parameters for valuation of subject property. These parameters should reflect a natural specificity of resources forming and the pattern of their utilization. Nowadays, land valuation is used only as a fiscal instrument in the land policy of the government, but valuation functions should not be limited by it. In order to make any managerial decision we need to have reliable data about one or another object and its degree of impact, i.e. functional valuation of territories is required, the aim of which is long-term social and economic well-being with a minimal environmental damage in relation to territories. This approach, in its turn, influences the environment and life quality of population and ecological balance with nature. Within the boundaries of the Tom lower course ecofriendly metabolic processes should be combined with functional stability of operating engineering systems for water resource utilization and military-industrial complexes, as well as for water-land property matters.

Taking into account the fact that the results of government cadastral land valuation (GCLV) influence the volume of payments for corporate and natural entities for the right to use the land, as well as on the revenue part of budget, in order to respect the interests of land proprietors and exclude the possible facts of budget mistreatment the following solution is offered. Such a group of factors as natural ones (especially, a water ecological factor), including hydrological features of the territory (ground water dynamics, presence of a depression cone), geological (geomorphologic) territory conditions, should be reflected in the GCLV results. In particular, these factors determine the size of missed profit speaking about agriculture, construction appreciation, repair costs at capital facilities operation, engineering and transport infrastructure. Also, they appear in increase or decrease in real property costs, the example of which may be found in carrying out valuation in Tomsk and Tomsk region, what, in its turn, influences the decrease in investment prospects in particular territories.

Problems of the underground water supply situated on the urbanized territory of the Tom lower course have already fallen beyond the scope of effortless speculations. It is necessary to make a decision on applying to these problems, but this issue is complicated by the absence of a unified managerial authority. Different interests of water consumers may be united with the help of establishing Water Consumers Association. This model of water resource management has been used in the USA (California) since 1960, in Spain – since 1985, in Uzbekistan – since 2002. Practices of these countries demonstrate that organizations of water consumers themselves allow making the most efficient decisions in water resource management, because they know the root of the problem firsthand. At establishing the organization of this type, it is possible to obtain constructive relations of water consumers in the task of water resource management. In order this Association to be successful in Russia, we need to start with forming environmental world outlook among population, because a great number of water resources has led to the declension of moral upbringing. All the problems connected with water are perceived as a natural inevitable phenomenon. But that is not good enough. It is necessary to assimilate the standards of other countries in the system «water-community», which are successful at using the managerial model of Water Consumers Association. Association should take into consideration the interests of all water consumers. Its policy should not be temporary and it should not depend on a current situation. Its establishment is required for environment forecasting and problems elimination. In general, this organization should think forward and it should not be commercial, otherwise, it will lead to the development of corrupt practices. To provide the development of Water Consumers Association without breaking Legislation of the Russian Federation, special legislative and statutory acts need to be drafted, which regulate the activity of Water Consumers Association. Activity and services should not be taxed, because all the cash flow must be directed to maintenance of water supply systems and, if necessary, to accident response and water systems reconstruction.

References

[1] Popov V K, O D Lukashevich and V A Korobkin et al. 2003 *Ecological and economic aspects of ground waters operation in the Ob'-Tomsk interstream* (Tomsk: TGASU) p 174

[2] Savichev O G, Tokarenko O G, Pasechnik E Yu et al. Microbiological composition of river waters in the Ob' basin (West Siberia) and its connection with hydrochemical indices *IOP Conf. Ser.: Earth Environ. Sci.* **27** 12035

[3] Popov V K and Kozina M V 2015 Ecologization of cadastral valuation for sustainable development of urbanized territories *Bulletin of Tomsk Polytechnic University. Geo Assets Engineering* **326** (11) 98-105

Remote magnetic resonance sounding for exploration of pore space microstructure and aquifer macrostructure

O A Shushakov [1,2], A G Maryasov [3] and A B Strelnikova[4]

[1, 3] Voevodsky Institute of Chemical Kinetics and Combustion, Siberian Branch of the Russian Academy of Sciences, 3 Institutskaya Str., Novosibirsk, 630090, Russia
[2] Novosibirsk State University, 2 Pirogova Str., Novosibirsk, 630090, Russia
[4] National Research Tomsk Polytechnic University, 30 Lenina Ave., Tomsk, 634050, Russia Novosibirsk State University, Russia

E-mails: [1]shushako@kinetics.nsc.ru, [2]maryasov@kinetics.nsc.ru

Abstract: Proton nuclear magnetic resonance is a type of nuclear magnetic resonance (NMR), which is widely used to detect hydrogen containing liquids, like water or hydrocarbons. The use of the Earth's magnetic field allows remote detection of fluids sub surface without generating an independent magnetic field that is expensive and complex enough to apply. An important feature of NMR in the Earth's field is that the signals produced by snow and ice are generally unobservable under experimental conditions created for liquid water prospecting. The characteristics of the RF response are associated with the molecular environment of the nuclei. This allows detecting NMR signals of water in pores of various size.

1. Introduction

Magnetic resonance sounding (MRS) is a direct method to detect aquifers, in contrast to conventional indirect geophysical methods, such as electrical exploration, seismic prospecting, etc. It has been proved that MRS in geomagnetic field can be applied to detect aquifers at depths up to 100 m and even deeper depending on the electromagnetic shielding provided by formations and the intensity of natural and anthropogenic electromagnetic noises [1-3]. The physical principle of this method implies detecting a resonance signal emitted as a result of the water proton spin magnetization. Macroscopic zones of aquifers are investigated through measurement of water proton nuclear magnetic relaxation (NMR) in geomagnetic field within rock pores and fractures. To excite the spins and receive the magnetic resonance signal, the antenna loop is laid out on the ground, in a circle of 100 m in diameter or in a figure-of-eight shape to minimize the effects of extraneous noises. A radiofrequency (RF) current $I(t) = I_0 \cos \omega_0 t$ is applied to the antenna thus producing alternating RF magnetic field in the space nearby. The frequency of oscillation is equal to the Larmor frequency of protons in the geomagnetic field. Due to resonance effect, nuclear magnetization vector is tilted away from its equilibrium direction alond the geomagnetic field and rotates around both the direction of the RF magnetic field and the geomagnetic field. After switching off of the excitation RF pulse, the magnetization precesses freely around the geomagnetic field direction. This precession induces alternating voltage in the same antenna thus producing the so called free nuclear induction signal. Magnetic resonance frequency in the geomagnetic field is several kilohertz, deadtime of measurement system is several milliseconds (Figure1.). The signal detected is produced by hydrodynamically mobile water molecules only. The water in very small pores of water-resisting rocks, e.g. clays, as well as crystallized, chemically bound, or frozen water, is characterized by shorter relaxation time, therefore, the signal is not detected after the system deadtime. The NMR amplitude is proportional to free-fluid index (i.e. movable fluid).

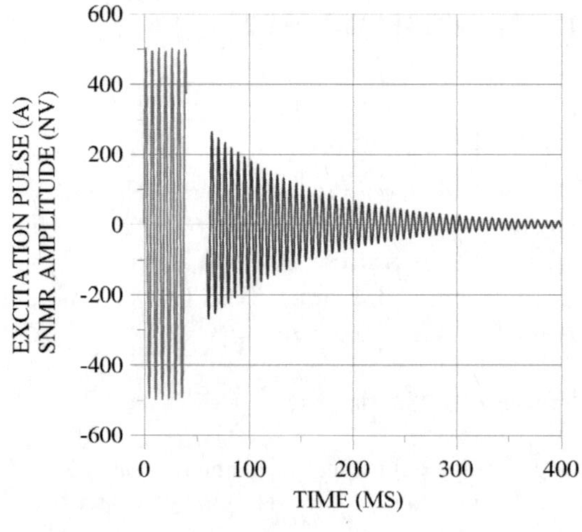

Figure 1. The principles of MRS signal detection in geomagnetic field. After the pulse of the field oscillating with the frequency equal to Larmor frequency the MRS signal (free-induction decay or FID) is induced in the antenna by nuclear magnetization freely precessing in geomagnetic field. The FID oscillates with the Larmor frequency and decays with the time constant of inhomogeneous spin-spin relaxation time. The MRS signal is detected with delay ("dead" time).

2. Bloch-Siegert effect in magnetic resonance sounding

If the water temperature is T = 293 K, and geomagnetic field strength is $B_0 = 6\cdot 10^{-5}$ T, then integral equilibrium magnetization, $M_0 = 1.93\cdot 10^{-7}$ J/(T*m³), is induced. If this magnetization is deviated from its equilibrium direction, it precesses around direction of geomagnetic field $\vec{B}_0$ with Larmor frequency of $\omega_0 = \gamma_H \cdot B_0$, where γ_H = 2.6753·10⁸ radian/(s∗T) is proton gyromagnetic ratio. In magnetic resonance experiments, RF magnetic field, $\vec{B}_1(t) = 2\vec{B}_1 \cos\omega_0 t$, perpendicular to $\vec{B}_0$, is applied. Interaction of nuclei (protons of water molecules) with external magnetic field, which is a sum of static geomagnetic field and RF field, $\vec{B} = \vec{B}_0 + 2\vec{B}_1 \cos\omega_0 t$, may be expressed in terms of spin Hamiltonian

$$\hat{H} = \hat{H}_0 + \hat{H}_1(t) = -\omega_0 \hat{I}_z - 2\omega_1 \hat{I}_x \cos\omega_0 t\,, \tag{1}$$

where $\vec{B}_0 \parallel$ z, $\vec{B}_1 \parallel$ x, and $\hat{I}_x$, $\hat{I}_z$ are components of spin operator of nucleus, $\omega_1 = \gamma B_1$ is frequency of spin precession in magnetic field with the strength of B_1 (Rabi's frequency).

For more accurate description of spin (and magnetization) motion within the frame rotating around $\vec{B}_0$ with frequency ω_0, let us calculate two first terms of the average Hamiltonian,

$$\bar{\tilde{H}}_1^{(0)}(t) = -\omega_1 \hat{I}_x\,, \tag{2}$$

$$\bar{\tilde{H}}_1^{(1)}(t) = -(i\omega_0/4\pi)\int_0^{2\pi/\omega_0} dt_2 \int_0^{t_2} dt_1 \left[\tilde{H}_1(t_2), \tilde{H}_1(t_1)\right] = (\omega_1^2/4\omega_0)\hat{I}_z\,, \tag{3}$$

where [A,B]=AB-BA is a commutator of operators A and B.

Eq. (3) shows that there is resonance frequency shift (so called Bloch-Siegert shift) [4]:

$$\Delta\omega = -\omega_1^2 / 4\omega_0. \tag{4}$$

The Bloch-Siegert shift (Eq. 4) provides an additional rotation of nuclear magnetization around z axis within the rotating ($\tilde{x}, \tilde{y}, z$) frame during pulse duration t_p:

$$\tilde{M}_x = \frac{\omega_1 \sin\Theta \cdot \Delta\omega}{\omega_{eff}^2} \cdot (1 - \cos\omega_{eff} t_p) \cdot M_0$$

$$\tilde{M}_y = \frac{\omega_1 \sin\Theta}{\omega_{eff}} \cdot \sin\omega_{eff} t_p \cdot M_0 \quad , \tag{5}$$

$$\omega_{eff} = \sqrt{\omega_1^2 \sin^2\Theta + \Delta\omega^2}$$

where Θ is the angle between static Earth's $\vec{B}_0$ and linearly polarized $\vec{B}_1$ RF magnetic fields in the lab frame.

The MRS signal immediately after the transmitting pulse (t=0) is

$$e_{t=0} = (\omega / I_0) \int_V \left\{ \sqrt{\tilde{M}_x^2(\mathbf{r}) + \tilde{M}_y^2(\mathbf{r})}\, B_1^2(\mathbf{r}) \sin\Theta(\mathbf{r}) / |B_1(\mathbf{r})| \right\} dV(\mathbf{r}) \quad . \tag{6}$$

MRS method calibration with regard to Bloch-Siegert effect was conducted via experiments on ice covered Novosibirsk Reservoir (Figure 2). Ice thickness and water depth were measured directly via drilling holes in the ice, and were 1±0.05m and 11±0.5m, respectively. Antenna of 50 m radius was used both for RF field generation and for the signal detection. The geomagnetic field α was 74°, proton resonance frequency was 2514 Hz ($B_0 = 5.9044 \cdot 10^{-5}$T).

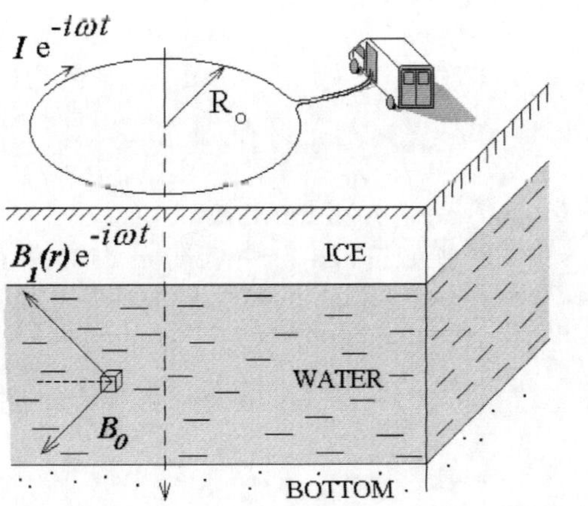

Figure 2. Experimental scheme of Bloch-Siegert effect detection on ice covered Novosibirsk Reservoir.

Figure3 and 4 demonstrate that the experimental MRS data fit particularly well with model calculations of the signal amplitude and phase with regard to the Bloch-Siegert effect.

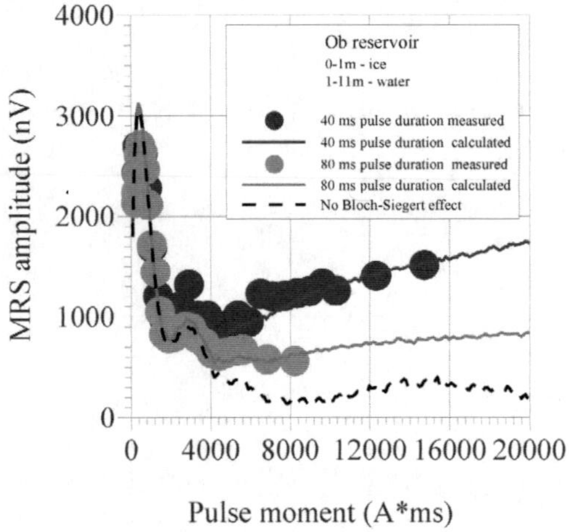

Figure 3. Amplitude of MRS signal versus the RF pulse intensity, for pulse lengths of 40 ms and 80 ms.

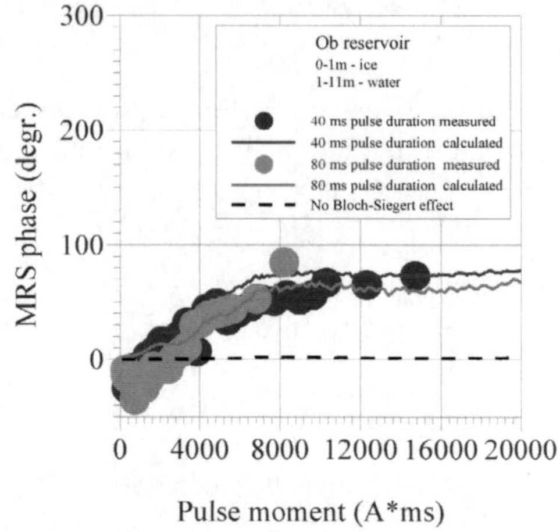

Figure 4. Phase of MRS signal versus the RF pulse intensity, for pulse lengths of 40 ms and 80 ms.

3. MRS relaxation and screening mechanisms

The investigation of double dielectric screening of the MRS signal allows estimating the total dissolved diamagnetic solids content in groundwater without drilling.

For example, in the half-space $z > 0$ with the uniform conductivity σ the magnetic field of the loop in the cylindrical frame has the form [1-3]:

$$B_{1z}(\mathbf{r}) = I_0 \cdot R_0 \int_0^{\infty} \frac{m^2}{m+u} e^{-uz} J_1(R_0 m) \cdot J_0(rm)\, dm \quad (7),$$

$$B_{1r}(\mathbf{r}) = I_0 \cdot R_0 \int_0^{\infty} \frac{mu}{m+u} e^{-uz} J_1(R_0 m) \cdot J_1(rm)\, dm \quad (8),$$

where R_0 is the antenna radius, $u = (m^2 - i\sigma\mu\omega)^{1/2}$, J_0 and J_1 are Bessel functions.

Distribution of water concentration with depth may be determined by inversion of the integral equation with experimentally measured and modeled NMR signal dependent on excitation intensity [3]. Figure 5 shows a comparison between MRS results in geomagnetic field and drilling and logging data of borehole 37, Novosibirsk.

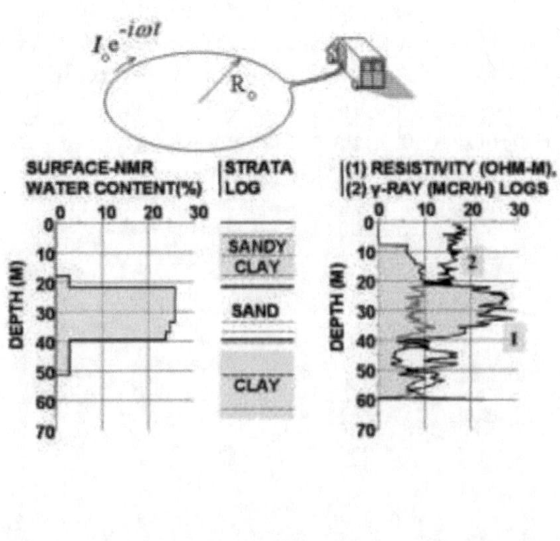

Figure 5. Comparison between geomagnetic MRS results and drilling and logging data of borehole 37, Novosibirsk.

The relaxation of the magnetization **M** that is proportional to the NMR signal is described using the Bloch-Torrey equations [5]:

$$\partial M_x/\partial t = D\,\Delta(M_x - M_{x0}) - M_x/T_{2bulk} + \gamma\,(\mathbf{M} \times \mathbf{B})_x \quad (9),$$

$$\partial M_y/\partial t = D\,\Delta(M_y - M_{y0}) - M_y/T_{2bulk} + \gamma\,(\mathbf{M} \times \mathbf{B})_y \quad (10),$$

$$\partial M_z/\partial t = D\,\Delta(M_z - M_{z0}) - M_z/T_{1bulk} + \gamma\,(\mathbf{M} \times \mathbf{B})_z \quad (11),$$

where D is the self-diffusion coefficient, T_{1bulk} and T_{2bulk} are the longitudinal and transverse relaxation times of bulk water.

The boundary conditions characterize the longitudinal (ρ_1) and the transverse (ρ_2) relaxivity on pore surface.

$$D\,(\mathbf{n}\cdot \nabla M'_x) + \rho_2\, M'_x = 0 \quad (12),$$

$$D\,(\mathbf{n}\cdot \nabla M'_y) + \rho_2\, M'_y = 0 \quad (13),$$

$$D\,(\mathbf{n}\cdot \nabla M'_z) + \rho_1\, M'_z = 0 \quad (14).$$

In resonance within rotating frame between RF-field pulses

$$\partial U/\partial t = D\,\Delta U - i\,\gamma\, G\, z\, U \quad (15),$$

where G is the magnetic field gradient, $U = M'_x + i\,M'_y$.

If we assume the uniformity of fluid in pore space and fast diffusion, the solution for the relaxation times is as follows [6]:

$$1/T_1 = 1/T_{1bulk} + \rho_1\, S/V \quad (16)$$

$$1/T_2 = 1/T_{2bulk} + \rho_2\, S/V + (\gamma\, G\, t)^2\, D/12 \quad (17)$$

$$1/T^*_2 = 1/T^*_{2bulk} + \rho_2\, S/V + (\gamma\, G\, t)^2\, D/3 + \gamma\, G\, a, \quad (18)$$

where S/V is the surface-to-volume ratio of pores, a is a core grain radius.

By the example of borehole 37 in Novosibirsk the spin-relaxation times were studied using two-pulse sequences (Figure 6, 7) [7].

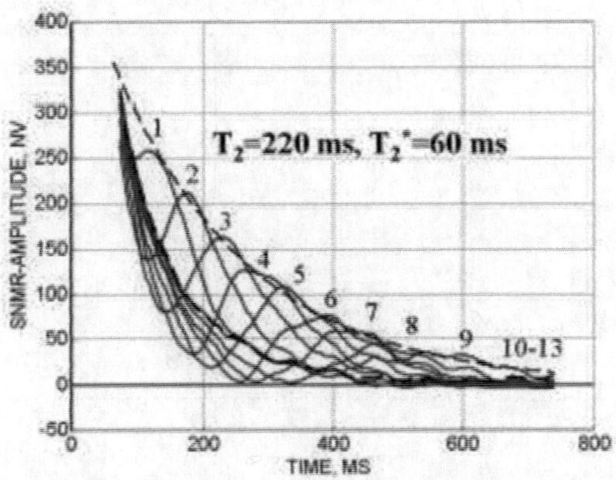

Figure 6. FID and spin-echo amplitudes measured via MRS (borehole 37, Novosibirsk), with sequence of pulses of 16 and 34 ms duration and delays between pulses from 100 to 700 ms (1-13).

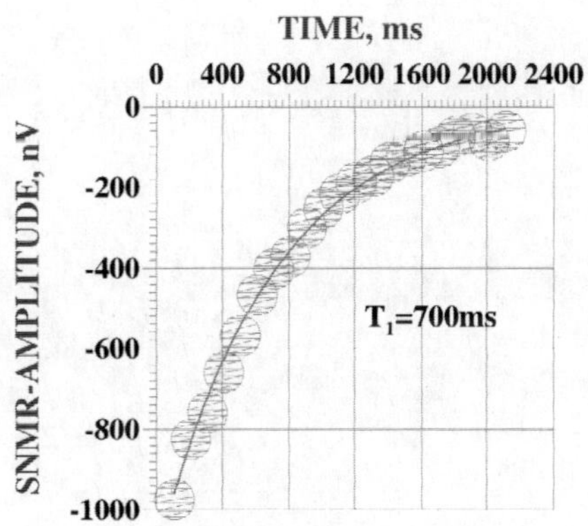

Figure 7. FID amplitude versus inversion time measured via MRS (borehole 37, Novosibirsk), with sequence of pulses of 32 and 18 ms duration and delays between pulses from 100 to 2100 ms.

Figure 6 illustrates the measurement of the homogenous spin-spin relaxation time T_2 equal to 220 ms via spin-echo method (the first pulse rotates the magnetization by 90°, the second one – by 180°). In this case the free-induction decay time T_2^* is 60 ms. Figure 7 exemplifies the measurement of the spin-lattice relaxation time T_1 equal to 700 ms via inversion-recovery method (the first pulse rotates the magnetization by 180°, the second one – by90°).

With regard to the longitudinal and transverse relaxation times for free water T_{1bulk} = 1.4 s and T_{2bulk} = 1 s measured on the ice cover of Novosibirsk reservoir, the core grain size a = 2.5 10^{-2} cm determined using drilling data, and diffusion coefficient of water D = 1.3 10^{-5} cm^2/s at 277 K, the surface relaxivities ρ_1 = 7 10^{-3} cm/s, ρ_2 = 3.5 10^{-2} cm/s and the field gradient G = 2 10^{-2} Gauss/cm have been determined.

References

[1] Shushakov O A and Legchenko A V 1994 *Geol. and geophysics* **35** 161 (in Russian)
[2] Shushakov O A 1996 *Geophysics* **61** 998
[3] Legchenko A V and Shushakov O A 1998 *Geophysics* **63** 75
[4] Bloch F and Siegert A 1940 *Phys.Rev.* **57** 522
[5] Torrey R C 1956 *Phys. Rev.* **104** 563
[6] Korringa 1956 *J.: Bull. Amer. Phys. Soc. (Ser. II)* **1** 216
[7] Shushakov O A and Fomenko V M 2004 *Appl. Magn. Res.* **25** 599

Groundwater compatibility with formation water and pay zone rocks in Pervomaysk oil-gas-condensate field to maintain formation pressure

N Trifonov[1,2], A Nazarov[2] and S Alekseev[3]

[1] A.A. Trofimuk Institute of Petroleum Geology and Geophysics Siberian Branch RAS, Tomsk, Russia
[2]National Research Tomsk Polytechnic University, Tomsk, Russia
[3]Institute of the Earth's Crust, Siberian Branch RAS, Irkutsk, Russia

E-mail: [1]Trifonovnik@mail.ru

Abstract. The paper describes the research results in determining the compatibility of groundwater from Aptain-Albian-Cenomanian aquifer with formation water and pay zone rocks in U_1 layer sediments, Pervomaysk oil field.

1. Introduction

Pervomaysk oil field is located within two territorial areas of the Russian Federation: southern area (the largest) in Kargasoksk region, Tomsk Oblast and northern area (the smallest) in Surgut region, Khanty-Mansi Autonomous District – Ugri. This oil field was discovered in 1969 as a result of drilled-in Upper Jurassic formation (well № 260) and further developed in 1981. Commercially, oil-bearing field is attributed to terrigenous Upper Jurassic sediments, Vasugan suite (U_1 layer embracing sandstones with interbedded aleurites and clays) [1].

Currently, Pervomaysk oil field is in the III development stage, which, in most cases, is usually marked by the production decline under conditions of water cutting. Since 1982 formation pressure maintenance (FPM) system by injecting groundwater from Aptain-Albian-Cenomanian aquifer has been applied in the southern oil field area, and in 1990- in the northern area [2, 3]. This aquifer, upper level of lower West Siberian artesian basin hydrostratigraphic unit, embraces significant groundwater supplies within the Western Siberian territory which are applied in FPM.

According to existing regulatory documents water used in FPM systems should be assessed to determine its compatibility to formation water and reservoir rocks. In this case, the authors have applied the numerical physic-chemical modeling based on software system HydroGeo [4, 5]. This software system is used to solve a wide spectrum of problems [6].

2. Modeling design

To model geochemical terrigenous processes applicable to "water-rock" system in U_1 layer the following method was used.

Modeling hydrogeochemical processes included (1) base particles - "independent" components: H^+, Na^+, K^+, Ca^{2+}, Mg^{2+}, NH_4^+, Fe^{2+}, Sr^{2+}, Al^{3+}, OH^-, Cl^-, HCO_3^-, SO_4^{2-}, H_2O, SiO_2; (2) dependent components and associates: $(CO_3)^{2-}$, H_2CO_3, CO_2, NH_4HCO_3, $(NH_4)_2CO_3$, NH_4Cl, NH_3, NH_4OH, $NaHCO_3$, $(NaCO_3)^-$, $(NaSO_4)^-$, $NaCl$, $Mg(HCO_3)_2$, $(MgHCO_3)^+$, $MgCO_3$, $MgSO_4$, $MgCl^+$, $MgCl_2$,

$(HSi_2O_6)^{3-}$, $KHCO_3$, KCl, $(CaHCO_3)^+$, $Ca(HCO_3)_2$, $CaCO_3$, $CaSO_4$, $CaCl^+$, $CaCl_2$, $(SrHCO_3)^+$, $SrCO_3$, $SrCl^+$, $SrCl_2$, $(FeHCO_3)^+$, $FeCO_3$, $FeCl^{+,}$ and (3) minerals (which could be dependent components): SiO_2 -calcedony, SiO_2 – quartz, $Al(OH)_3$ – hydrargillite, $Al_2Si_2O_5(OH)_4$ -kaolinite, $Al_2Si_4O_{10}(OH)_2$ –, montmorillonite, $Al_4Si_4O_{10}(OH)_8$ – chlorite (Al), $NaAlSi_3O_8$– albite, $KAlSi_3O_8$ -feldspar (K), $CaAl_2Si_2O_8$ – anorthite, $MgAl_2SiO_6$ – monocline pyroxene, $CaCO_3$ -calcite, $FeCO_3$ – siderite, $(Fe^{2+})(OH)_2$ – amakinite, $CaSO_4$ – anhydrite, $CaSO_4(H_2O)_2$ – gypsum [7]. The first (independent components) and third (minerals) series were determined by the water analysis of investigated components, mineral composition identification of formation rocks and expected composition of secondary solid phase. Applying Pitzer model in calculating solution particle activity coefficient the required ionic associate composition was determined by a series of preliminary calculations including their maximum number. Furthermore, those associates that were of insignificant concentrations were excluded. Analogous selection was applied relevant to the minerals.

The calculations were performed relevant to two different conditions: (1) at standard conditions (22°C and 0.1 MPa), approximate to reproduced storage conditions and laboratory sampling, and (2) at formation (reservoir) conditions embracing thermodynamic conditions of Vasugan suite sediments (U_1 layer -90.7°C and 25.8 MPa), where groundwater injection into Pokurskoe suite sediments was planned.

The first stage included importing initial lab analysis data of formation water from Vasugan suite sediments (U_1 layer), Pervomaysk oil field into the software system HydroGeo (Table 1). This included the recalculations for solution electroneutrality reduction at standard conditions: t=22°C, P=0.1 MPa (electroneutrality reduction was performed to Cl ion). Then, the water composition in Pervomaysk oil field was reduced to thermodynamic formation conditions (90.7°C and 25.8 MPa) involving CO_2 invasion, which, in its turn, changed the acid-alkali properties of the solution itself (Table 1) and being normalized to the conventional reservoir formation condition – equilibrium via selecting standard Gibbs free energy values relevant to the equilibrium of both flowing solution composition and rocks [8].

Table 1. Groundwater composition changes in Vasugan suite (U_1 layer), Pervomaysk oil field during modeling preparation.

Component	Ca^{2+}	Mg^{2+}	Na^+	K^+	Fe^{2+}	Al^{3+}	HCO_3	Cl^-	$(SO_4)^{2-}$	SiO_2	$(CO_3)^{2-}$	CO_2	pH	M (gr/lit)	Density(kg/m^3)	Ionic strength (kg/m^3)	T° (C)	P (MPa)
Initial electro-neutrality	1160.8	300.1	12961.9	12.8	54.5	5.0E^{-05}	319.8	23427.0	0.05	3.0E^{-06}	2.1	5.5	7.0	41.2	1027.7	0.7	22.0	0.1
Balanced to CO_2 in layer	1169.4	317.2	13166.1	11.9	52.7	5.0E^{-05}	0.0	23999.2	0.04	3.3E^{-05}	0.0	6.5	3.1	40.2	1000.7	0.8	90.7	25.8

Reference rock composition was based on the lithologic description data of U_1 layer: quartz- 48%, calcite-10.5%, albite and K-feldsdpar- 10%, respectively, anorthite- 9%, kaolinite and monocline pyroxene- 6% respectively, siderite- 0.3%, chlorite- 0.2%. In addition, the following accessory mineral impurities were included: hydrargillite, montmorillonite, dolomite, anhydrite and gypsum as expected from the overall lithological observation. Total porosity is 16.5%.

The second modeling stage involved preparing groundwater within Aptain-Albian-Cenomanian sediments after the same pattern as for formation water.

The changes, conditioned by possible technogenic alterations of the solution composition itself, occurring throughout the groundwater flow path within Aptain-Albian-Cenomanian sediments from water production wellhead to water injection well bottom hole are associated with thermodynamic condition shifting. Due to the lack of information on injected water flow rate and discharge, the calculations for pipelines and field equipment are performed excluding the residence time. In this case, these results are not actual but only maximum possible salt precipitation (Figure 1).

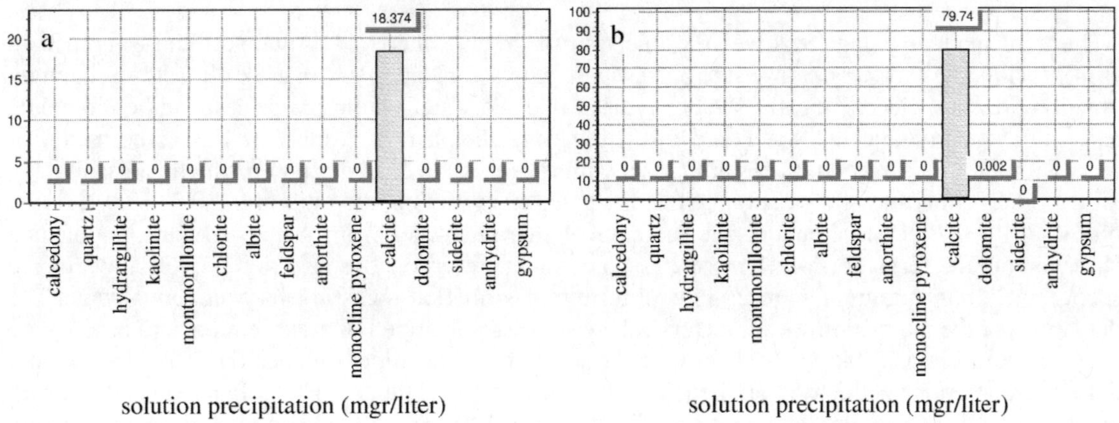

Figure 1. Possible salification; (a) in gathering facilities and processing systems for groundwater within Aptain-Albian-Cenomanian sediments under standard thermodynamic conditions; (b) in wellbore during groundwater injection into pay zone

The changes in solution composition are depicted in Table 2. Modeling data of possible chemical processes, occurring during technogenic mineral formation, in both subsurface equipment and injected wellbores indicate insignificant solution composition changes even if the following fact is regarded, i.e. calcite precipitation of 18.4 mgr. in one litre in subsurface equipment and 79.7mgr. - in injection wells. Moreover, gradual accumulation of carbonates in subsurface equipment and water injection well bottomhole could result in certain production decline.

Table 2. Groundwater composition changes within Aptain-Albian-Cenomanian sediments, Pervomaysk oil field during modeling preparation.

Component mg/dm³	Ca^{2+}	Mg^{2+}	Na^{+}	K^{+}	Fe^{2+}	Al^{3+}	HCO_3	Cl^{-}	$(SO_{4})^{2-}$	SiO_2	$(CO_3)^{2-}$	CO_2	pH	M (gr/lit)	Density(kg/m^3)	Ionic strength (kg/m^3)	T° (C)	P (MPa)
Initial electro-neutrality	549.7	89.6	6674.8	80.5	1.60	4,0E-06	109.2	11639.2	0.83	1,2E-06	1.3	0.5	7.6	19.9	1012.8	0.4	22.0	0.1

Settling simulation and behavior in subsurface equipment	543.4	89.6	6675.7	80.5	1.61	1,3E-09	101.1	11639,1	0.83	1,8E-06	0.5	1.0	7.2	19.8	1012.2	0.4	22.0	0.1
At bottomhole (reduction to formation conditions)	528.1	84.5	6794.9	78.8	1.65	3.2E-10	8.9	11870.1	0.86	2,2E-04	0.2	0.6	6.5	19.8	987.7	0,4	90.7	25.8

The third modeling stage involves the injection of groundwater from Aptain-Albian-Cenomanian sediments directly into formation water of Vasugan suite sediments, Pervomaysk oil field to maintain formation pressure. Geochemically, the behavior of injected water within pay zones is rather complex. This is the result of injected water and formation water being mixed and / or this combination of waters within the reservoir could possibly react with rock minerals and gas phase which is dissolved in residual oil. The mixture of formation (A) and injected (B) waters in the ratio of 0.8:0.2 – 0.0: 1.0 (i.e. from 80:20 to 0:100 %), simulates the increase of injected water from Aptain-Albian-Cenomanian sediments into the formation water from 20% to complete reservoir filling. Progressively towards the injection well, this mixture is reduced in equilibrium to formation rocks in every modeling stage. The calculations of the interaction within water-rock system are performed without reference to time, i.e. to complete equilibrium (Table 3). To determine the solution composition of specified mixed waters the Mixture/Evaporation model was applied; and to calculate the interaction of obtained mixture with formation rocks the Solution/Precipitation model was used (based on Hydrogeochemistry of software system HydroGeo).

Table 3. Solution composition at the final modeling stage, mgr/litre (pay zone after the interaction with rocks).

Component	U_1 layer	A:B=0.8:0.2	A:B=0.6:0.4	A:B=0.4:0.6	A:B=0.2:0.8	A:B=0.0:1.0	Aptain-Albian-Cenomanian
Ca^{2+}	1169.38	941.08	784.47	628.78	499.51	455.55	528.05
Mg^{2+}	317.18	259.57	218.56	175.16	131.92	90.22	94.52
Na^{+}	13166.10	11792.40	10538.50	9279.58	8019.41	6720.32	6794.90
K^{+}	11.88	27.47	38.92	49.95	63.67	76.92	78.81
Fe^{2+}	52.69	19.80	5.77	2.24	0.16	0.59	1.65
Al^{3+}	5.00E-05	8.06E-05	3.25E-06	4.74E-04	2.41E-05	2.45E-06	3.18E-10
$(HCO_3)^{-}$	0.04	0.00	0.00	0.00	0.00	0.00	8.90
Cl^{-}	23999.20	21303.40	18931.10	16554.20	14172.10	11707.60	11870.10
$(SO_4)^{2-}$	0,04	0.04	0.03	0.03	0.04	0.03	0.86
SiO_2	3.33E-05	3.99E-12	2.11E-05	1.78E-05	1.92E-05	1.81E-05	0.00
$(CO_3)^{2-}$	5.36E-07	7.84E-12	7.60E-12	7.28E-12	7.25E-12	5.54E-12	0.16
H_2CO_3	427.14	0.12	0.23	0.39	0.46	0.05	36.60
CO_2	6.50	0.00	0.00	0.01	0.01	0.00	0.55
pH	3.05	2.45	2.33	2.24	2.25	2.71	6.50
M(gr/litre)	40.20	35.19	31.20	27.21	23.26	19.27	19.76
Density.(kg/m^3)	1000.68	997.92	995.29	992.65	990.04	987.45	987.68
Ionic strength (mole/litre)	0.76	0.66	0.59	0.51	0.43	0.36	0.36

3. Results of modeling

Technogenic geochemical changes of the water and rock compositions, based on modeling data, provide evidence of the fact that secondary mineral formation processes predominate over formation rock solution processes (in 1 liter of solution the amount of newly-formed minerals is 120-380 mgr more than amount of dissolved minerals). This, in its turn, indicates the possible decrease of the reservoir properties and injection well rate.

Basically, technogenic reservoir changes are associated not only with the dissolution of feldspar, chlorite, kaolinite and quartz but also the formation of secondary sulphates, clay (montmorillonite) and carbonates (calcite and siderite) (Figure 2). Every washout cycle initiates comparatively insignificant rock composition changes and frequent washout could result in a rather significant alteration of the mineral composition and porosity [9], while the latter gradually decreases causing relevant changes of the reservoir properties.

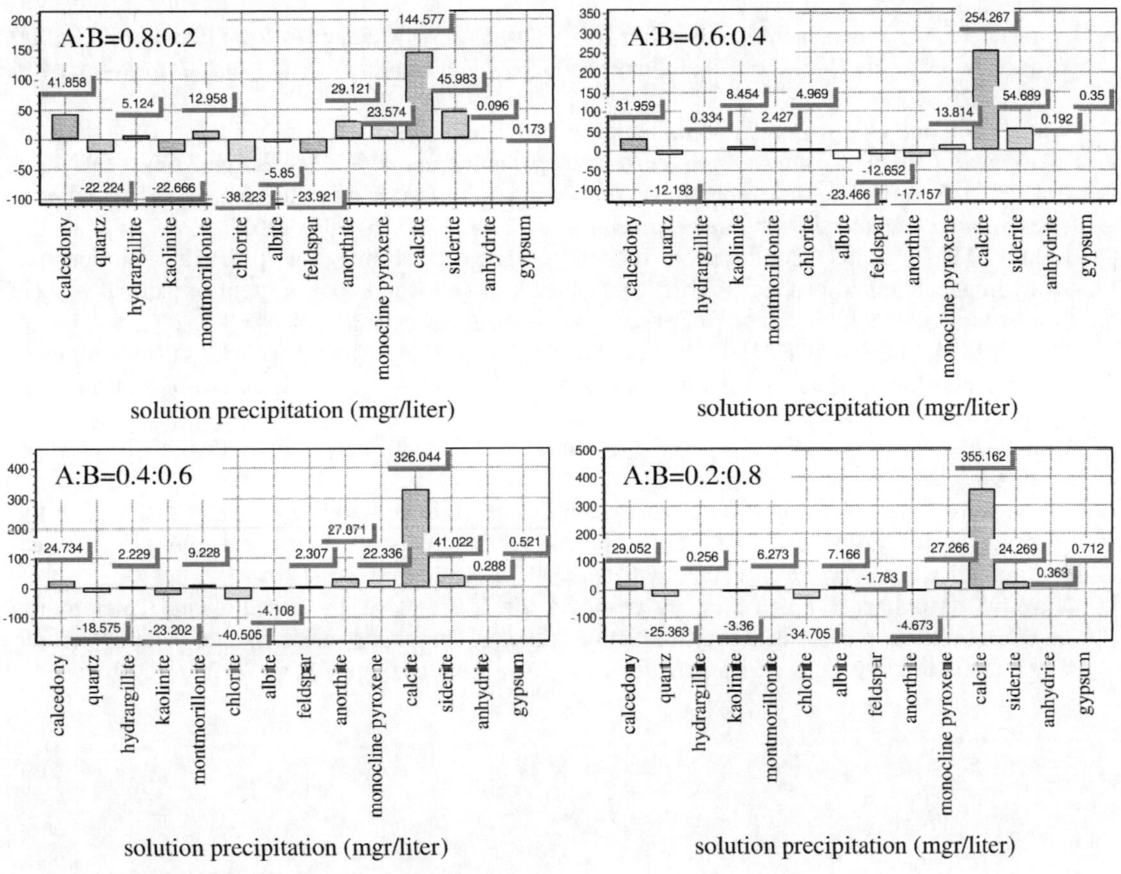

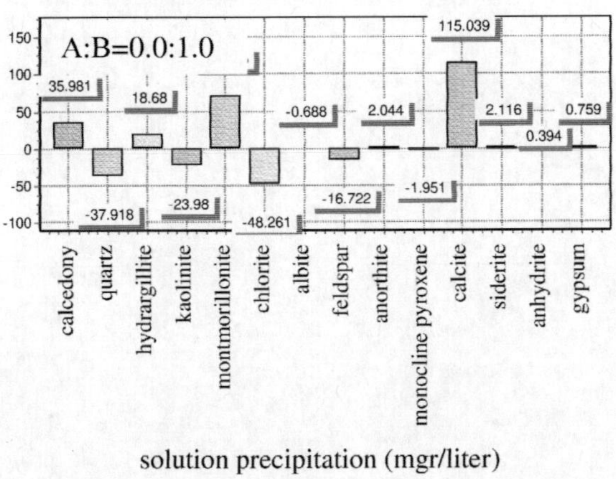

Figure 2. Precipitation (positive values) and solution (negative values) of minerals in the pay zone for different portions of mixed formation water and groundwater water from Aptain-Albian-Cenomanian sediments to maintain formation pressure.

4. Conclusion

Modeling data results of Aptain-Albian-Cenomanian water injection into pay zone (U_1 layer) indicate a relatively good compatibility of above-mentioned waters. It should be noted that the quantitative composition changes of solutions and rocks are the possible maximum changes which actually could be lower (if considering interaction time in the water-rock system).

References

[1] *Geology and mineral resources of Russia* 2000 Ed. Kontorovich A E, Surkov V S (Saint-Petersburg, VSEGEI) p 477

[2] Ahmadun F L , Pendashteh A, Abdullah L C, Biak D R A, Madaeni S S and Abidin Z Z 2009 Review of technologies for oil and gas produced water treatment *J. Hazardous Materials* **170** (2-3) 530 – 551

[3] Warner D L and Orcutt D H 1973 Industrial wastewater – injection wells in the USA – status of use and regulation *Underground waste management and artificial recharge* **2** 687 – 697

[4] Bukaty M B 2002 The development of software for the issues on hydrogeology TPU bulletin. *Geology and mineral resources exploration in Siberia* 305 (6) 348 – 365

[5] Bukaty M B 2005 Geomigration modeling. The opportunity of practical solutions of hydrogeochemical problems. Quality problems of the water-rock system numerical models *Geological evolution and self-organizing of the water-rock system* **1** 243 – 332

[6] Dutova E M, Bukaty M B, Nevolko A I, Pokrovsky D S and Shvartsev S L 2006 Hydrogenic concentration of gold in alluvial placers of the Egorevskoe area (Salair) *Russian Geology and Geophysics* **47** (3) 364 – 376

[7] Collins A G 1975 *Geochemistry of oilfield waters (*Elsevier Amsterdam–Oxford–New York) p 496

[8] Trifonov N S and Nazarov A D 2015 Possibility of produced and wastewater disposal in the process of Yurubcheno field development by methods of numerical physico-chemical simulation *Int. Multidiscip. Sci. GeoConf. (*SGEM, Bulgaria) **2** 541 – 549

[9] Naseri S, Moghadasi J and Jamialahmadi M 2015 Effect of temperature and calcium ion concentration on permeability reduction due to composite barium and calcium sulfate precipitation in porous media *J. Natural Gas Science and Engineering* **22** 299 – 312

System-integrated GIS-based approach to estimating hydrogeological conditions of oil-and-gas fields in Eastern Siberia

L I Auzina[1] and A V Parshin[1,2]

[1]Irkutsk National Research Technical University,
[2] A.P. Vinogradov Institute of Geochemistry SB RAS

E-mail: [1]lauzina@mail.ru

Abstract. The article describes the basic methodological elements of system-integrated approach to estimating water abundance in the upper hydrodynamic zone of Eastern Siberia territory to optimize water exploration. The technique is based on materials available at the pre-survey stages of exploration. When processing the information, the integrated geoinformation analysis has been used.

1. Introduction

At present in the territory of Eastern Siberia hydrocarbon fields with low reservoir pressure and temperature (6-20^{0}C) are being developed and planned to be developed, which, in turn, implies significant expendetures with less economic benefits than, for instance, in Western Siberia. Hence, the issues of investment optimization at different types of exploration is more topical than ever before.

One of the most urgent problems faced by production engineers in the oil and gas fields is groundwater prospecting for multi-purpose water supply and optimization of water well location.

The maximum geological and economic effect is possibe to achieve by means of developing the estimating system of the most perspective water sites that would be applied at the early stages of exploration. To solve the geological problem, such an estimating system should include minimal amount of field information due to maximum use of prior information or information provided by: archives and literature sources as well as the Earth remote sensing data. In this case, the estimate of water abandunce perspectives of study areas should be scientifically based on the used data.

At present at the early stages of exploration the pre-survey information analysis is performed, however, cartography developed at this stage is not used for complete forecast information to full extent. At the contemporary stage of studying perspective oil fields in Eastern Siberia, the information obtained at the pre-survey stage cannot be directly used for exploration due to the lack of information on study areas. Thus, identification of high water abandunce zones for subsequent drilling with the help of direct hydrogeological parameters is virtually impossible.

The purpose of the given article is to determine a number of indirect parameters that are possible to estimate at the initial exploration stage and to develop integrated index of water abandunce, which is a basis for exploration optimization involving groundwater reserve estimates, that is collectively defined as a system-integrated approach.

2. Objects and methods

The integrated approach based on integrated-model analysis has been used before for estimating sustainability of underground hydrosphere in the urban territories, solution of optimization problems of geological exploration of Lenskoye gold fields (Auzina, 1997). The current article considers modern GIS-based technique.

Development of the described technique was performed within the exploration project at oil and gas condensate fields (OGCF) of Chonsk group (Eastern Siberia) located in the extemely complex geological-structural and permofrost-hydrogeological conditions. The geological structure of the study area is defined by Archean, Proterozoic, Paleozoic formations, Mesozoic and Cenozoic deposits, as well as trias intrusions.

Most fields are confined by Nepsko-Botuobin anteclise, within which, as a result of several stages of tectogenesis, multi-block structure has been conditioned by fractures of different ages of north-eastern, north-western and submeridional directions. Sufficient deformations are conditioned by intrusion of trappean rock into Early-Triassic period of magmatism. The most significant result of neotectonic uplift was development of contemporary relief and identification of zones with different water exchange regimes in the cross section: free, confined by the bottom of Middle and Upper Cambrian deposits, and impeded, beginning from the aquifer of Lower and Middle Cambrian rocks.

Spatial distribution of permafrost rocks is of rather a complex pattern and not confined by definite relief features.

Such a diversity of natural conditions determined the extreme complexity in pre-survey estimate of hydrogeological features of study areas and the necessity to develop integrated index including a number of parameters influencing water abandunce of prospective hydrogeological bodies in the upper hydrodynamic zone and identified at the initial survey stages.

The technique was implemented using geoinformation approach, some elements of which have been used for exploration of ore fields (Parshin, Spiridonov, 2014) and geoecological monitoring (Filimonova et al, 2015, Pokrovsky et al, 2014, Pokrovsky et al, 2015). In this case the technique was applied to explored and operated groundwater deposits, and then it was used for new oil and gas fields.

At the initial stage of pre-surver operation, marking and vectoring of archieve cartography is performed for study area and the adjacent sites including topographic and geological mapping, accessible satellite information, and results of aerogeophysical survey. This information allows the initial digital base and database (DB) development added by the results of the field hydrometric and filtration opertions including hydrogeological, geocryological maps, patterns of head distribution, groundwater level, filtration parameters of water-bearing material and some other information. One of the most important parameters from the list above is a consistent digital elevation model (DEM) that allows calculating a number of morpho-structural parameters.

Based on the initial information, the additional layers of morpho-structural parameters are calculated:

- fracture index defined by relationship of river bed length not confined by Quaternary deposits and crossing the square of statistical subsqure to its size, which depends on research details (Auzina, 2014), (Figûe 1a);
- proximity of target point to water stream crossing of the 3-d and 4-th orders;
- height and distance to the nearest drain;
- height and distance to the main drain;
- terrian ruggedness index (Riley et al, 1999), (Figûe 1b);
- downward gradient, exposure;
- elevation of target point.

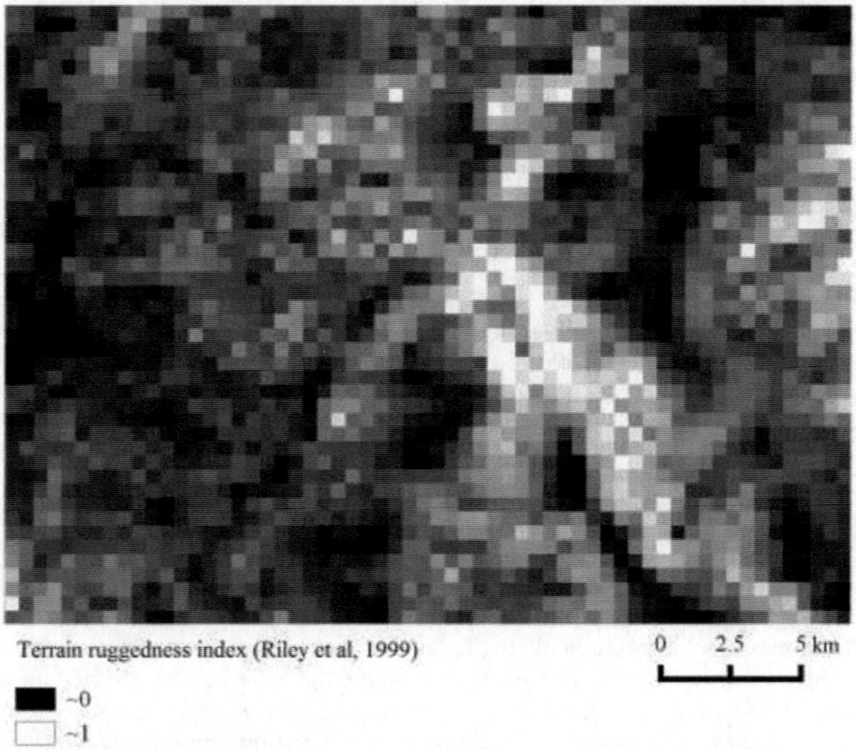

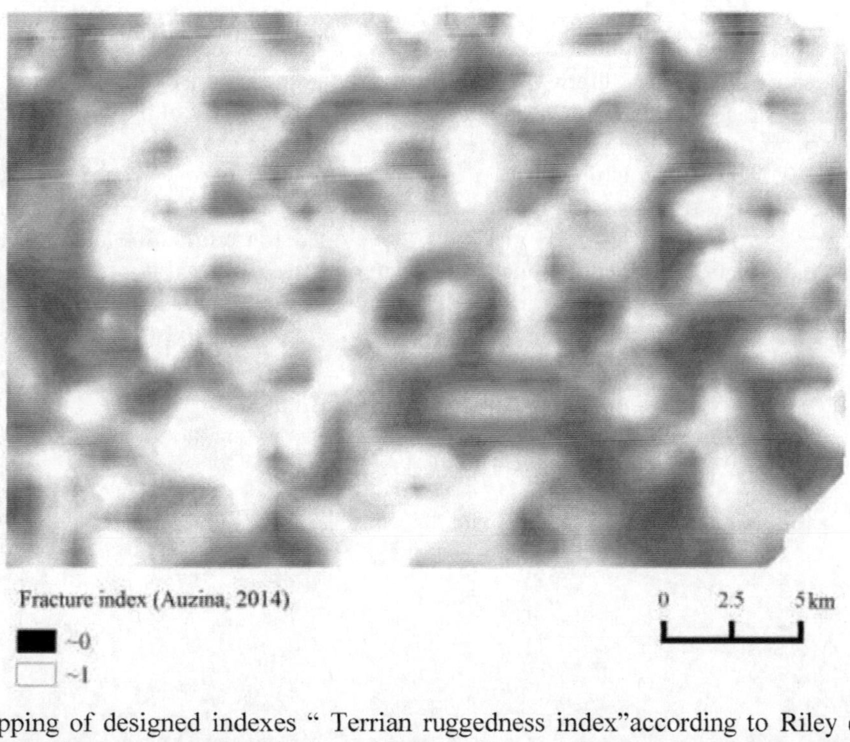

Figure 1. Mapping of designed indexes " Terrian ruggedness index"according to Riley et al. 1999 (above) and "Specific fracture index" (below). Mapping is normalized and scaled in such a way that black colour means minimum index (values close to 0), white – maximum index (values close to 1).

All mentioned indexes are determined by map layers, digital layers of drainage system and DEM, i.e. they are available at the pre-survey exploration stage.

Thus, the database was generated consisting of more than 50 predictors. They were independent vector layers in the general geoinformation project "Hydropoisk". As is known, a large number of GIS-project layers is not convenient for visual analysis, therefore, it is appropriate to generalize it to a less number of informative integrated indexes by means of GIS mapping (Budyak, Parshin, 2013; Filimonova et al, 2015).

Reduction in information size without loss of information was performed by means of development of integrated index of water abundance (IIWA). To identify the key factors concerned with water abundance, the factor analysis was used for geodata of the analogous site by means of robust principal components method. As a result, among the whole set of primary and designed indexes a number of parameters were distinguished that reflected water abandunce of the perspective sites and, at the same time, did not involve any difficulty to obtain them at the early exploration stage. Despite the significant difference in geological-hydrogeological and permafrost conditions of the license areas, the list of universal and efficient parameters included fracture index, proximity to water stream crossing of the 3-d and 4-th orders, height above the main and nearest drain, distance to the nearest drain, terrian ruggedness index, seldom - downward gradient, elevation of target point, distance to fault zones of definite types.

The values of indexes were brought to uniform dimention by means of normalizing, scaling, and/or classification by Delphi method, after that corresponding raster layers were developed with unified grid step. The potential hydrogeological significance of every parameter and, consequently, the raster values were ranged within the interval from 0 to 1, where the least values reflected unproductive sites from the point of view of water exploration. Integrated index of water abandunce means a sum of all significant parameters. It was plotted in conditional units from minimum to maximum values, where the latter denoted the most perspective sites. All calculations can be performed both in rasters (as in the above example) and in tables of database. The suggested technique is based on open GIS technology, mostly DBMS PostgreSQL/PostGIS and QuantumGIS system, therefore, it is in full conformity with the latest government decision on refusal from public procurement of propriety foreign software programms.

3. Results and Discussion

Integrated index designed on the basis of OGCF developed for production to a different extent or being currently developed has been applied to the sites of the early exploration stage. The example below shows (Figûe 2) maximum values denoting the most perspective sites which are in red colour.

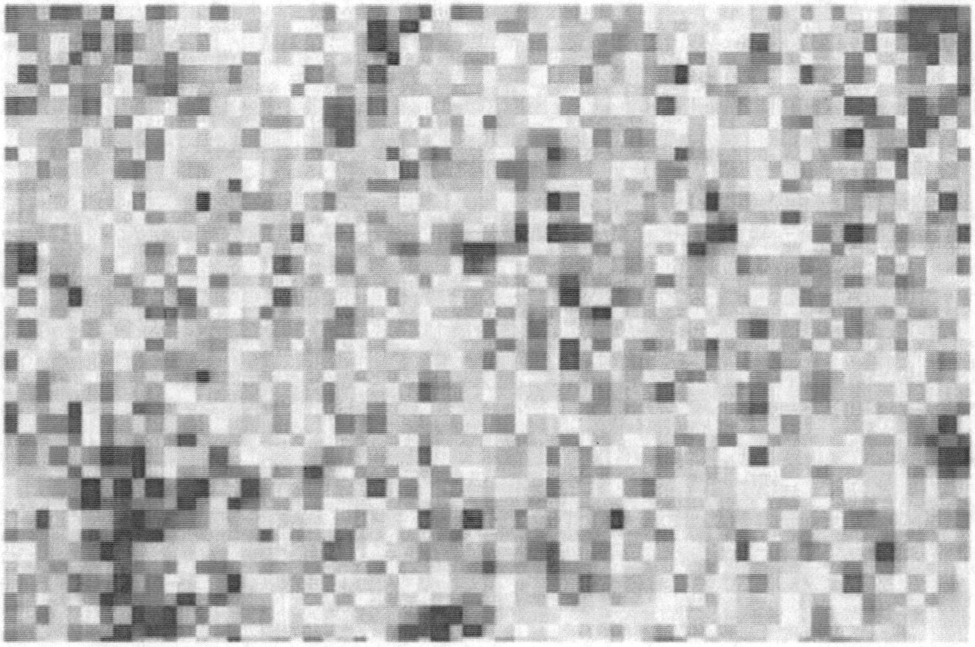

Figure 2. Mapping of model calculation of integrated water abundance index.

4. Conclusion

In some sites, the drilling results have shown that, on the whole, the suggested mapping relibly determines the expected hydrogeological conditions. Depending on the task and a set of available geodata, the integrated index can be calculated with different detail level.

The technique was tested in one of the license sites of Chonsk group deposits. The initial results of implementation of suggested approach has proved its high efficiency: more relible forecasting allowed solving geological problem, having reduced the number of exploration wells by 30%, in this case the possibility of well inclusion into perspective ones increased 60-80%. At the moment, testing of its efficiency is underway at other license sites.

The suggested approach is universal to a great extent. Depending on the set target and exploration scale, one can only change a number of parameters, but the basic indexes of geodata processing technique remain the same.

References

[1] Auzina L I 1997 Assessment of Hydrogeological Conditions of Some Solid Mineral Deposits of Eastern Siberia Using Integrated-Modeling Analysis *Methods of exploration, prospecting and evaluation of ore deposits* **21** 86 – 94 [in Russian]

[2] Auzina L I 2014 One of the approaches to optimization of water exploration at the oil and gas fields of Eastern Siberia *Izvestiya Sibirskogo otdeleniya sektsii nauk o zemle Rossiyskoi Academii Estestvennykh nauk. Geologiya, Poiski i Razvedka Rudnykh Mestorozhdeniy* **6** 82 – 88 [in Russian]

[3] Parshin AV and Spiridonov A M 2014 Methodical and engineering solutions of geological and geochemical GIS to explore gold-ore deposits *Geologiya i mineralno-syr'evye resursy Sibiri* **3 – 2** 72 – 76 [in Russian]

[4] Filimonova L M, Parshin A V and Bychinsky V A 2015 Assessment of Air Pollution in the Area of Aluminum Production by means of Geochemical Survey of Snow *Russian Meteorology and Hydrology* **10** 75 – 84 [in Russian]

[5] Budyak A and Parshin A 2013 GIS mapping of geological features of the Baikal mountain region based on integrated geochemical indicators *Goldschmidt Conference Abstracts Mineralogical Magazine* **B** 154

[6] Riley S, DeGloria S and Elliot R 1999 A Terrain Ruggedness Index That Quantifies Topographic Heterogeneity *Intermountain Journal of Sciences* **5** 23 – 27

[7] Pokrovsky V, Pokrovsky D, Dutova E, Nikitenkov A and Nazarov A 2014 Degree of Areal Drainage Assessment Using Digital Elevation Models *IOP Conf. Ser.: Earth Environ. Sci.* **21(1)**

[8] Pokrovsky V, Dutova E, Kuzevanov K, Pokrovsky D and Nalivaiko N 2015 Hydrogeological Conditions Changes of Tomsk, Russia *IOP Conf. Ser.: Earth Environ. Sci.* 012031 URL: http://iopscience.iop.org/1755-1315/27/1/012031) **27** 7

Institute of Physics
Dirac House, Temple Back
Bristol BS1 6BE UK

ISSN: 1755-1307
ISBN 978-1-5108-2274-0

CURRAN ASSOCIATES INC.
proceedings.com